TRANSLATIONS SERIES IN MATHEMATICS AND ENGINEERING

A.V. Balakrishnan
General Editor

M.I. Yadrenko
Spectral Theory of Random Fields
1983, viii + 259 pp.
ISBN 0-911575-00-6 Optimization Software, Inc.
ISBN 0-387-90823-4 Springer-Verlag New York Berlin Heidelberg Tokyo
ISBN 3-540-90823-4 Springer-Verlag Berlin Heidelberg New York Tokyo

G.I. Marchuk
Mathematical Models In Immunology
1983, xxv + 353 pp.
ISBN 0-911575-01-4 Optimization Software, Inc.
ISBN 0-387-90901-X Springer-Verlag New York Berlin Heidelberg Tokyo
ISBN 3-540-90901-X Springer-Verlag Berlin Heidelberg New York Tokyo

A.A. Borovkov, Ed.
Advances In Probability Theory: Limit Theorems and Related Problems
1984, xiv + 378 pp.
ISBN 0-911575-03-0 Optimization Software, Inc.
ISBN 0-387-90945-1 Springer-Verlag New York Berlin Heidelberg Tokyo
ISBN 3-540-90945-1 Springer-Verlag Berlin Heidelberg New York Tokyo

V.A. Dubovitskij
The Ulam Problem of Optimal Motion of Line Segments
1985, xiv + 114 pp.
ISBN 0-911575-04-9 Optimization Software, Inc.
ISBN 0-387-90946-X Springer-Verlag New York Berlin Heidelberg Tokyo
ISBN 3-540-90946-X Springer-Verlag Berlin Heidelberg New York Tokyo

N.V. Krylov, R.S. Liptser, and A.A. Novikov, Eds.
Statistics and Control of Stochastic Processes
1985, xiv + 507 pp.
ISBN 0-911575-18-9 Optimization Software, Inc.
ISBN 0-387-96101-1 Springer-Verlag New York Berlin Heidelberg Tokyo
ISBN 0-540-96101-1 Springer-Verlag Berlin Heidelberg New York Tokyo

Yu. G. Evtushenko
Numerical Optimization Techniques
1985, xiv + 561 pp.
ISBN 0-911575-07-3 Optimization Software, Inc.
ISBN 0-387-90949-4 Springer-Verlag New York Berlin Heidelberg Tokyo
ISBN 3-540-90949-4 Springer-Verlag Berlin Heidelberg New York Tokyo

Continued on page 297

Siberian Branch of
the USSR Academy of Sciences

PROCEEDINGS OF THE INSTITUTE OF MATHEMATICS
Novosibirsk

ADVANCES IN PROBABILITY THEORY

LIMIT THEOREMS FOR SUMS OF RANDOM VARIABLES

Edited by A.A. Borovkov

Optimization Software, Inc.
Publications Division, New York

Editor
A.A. Borovkov
Institute of Mathematics
Siberian Branch
USSR Academy of Sciences
Novosibirsk-90
U.S.S.R.

Editor
A.V. Balakrishnan
School of Engineering
University of California
Los Angeles
California 90024
U.S.A.

Library of Congress Cataloging in Publication Data

Predel´nye teoremy dlia summ sluchainykh velichin.
English.
Advances in probability theory.

(Translations series in mathematics and engineering)
Translation of: Predel´nye teoremy dlia summ
sluchainykh velichin.
Includes bibliographies.
1. Limit theorems (Probability theory)
2. Stochastic processes. I. Borovkov, Aleksandr
Alekseevich. II. Balakrishnan, A.V. III. Title:
Limit theorems for sums of random variables.
IV. Series.
QA273.67.P72513 1985 519.2 85–27931
ISBN 0–911575–17–0
ISBN 0–387–96100–3 (Springer–Verlag)

Exclusively authorized English translation of the original Russian edition of *Predel'nye teoremy dlya summ sluchajnykh velichin*, Volume 3 of the Proceedings of the Institute of Mathematics, the Siberian Branch of the USSR Academy of Sciences, edited by Academician *Sergej L'vovich SOBOLEV*, published in 1984 by the Nauka Publishing House, Novosibirsk, U.S.S.R.

Printed in the United States of America.

Worldwide Distribution Rights by Springer-Verlag New York Inc., 175 Fifth Avenue, New York, New York 10010, USA and Springer-Verlag Berlin Heidelberg New York Tokyo, Heidelberg Platz 3, Berlin-Wilmersdorf-33, The Federal Republic of Germany.

ISBN 0-911575-17-0 Optimization Software, Inc.
ISBN 0-387-96100-3 Springer-Verlag New York Berlin Heidelberg Tokyo
ISBN 3-540-96100-3 Springer-Verlag New York Berlin Heidelberg Tokyo

ABOUT THE AUTHORS

BORISOV, Igor S. Member, Institute of Mathematics, Novosibirsk, Siberian Branch of the USSR Academy of Sciences. Associate Professor, Novosibirsk State University. Novosibirsk State University Graduate (1974). Kandidat of Science (1978). Main scientific interests: Limit Theorems for Sums of Random Variables in Linear Spaces.

KHODZHIBAEV, Valy R. Instructor, Polytechnic Institute, Namangan. Novosibirsk State University Graduate (1978). Kandidat of Science (1983). Main scientific interests: Boundary–value Problems for Random Processes with Independent Increments.

LUGAVOV, Vyacheslav S. Instructor, Teachers Institute, Kurgan. Novosibirsk State University Graduate (1976). Main scientific interests: Random Processes with Independent Increments.

MIROSHNIKOV, Andrej L. Instructor, Novosibirsk Institute of Geodesy, Air Photography and Cartography. Novosibirsk State University Graduate (1977). Kandidat of Science (1983). Main scientific interests: Probability Inequalities for Sums of Independent Random Variables.

MOGUL'SKIJ, Anatolij A. Member, Institute of Mathematics, Novosibirsk, Siberian Branch of the USSR Academy of Sciences. Novosibirsk State University Graduate (1969). Doctor of Science (1983). Professor, Novosibirsk Agricultural Institute. Main scientific interests: Limit Theorems for Random Processes.

POZHIDAEV, Alexander V. Instructor, Novosibirsk Institute of Railroad Engineering. Novosibirsk State University Graduate (1977). Kandidat of Science (1983). Main scientific interests: Differential Equations with Random Coefficients.

SAKHANENKO, Alexander I. Member, Institute of Mathematics, Novosibirsk, Siberian Branch of the USSR Academy of Sciences. Associate Professor, Novosibirsk State University. Novosibirsk State University Graduate (1972). Kandidat of Science (1975). Main scientific interests: Limit Theorems for Sums of Independent Random Variables.

TOPCHIJ, Valentin A. Member, Computing Center, Omsk, Siberian Branch of the USSR Academy of Sciences. Associate Professor, Omsk State University. Novosibirsk State University Graduate (1972). Kandidat of Science (1978). Main scientific interests: Limit Theorems for Branching Random Processes.

UTEV, Sergej A. Member, Institute of Mathematics, Novosibirsk, Siberian Branch of the USSR Academy of Sciences. Novosibirsk State University Graduate (1979). Kandidat of Science (1984). Main scientific interests: Limit Theorems for Sums of Weakly Dependent Random Variables.

Preface

This volume features one of the main current trends in Probability Theory, succinctly expressed in the title "Limit Theorems for Sums of Random Variables".

Articles on convergence–rate bounds for sums of independent random variables in limit theorems make up the bulk of the volume. A.I. Sakhanenko's work involves a new estimate in the classical invariance principle, while I.S. Borisov and S.A. Utev study convergence rates for empirical measures and mixing sequences. Sharp asymptotics are derived by A.A. Mogul'skij for the sojourn probability of a random walk in a receding strip. V.R. Khodzhibaev develops asymptotic expansions in a two–boundary problem with continuous time.

The papers in the second group are more diverse and are mainly devoted to investigating the limiting behavior of random sequences and processes. A.V. Pozhidaev examines asymptotic normality of solutions of parabolic equations with random coefficients. V.S. Lugavov analyzes distributions of boundary–value functionals for controlled processes. V.A. Topchij focuses his attention on the asymptotics of the probability of extending branching processes. A.L. Miroshnikov derives concentration–function inequalities.

A.A. Borovkov

Contents

Part 1. THE INVARIANCE PRINCIPLE. CONVERGENCE–RATE ESTIMATES AND LARGE DEVIATIONS 1

A.I. Sakhanenko
Convergence Rate in the Invariance Principle for Non-identically Distributed Variables with Exponential Moments 2

S.S. Utev
Inequalities for Sums of Weakly Dependent Random Variables and Estimates of the Convergence Rate in the Invariance Principle 73

V.R. Khodzhibaev
Asymptotic Analysis of Distributions in Two–boundary Problems for Continuous–time Random Walks 114

A.A. Mogul'skij
Probabilities of Large Deviations for Trajectories of Random Walks 140

Part 2. LIMIT THEOREMS FOR RANDOM PROCESSES OF PARTICULAR TYPES 185

I.S. Borisov
On the Convergence Rate in the Central Limit Theorem 186

V.S. Lugavov
On the Distribution of the Sojourn Time on a Half–axis and the Final Position of a Process with Independent Increments Controlled by a Markov Chain 214

A.L. Miroshnikov
Inequalities for the Integral Concentration Function 236

A.V. Pozhidaev
Asymptotic Normality of Solutions of Parabolic Equations with Random Coefficients 253

V.A. Topchij
Asymptotics of the Probability of Continued Critical General Branching Processes with No Second Moment for the Number of Decendants 269

List of forthcoming publications 297
Transliteration table 299

Part 1

The Invariance Principle. Convergence-Rate Estimates and Large Deviations

A. I. Sakhanenko

CONVERGENCE RATE IN THE INVARIANCE PRINCIPLE FOR NON-IDENTICALLY DISTRIBUTED VARIABLES WITH EXPONENTIAL MOMENTS

1. Introduction and the main theorem

Given $\eta_1,\ldots,\eta_n$, a sequence of independent normal random variables; we are required to construct a sequence of independent *random variables* $\xi_1,\ldots,\xi_n$ with *a priori* given distribution, making

$$\Delta = \Delta_n \equiv \max_{m\leqslant n}\left|\sum_{j\leqslant m}\xi_j - \sum_{j\leqslant m}\eta_j\right|$$

as small as possible.

Solution of this problem is the chief obstacle in obtaining estimates both in the Donsker-Prokhorov Invariance Principle, when the rate of convergence of Δ_n to zero is determined by ε_n, where $P(\Delta_n > \varepsilon_n) \le \varepsilon_n$ (see, e.g., [1]-[3]) and in the Von Strassen Invariance Principle, where this convergence rate is characterized by a sequence δ_n such that $\Delta_n/\delta_n \to 0$ a.s. (see [4]-[6]). Therefore, following [5], [6], our objective is to obtain estimates for Δ_n. Of course we assume that the distributions of the $\xi_1,\ldots,\xi_n$ satisfy the condition

$$\mathbf{M}\xi_j = \mathbf{M}\eta_j = 0,\ \mathbf{D}\xi_j = \mathbf{D}\eta_j > 0 \quad \forall j. \tag{1}$$

For $P(\Delta > x)$, it is not difficult to construct lower bounds (see [3], [7], [8]), which show that (at least for identically distributed variables) $P(\Delta > x)$ decreases, as $x \to \infty$, not slower than

$$\mathbf{P}(\max_j|\xi_i| > x) \sim \sum_j \mathbf{P}(|\xi_j| > x).$$

For identically distributed random variables $\xi_1,\ldots,\xi_n$, Komlós, Major, and Tusnády [6] have derived upper bounds for $P(\Delta_n > x)$ having thus a regular order of dependence on n and x. In particular, if

$$\mathbf{M}\exp(t\xi) < \infty \ \text{ for } \ |t| \leqslant t_0,\ t_0 > 0, \tag{2}$$

then (see [6]) $\xi_1,\dots,\xi_n$ can be given on a common probability space along with $\eta_1,\dots,\eta_n$ so that

$$\mathbf{P}(\Delta_n > x) \leqslant C(\mathcal{F})\, n^{K(\mathcal{F})} e^{-c(\mathcal{F})x}, \tag{3}$$

where the constants $C(F) < \infty$, $K(F) < \infty$ and $c(F) > 0$ depend only on the common distribution F of the random variables $\xi_1,\dots,\xi_n$.

However, the method of a common probability space used in [5], [6] is not directly extendable to the case of non-identically distributed summands, since for the distribution of $\xi_1,\dots,\xi_n$ at least one of the following two conditions has to be satisfied: either $\xi_1,\dots,\xi_n$ have sufficiently smooth (or lattice) distributions, or they are identically distributed. The well-known method of Prokhorov [1] and of Skorokhod [9] are applicable to non-identically distributed summands as well, but the estimates obtained when $\{\xi_j\}$ have higher-order moments are inferior to Komlós-Major-Tusnády estimates.

In this paper, we suggest yet another method for constructing $\xi_1,\dots,\xi_n$. Its main difference from the Komlós-Major-Tusnády method is the use of a technique of smoothing, which although it significantly complicates the problem, still allows one to obtain good estimates in the case of non-identically distributed variables $\xi_1,\dots,\xi_n$.

Our main result is the following theorem.

T h e o r e m 1. If condition (1) is satisfied and for some $\lambda > 0$ the inequalities

$$\lambda \mathbf{M}|\xi_j|^3 \exp(\lambda|\xi_j|) \leqslant \mathbf{D}\xi_j \qquad \forall j \tag{4}$$

hold, the random variables $\xi_1,\dots,\xi_n$ can be constructed with respect to $\eta_1,\dots,\eta_n$ so that

$$\mathbf{M}e^{c\lambda\Delta} \leqslant 1 + \lambda B, \tag{5}$$

where $c > 0$ is some absolute constant and $B^2 = \sum_{j\leq n} D\xi_j$.

The Komlós-Major-Tusnády estimate (3) follows from (5) for $B^2 = nD\xi_1$ since conditions (2) and (4) are equivalent.

Note the characteristic

$$B^{-1} \max_j (\mathbf{M}|\xi_j|^3 \exp(\lambda|\xi_j|))/\mathbf{D}\xi_j$$

used in Theorem 1 which is a simple majorant of the Lyapunov ratio for conjugate distributions. This ratio is inevitable since we need estimates of the convergence rate in the central limit theorem (for quantiles of conditional distributions), with large deviations taken into account.

We emphasize the fact that estimate (5), in contrast to (3), shows explicit dependence on the distributions of $\xi_1,\ldots,\xi_n$. Hence we can apply (appropriately choosing λ in (4)) this inequality for truncations of arbitrary random variables, in order to obtain estimates in the invariance principle even when there are no exponential moments. In particular, we have

$$\mathbf{M}\Delta^s \leqslant C(s) \sum_{j\leqslant n} \mathbf{M}|\xi_j|^s,$$

where the constant $C(s)$ depends only on s, as will be proved in a future work of this author.

Section 3 contains the proof of Theorem 1 as well as a description of the method for constructing $\xi_1,\ldots,\xi_n$. The proof requires estimates for the quantiles of conditional distributions of sums of non-identically distributed summands; the estimates need to be more exact than the ones used for identically distributed random variables in a simpler method -- the Komlós-Major-Tusnády method. These estimates are given in Section 2. Note that Theorem 3 of Section 2 containing a new estimate in the local limit theorem for densities, is of independent interest.

This paper consists of three sections. Section 1 is introductory. Sections 2 and 3 have subsections with a double enumeration: the first digit denotes the section, the second the subsection. The numbering of theorems and remarks is continuous; moreover, each subsection has its own numbering of lemmas and formulas. In references to lemmas and formulas of a given subsection we shall use a single digit, while for citing assertions from another subsection of the same section, we shall use double digits, the first digit referring to the subsection. In Section 3 there are several references to lemmas from Section 2 -- then a triple enumeration is used.

We use the symbol O only when the constant is absolute, and the symbol θ stands for any value of magnitude not exceeding 1. Furthermore, C_1, C_2, ... are positive absolute constants, but in different subsections the C_i can designate different values. The symbols $C < \infty$ and $c > 0$ stand for any absolute constants retaining their meaning only within a few paragraphs. We use $\Phi(x)$ and $\phi(x) = (2\pi)^{\frac{1}{2}} \exp(-x^2/2)$ to denote standard normal distribution and its density, respectively.

2. Estimates of the densities and quantiles of smooth distributions

2.1. BASIC RESULTS

Let $\xi_1, \ldots, \xi_m$ be a sequence of independent random variables satisfying for some $R \neq 0$ the condition

$$\mathbf{M}\xi_j = 0, \ \mathbf{M}|\xi_j|^4 \exp(R\xi_j) < \infty \quad \forall j. \tag{1}$$

Let $S = \xi_1 + \cdots + \xi_m$, with

$$0 < \mathbf{D}S \equiv B^2 < \infty. \tag{2}$$

Clearly, the second inequality in (1) for $h \neq R$ coincides with Cramér's condition: $\mathrm{M} \exp(h\xi_j) < \infty$ for all j and h belonging to the interval containing 0 and R, i.e., for

$$0 \leqslant h/R \leqslant 1. \tag{3}$$

Waiving the assumption $B = 1$ in (2) does not increase the generality, but is very burdensome in Subsection 2.9. We shall assume that in the sequel the parameter h satisfies condition (3) and the sums are taken over j ranging from 1 to m.

Next we consider the sequence of independent random variables $\xi_1(h), \ldots, \xi_m(h)$ having "conjugate" distributions (or distributions that are Cramér's transformation of distributions of the $\xi_1, \ldots, \xi_m$), i.e., for $\xi = \xi_j$ having the distribution functions

$$\mathbf{P}(\xi(h) < x) = \int_{z<x} e^{hz} d\mathbf{P}(\xi < z)/\mathbf{M}e^{h\xi}. \tag{4}$$

We assume without loss of generality that $\xi_j(0) = \xi_j$. Using the notation $\xi_{j,h} = \xi_j(h) - \mathrm{M}\xi_j(h)$, we put

$$S(h) = \Sigma\xi_j, \quad A(h) = \mathbf{M}S(h),$$

$$B^2(h) = \mathbf{D}S(h), \quad \Gamma(h) = \Sigma\mathbf{M}(\xi_{j,h})^3,$$

$$L^*(h) = B^{-3}(h) \sum \mathbf{M}|\xi_{j,h}|^3, \quad L(R) = \max_{0 \leqslant h/R \leqslant 1} \sum \mathbf{M}|\xi_{j,h}|^3,$$

$$K^*(h) = B^{-4}(h) \sum \mathbf{M}|\xi_{j,h}|^4, \quad K(R) = \max_{0 \leqslant h/R \leqslant 1} \sum \mathbf{M}|\xi_{j,h}|^4.$$

Since $L^*(h)$ and $K^*(h)$ are the Lyapunov ratios of orders three and four, we have (see [10]):

$$B^{-6}(h)\Gamma^2(h) \leqslant (L^*(h))^2 \leqslant K^*(h). \tag{5}$$

Now let $\Lambda(h) \equiv \ln \mathbf{M}e^{hS}$ be the generating function of the cumulants. By the properties of the latter, on the interval (3) we obviously have

$$\Lambda'(h) = A(h),\ \Lambda''(h) = B^2(h) > 0,\ \Lambda'''(h) = \Gamma(h), \tag{6}$$

$$\Lambda(0) = A(0) = 0,\ B(0) = B,\ \Gamma(0) = \mathbf{M}S^3 = \Gamma, \tag{7}$$

$$\Lambda^{IV}(h) = \Sigma[\mathbf{M}(\xi_{j,h})^4 - 3\mathbf{D}^2\xi_{j,h}]. \tag{8}$$

It follows immediately from (6) that for h satisfying (3) we have the estimate

$$|\Lambda'''(h)| \leqslant L(h) \leqslant L(R), \tag{9}$$

and from (8) and the inequality

$$\mathbf{D}^2\xi_{j,h} = \mathbf{M}^2(\xi_{j,h})^2 \leqslant \mathbf{M}(\xi_{j,h})^4$$

for these h we have also:

$$|\Lambda^{IV}(h)| \leqslant 2K(h) \leqslant 2K(R). \tag{10}$$

Since $L(R)$ and $K(R)$ will be used in the sequel as convenient characteristics, making it possible to estimate simultaneously both the Lyapunov ratios $L^*(R)$ and $K^*(R)$ and the derivatives of $\Lambda(h)$, as well as (see formula (15)) the derivatives of $\lambda(x)$ and $\beta(x)$, it is appropriate to state the following useful estimates:

$$L(R) \leqslant 8\sum \mathbf{M}|\xi_j|^3 \max\{\exp R\xi_j, 1\}, \tag{11}$$

$$K(R) \leqslant 16\sum \mathbf{M}|\xi_j|^4 \max\{\exp R\xi_j, 1\}, \tag{12}$$

which will be proved in Lemma 6.1. The inverse inequalities also hold with other constants, however, only when there are no "distinguished" summands, i.e., for $R^2 \max_{j,h} \mathbf{D}\xi_j(h) = O(1)$.

Let

$$f(t, h) = \mathbf{M}e^{(h+it)S}/\mathbf{M}e^{hS},\ T(h) = 1/(4L^*(h)B(h)),$$

$$U^*(h) = B(h) \int_{|t|\geqslant T(h)} |f(t, h)|\, dt, \quad U(R) = \max_{|h|<|R|} \int_{|t|\geqslant T(h)} |f(t, h)|\, dt.$$

We note (see [10]) that by the independence of $\xi_1, \ldots, \xi_m$ the random variable $S(h)$ has conjugate distribution (i.e., (4) holds for $\xi = S$), and therefore $f(t,h)$ is the characteristic function of $S(h)$. It follows from the inversion formula for characteristic functions that if $U^*(h)$ is finite for some h, we are assured of the existence of a bounded distribution density $p(x,h)$ of $S(h)$. It follows from (4) that for $\xi = S$ one then has

$$p(x) = e^{\Lambda(h)-hx}p(x,\ h), \tag{13}$$

where $p(x) = p(x,0)$ is the density of S.

If we now try to obtain estimates for the density $p(x)$, using (13), and estimates in the local limit theorem for the density $p(x,h)$, we get the best accuracy if we set $h = H(x)$ in (13), where $H(x)$ is the solution of the equation

$$A(H(x)) = x. \tag{14}$$

Let

$$\lambda(x) = xH(x) - \Lambda(H(x)),\ \ \beta(x) = \ln B(H(x)). \tag{15}$$

T h e o r e m 2. If $H(x)$ satisfies the condition

$$0 \leqslant H(x)/R \leqslant 1, \tag{16}$$

we have

$$p(x) = (2\pi)^{-1/2}e^{-\lambda(x)-\beta(x)}(1+\theta\delta_L(x)), \tag{17}$$

$$p(x+z) \leqslant (2\pi)^{-1/2}e^{-\lambda(x)-\beta(x)-zH(x)}(1+\delta_L(x)), \tag{18}$$

where

$$\delta_L(x) = (2/5)U^*(H(x)) + 6L^*(H(x)). \tag{19}$$

If the condition

$$2|R|L(R) \leqslant B^2 \tag{20}$$

is satisfied, (16)-(18) hold for x satisfying the inequalities

$$0 \leqslant x/R \leqslant (2/3)B^2, \tag{21}$$

and we also have the following estimates:

$$3/4 \leqslant H(x)/(xB^{-2}) \leqslant 3/2,\ \ x \neq 0, \tag{22}$$

$$1/2 \leqslant B^2(H(x))/B^2 \leqslant 2, \tag{23}$$

$$\lambda(x) = (x/B)^2/2 + \theta x^3 L(R)B^{-6}, \tag{24}$$

$$\beta(x) - \ln B = 2\theta|x|L(R)B^{-4}, \tag{25}$$

$$\delta_L(x) \leqslant BU(R) + 15B^{-3}L(R). \tag{26}$$

Even more precise results are given, as a rule, by the following

Theorem 3. If x satisfies condition (16), then

$$p(x) = (2\pi)^{-1/2}e^{-\lambda(x)-\beta(x)}(1+\theta\delta_K(x)), \tag{27}$$

where

$$\delta_K(x) = (2/5)U^*(H(x)) + 12K^*(H(x)). \tag{28}$$

If the condition

$$4R^2K(R) \leqslant B^2 \tag{29}$$

is satisfied, (16)-(18) and (26) hold, while for x satisfying (20) we have estimates (21)-(25) and

$$\lambda(x) = (x/B)^2/2 - (x/B)^3(\Gamma/B)^3/6 + 2\theta x^4K(R)/B^8, \tag{30}$$

$$\beta(x) - \ln B = (x/B)(\Gamma/B^3)/2 + 40x^2K(R)B^{-6}, \tag{31}$$

$$\delta_K(x) \leqslant BU(R) + 32B^{-4}K(R), \tag{32}$$

$$\delta_L(x) \leqslant BU(R) + 10(B^{-4}K(R))^{1/2}. \tag{33}$$

Relations (17)-(19) and (27)-(28) will be proved in Subsection 2.2 and the remaining assertions of Theorems 2 and 3 in Subsection 2.3.

Note that to improve the accuracy of the estimates for fixed x in Theorems 2-5, we should set R = (3/2)x in Theorems 2 and 3, R = 2x in Theorem 4 and R = 8x in Theorem 5.

The choice of the constant 4 in condition (29) (and 2 in condition (29)) is mostly arbitrary and dictated by our desire to obtain a "decent" constant 2/3 in (21). It is clear that if we replace the constant in (29) by a larger one (i.e., if we make this condition more stringent), we extend the regions of x for which Theorems 2-5 hold (i.e., in particular, we increase the constant 2/3 in (21) to some c, 2/3 < c < 1).

REMARK 1. Suppose that all the following conditions are satisfied for the sequence $\{\xi_j\}$:

$$\mathbf{M}\xi_j = 0,\ \mathbf{M}\exp(\pm R_0\xi_j) \leqslant R_1 < \infty \quad \forall j,\ R_0 > 0, \tag{34}$$

$$B^2 \geqslant \delta m,\ \delta > 0, \tag{35}$$

$$U(R_0) \leqslant R_2/m,\ R_2 < \infty. \tag{36}$$

In this case, from the inequality $x \leq e^x + e^{-x}$ we obtain that

$$h^k\mathbf{M}|\xi|^k e^{\pm h\xi} \leqslant \mathbf{M}e^{(k+1)h\xi} + \mathbf{M}^{-(k+1)h\xi} \quad \text{for} \quad h = R_0/(k+1).$$

This relation and (11), (12) and (34) imply the inequalities

$$L(R) \leqslant 2 \cdot 8 \cdot 4^3 R_1(R_0)^{-3} m \leqslant \delta m R_3/2, \tag{37}$$

$$K(R) \leqslant 2 \cdot 16 \cdot 5^4 R_1(R_0)^{-4} m \leqslant \delta m R_4/4, \tag{38}$$

$$B^2 = \Sigma \mathbf{M}(\xi_{lj})^2 \leqslant 2 \cdot 3^2 \cdot R_1(R_0)^{-2} m. \tag{39}$$

It follows from (35) and (37) that condition (20) of Theorems 2 and 3 is satisfied for $R = \pm 1/R_3$ and therefore for

$$|x| \leqslant (2/3) R\delta m \tag{40}$$

all the assertions of Theorem 2 hold. In particular, substituting (35), (36), (37) and (39) into (26) and (25), we obtain from (17) an estimate of the form

$$p(x) = (2\pi)^{-1/2} e^{-\lambda(x)}(1 + \theta R_5(1 + |xm^{-1/2}|)m^{-1/2}) \tag{41}$$

for x satisfying (40). Since conditions (35) and (36) coincide with conditions 2 and 3 in [11] and (34) is weaker than condition 1 of [11], Theorem 3 in [11] is a special case of our Theorem 2, the constants R_4 and R_5 in (40) and (41) being easily expressible in terms of R_0, R_1, R_2 and δ when Theorem 2 is used.

The fact that an assertion of the form (41) holds under a constraint of the form (40) (i.e., for $|x| \leqslant R_0(\{\xi_j\})$) instead of $x = o(m)$, as was in [11], has been noted in [5].

Note that if instead of (36) we use the slightly more stringent condition

$$U(R_0) \leqslant R_7/m^{3/2}, \ R_7 < \infty, \tag{42}$$

we can obtain by virtue of (38) and (39) that if conditions (34), (35) and (42) hold, all the assumptions of Theorem 3 for $R = \pm(1/R_4)^{\frac{1}{2}}$ are satisfied and, in particular, (27) can be rewritten in the form

$$p(x) = (2\pi)^{-1/2} e^{-\lambda(x)-\beta(x)}(1 + \theta R_8/m) \tag{43}$$

for $|x| \leqslant (2/3)\delta Rm$.

We note that, as follows from the proof (see Lemmas 2.2 and 2.3), higher accuracy than (41) can be achieved in (43) by using the following estimate for the conjugate distributions in the local limit theorem:

$$p(0) = \varphi(0) + \theta R_9 m^{-1}$$

instead of

$$p(x) = \varphi(x) + \theta R_{10} m^{-1/2}.$$

REMARK 2. Let $\xi, \xi_1, \xi_2, \ldots$ be a sequence of independent identically distributed random variables satisfying the following conditions: $\mathrm{M}\xi = 0$, $0 < \mathrm{D}\xi < \infty$ and

$$\int_{|t|<\infty} |\mathbf{M}e^{(h+it)}|^{k(\mathcal{F})}\,dt < C_1(\mathcal{F}) \quad \text{for} \quad |h| \leqslant c_0(\mathcal{F}), \tag{44}$$

where $k(F) > 0$, $c_0(F) > 0$, and F designates the common distribution of $\xi, \xi_1, \xi_2, \ldots$. It follows from (44) in particular that

$$\mathbf{M}e^{h\xi} \leqslant C_2(\mathcal{F}) < \infty \quad \text{for} \quad |h| \leqslant c_0(\mathcal{F}), \tag{45}$$

$$|f_0(t, h)| \leqslant e^{-c(\varepsilon,\mathcal{F})} \quad \text{for} \quad |t| \geqslant \varepsilon \quad \text{and} \quad |h| \leqslant c_0(\mathcal{F}), \tag{46}$$

where $c(\varepsilon, F) > 0$ $\forall\varepsilon > 0$, and

$$f_0(t, h) = \mathbf{M}e^{(h+it)\xi}/\mathbf{M}e^{h\xi}$$

is the characteristic function of ξ when ξ has conjugate distribution. It follows from (45) that condition (34) of Remark 1 is satisfied for $R_1 = C_2(F)$, $R_0 = c_0(F)$. Condition (35) holds automatically for $\delta = \mathrm{D}\xi$. It follows from (37) and (38) that (20) is satisfied for $R = \pm 1/R_3 = \pm C_3(F)$ and (29) for $R = \pm(1/R_4)^{\frac{1}{2}} = \pm C_4(F)$. Thus, from (44), (46) and the definition of $U(R)$ we have that

$$U(R) \leqslant \int_{|t|>\varepsilon(\mathcal{F})} |f_0(t, h)|^m\,dt \leqslant e^{-(m-k(\mathcal{F}))c(\varepsilon(\mathcal{F}),\mathcal{F})}C_1(\mathcal{F}) \tag{47}$$

since $\mathrm{M}e^{h\xi} \geqslant 1$. By (47), for $m > m_0(F)$ conditions (36) and (42) of Remark 1 are satisfied and, therefore, (41) and (43) hold true.

Note now that most of the notation introduced in this subsection for the variable S (except $L^*(h)$, $K^*(h)$, $T(h)$, $L(R)$ and $K(R)$) do not depend on the concrete representation of S in the form $S = \sum \xi_i$. Hence we can redefine S as follows. In doing so, we assume that S is an arbitrary variable satisfying (2) and the condition $\mathrm{M}|S|^4 e^{RS} < \infty$. We retain for S all the notation introduced above but agree to mean by $L^*(h)$, $K^*(h)$, $T^{-1}(h)$, $L(R)$ and $K(R)$ the infima in the corresponding definitions over all representations of S as $S = \xi_1 + \cdots + \xi_m$, i.e., the sum of several independent variables satisfying (1).

We say that the random variable S (more precisely, its distribution) belongs to the class $\mathcal{D}(R)$ for $R > 0$ if conditions (2) and

$$4R^2K(R) \leqslant B^2, \quad 4R^2K(-R) \leqslant B^2 \tag{48}$$

are satisfied. We assume that $S \in \mathcal{D}_0(R) \subset \mathcal{D}(R)$ if conditions (48) and

$$R^2U(R) \leqslant B^{-3},\ R^2U(-R) \leqslant B^{-3} \tag{49}$$

are satisfied. Note that substituting (48) and (49) into (32) and (33) yields

$$\delta_K(x) \leqslant 9(RB)^{-2} \leqslant 2/3, \quad \delta_L(x) \leqslant 6(RB)^{-1} \quad \text{for } RB \geqslant 4 \tag{50}$$

if $S \in \mathcal{D}_0(R)$ and $|x| \leqq (2/3)RB^2$.

Next we proceed to estimates of quantiles.

T h e o r e m 4. If $S \in \mathcal{D}_0(R)$ and $RB \geqq 4$, then

$$F(x) = \Phi(a(x) + \delta(x) + O((RB)^{-2})) \tag{51}$$

for $|x| \leqq RB^2/2$, where

$$a(x) = x(2x^{-2}\lambda(x))^{1/2},\ a(0) = 0,$$
$$\delta(x) = a^{-1}(x)[\beta(x) + \ln a'(x)].$$

In particular,

$$F(x) = \Phi(x/B - (\Gamma/B^3)((x/B)^2 - 1)/6 + O(|x/B|^3 + 1)(RB)^{-2})), \tag{52}$$

$$F(x) = \Phi(x/B + O((x/B)^2 + 1)(RB)^{-1}). \tag{53}$$

Note that if $\xi, \xi_1, \xi_2, \ldots$ are independent and identically distributed, $M\xi = 0$, $D\xi = 1$, $M\xi^3 = \gamma$ and satisfy condition (40), then, as follows from Remark 2, for $n \geqq m_0(F)$, formulas (52) and (53) can be rewritten as follows:

$$\mathbf{P}\Big(\sum_{1\leqslant j\leqslant n} \xi_j < xn^{1/2}\Big) = \Phi(x - (\gamma/6)(x^2 - 1)n^{-1/2} + \theta C_5(\mathcal{F}) \tag{54}$$
$$(|x|^3 + 1)n^{-1}) = \Phi(x + \theta C_6(\mathcal{F})(x^2 + 1)n^{-1/2}) \quad \text{for} \quad |x| \leqslant \delta(\mathcal{F})n^{1/2}.$$

The last representation in (54) is proved in [5].

REMARK 3. Indeed, equality (53) and the last relation in (54) hold even without condition (49), i.e., without the smoothness assumption. Furthermore, in this case we have

$$F(x) = \Phi(a(x) + O((RB)^{-1})),$$

a stronger version of which is equality (51). This will be proved in a future work of this author.

Now let $S^{(1)}$ and $S^{(2)}$ denote two random variables with distribution functions $F^{(1)}$ and $F^{(2)}$, respectively, and let

$$\mathbf{M}S^{(j)}=0,\ \mathbf{D}S^{(j)}=B^2,\ \mathbf{M}(S^{(j)})^3=\Gamma\quad \forall j. \tag{55}$$

We emphasize the fact that in (54), B and Γ do not depend on j. We retain for $S^{(j)}$ all the notation introduced above for S, endowing them with an additional superscript j.

Theorem 5. If $S^{(1)}$ and $S^{(2)}$ belong to the class $\mathcal{D}_0(R)$ and condition (55) is satisfied, then for $|x| \leq RB^2/10$ and $RB \geq 6$ we have

$$F^{(1)}(x)=F^{(2)}(x+O(|x/B|^3+1)(RB)^{-1}). \tag{56}$$

Theorems 4 and 5 will be proved in Subsection 2.4. In Subsection 2.5 we will prove analogs of these assertions for conditional distributions, and in Subsection 2.6 we shall give simple properties of the classes $\mathcal{D}(R)$ and $\mathcal{D}_0(R)$ which we will need later on.

2.2 ESTIMATES FOR THE DENSITIES OF CONJUGATE DISTRIBUTIONS

For fixed h satisfying condition (1.3) we put

$$\begin{gathered}\eta_j=\xi_{j,h}/B(h),\ L=L^*(h),\ K=K^*(h),\ U=U^*(h),\\ G=\Sigma\mathbf{M}(\eta_j)^3,\\ Z=Z(h)\equiv(S(h)-A(h))/B(h)=\Sigma\eta_j.\end{gathered} \tag{1}$$

Let $f(t)$ denote the characteristic function of the random variable Z. We note that

$$|f(t)|\leqslant\exp(-t^2/3)\quad\text{for}\quad |t|\leqslant 1/(4L) \tag{2}$$

and, also, that $|f(t)| = |f(t/B(h),\ h)|$ and therefore

$$U=\int_{|t|\geqslant 1/(4L)}|f(t)|\,dt. \tag{3}$$

Lemma 1. For t satisfying the inequality

$$Kt^4\leqslant 1, \tag{4}$$

the following relation holds:

$$\mu(t)\equiv f(t)-(1-iGt^3/6)\exp(-t^2/2)=\theta K(t^4+t^6/72)\exp(-t^2/2). \tag{5}$$

Proof. By inequality (4), for the characteristic function $v = v_j(t)$ of the random variable $\eta = \eta_j$ the following well-known relations hold (see, for example, [10]):

$$|v-1| \leqslant D\eta t^2/2 \leqslant (\mathbf{M}\eta^4 t^4)^{1/2}/2 \leqslant 1/2, \tag{6}$$
$$v = 1 - D\eta t^2/2 - i\mathbf{M}\eta^3 t^3/6 + \theta \mathbf{M}\eta^4 t^4/24. \tag{7}$$

Expanding the logarithm in a Taylor series and using (6) and (7), we obtain

$$\begin{aligned} \ln v &= (v-1) + (1/2)(v-1)^2/(1+\theta(v-1))^2 \\ &= -D\eta t^2/2 - i\mathbf{M}\eta^3 t^3/6 + \theta(1/24 + 1/2)\mathbf{M}\eta^4 t^4. \end{aligned} \tag{8}$$

Putting $v = v_j$ in (8) and summing the resulting expressions over j, we obtain

$$\ln f(t) = \sum \ln v_j(t) = -t^2/2 - iGt^3/6 + \theta K(13/24)t^4. \tag{9}$$

It follows from the expansion of the exponential in a Taylor series that

$$\mu_1 \equiv \exp(-iGt^3/6) = 1 - iGt^3/6 + \theta(Gt^3/6)^2/2, \tag{10}$$
$$\mu_2 \equiv \exp \theta K(13/24)t^4 = 1 + \theta K(13/24)t^4 \exp K(13/24)t^4. \tag{11}$$

Now from formulas (9)-(11) and (4) we have

$$\begin{aligned} f(t)\exp(t^2/2) &= \mu_1\mu_2 = \mu_1(\mu_2 - 1) + \mu_1 \\ &= \theta K(13/24)t^4 \exp(13/24) - iGt^3/6 + \theta G^2 t^6/72 \end{aligned} \tag{12}$$

implying (5) since $G^2 \leqq L^2 \leqq K$ by (1.5). ∎

Let

$$\psi(t) = \exp(-t^2/2), \quad \gamma_k = \int |t|^k \psi(t)\,dt,$$
$$\mu_0(t) = f(t) - \psi(t)$$

and let $q(x)$ denote the distribution density of the random variable Z.

In that case, by the inversion formula for characteristic fucntions we have

$$q(x) - \varphi(x) = (2\pi)^{-1}\int e^{-itx}\mu_0(t)\,dt. \tag{13}$$

L e m m a 2.

$$|q(0) - \varphi(0)| \leqslant 12(2\pi)^{-1/2}K + U/2\pi.$$

Proof. By (13) and the oddness of the function $t^3\psi(t)$, we have for any $T > 0$

$$\begin{aligned} 2\pi(q(0) - \varphi(0)) &= \int \mu_0(t)\,dt = \int_{|t|\leqslant T} \mu(t)\,dt + \int_{|t|>T} f(t)\,dt \\ &\quad - \int_{|t|>T} \psi(t)\,dt \equiv I_1(T) + I_2(T) - I_3(T). \end{aligned} \tag{14}$$

For $T = K^{-1/n}$, it follows from (2) and (3) that

$$|I_2(T)| \leqslant I_4(T) + U, \tag{15}$$

where

$$I_4(T) = \int_{|t|>T} \exp(-t^2/3)\,dt \leqslant \int Kt^4\psi((2/3)^{1/2}t)\,dt = (3/2)^{5/2}K\gamma_4. \tag{16}$$

Lemma 1 implies that

$$|I_1(T)| \leqslant \int_{|t|\leqslant T} (Kt^4 + Kt^6/72)\,\psi(t)\,dt \leqslant (K\gamma_4 - I_3(T)) + K\gamma_6/72. \tag{17}$$

Summing (14)-(17), we obtain the required assertion, provided we note that

$$\gamma_3 = 4,\ \gamma_4 = 3(2\pi)^{1/2},\ \gamma_6 = 15(2\pi)^{1/2}. \quad \blacksquare \tag{18}$$

Lemma 3.

$$|q(x) - \varphi(x)| \leqslant (13L + U)/2\pi.$$

Proof. We make use of the estimate (see [12])

$$|\mu_0(t)| \leqslant \psi(t)[e^{\alpha(t/T)} - 1] \ \text{ for } |t| \leqslant T \equiv L^{-1/3}, \tag{19}$$

where for $0 \leqq x \leqq 1$

$$\alpha(x) = -x^2/2 + x^3/6 - \ln(1 - x^2/2) = x^3/6 + \sum_{k\geqslant 2} k^{-1}(x^2/2)^k \leqslant x^3\alpha(1) \leqslant 0.36x^3 \leqslant 0.36. \tag{20}$$

From (19) and (20) we obtain that for $|t| \leqq T$

$$|\mu_0(t)| \leqslant \psi(t)\cdot 0.36L|t|^3e^{0.36} \leqslant 0.52L|t|^3\psi(t). \tag{21}$$

Note that by formula (13)

$$(2\pi)|q(x) - \varphi(x)| \leqslant \int |\mu_0(t)|\,dt \leqslant I_0(T) + U + I_4(T) + I_3(T), \tag{22}$$

where

$$I_0(T) \equiv \int_{|t|\leqslant T} |\mu_0(t)|\,dt.$$

It follows from (21) and the inequality $1 \leqq L|t|^3$ for $|t| \geqq T$ that

$$I_0(T) + I_3(T) \leqslant L_3\gamma_3. \tag{23}$$

Next, for the given T, estimate (2) implies that

$$I_4(T) \leqslant \int L|t|^3\exp(-t^2/3)\,dt = (3/2)^2L\gamma_3\ , \tag{24}$$

the required assertion following from (13) and (22)-(24). ■

Note now that $p(y,h) = B^{-1}(h)\,q((y - A(h)/B(h))$ by (1). It follows from this and inequality (1.10) that

$$p(y) = B^{-1}(h)e^{\Lambda(h)-hx-h(y-x)}q(y - A(h)/B(h)). \qquad (25)$$

Substituting $h = H(x)$ and $y = x$ into (25) and estimating $q(0)$, using Lemma 2, we prove assertions (1.27) and (1.28) of Theorem 3. From Lemma 3 we have (1.17) and (1.19) in Theorem 2. Inequality (1.18) for the same $\delta_L(x)$ obviously follows for $h = H(x)$ and $y = x + z$ from (25), Lemma 3 and the inequality $\phi(u) \leqq \phi(0)$.

REMARK 4. Since $(1 - iGt^3/6)\,\psi(t) \equiv \psi(t,G)$ is the Fourier transformation of the function $(1 - (G/6)(x^3 - 3x))\,\phi(x) \equiv \phi(x,G)$, one can prove in the same way as Lemmas 2 and 3 the relation

$$|q(y) - \varphi(y, G)| \leqslant (2\pi)^{-1}\int |f(t) - \psi(t, G)|\,dt = O(K + U),$$

which together with (25) for $y = x + z$ and $h = H(x)$ imply the estimate

$$\begin{aligned} p(x+z) &= (2\pi)^{-1/2}e^{-\lambda(x)-\beta(x)-zH(x)} \\ &\times [\varphi(z/B(H(x)),\ B^{-3}(H(x))\Gamma(H(x))) + O(\delta_K(x))] \end{aligned} \qquad (26)$$

which may be more precise that (1.18). For $z = 0$ formulas (26) and (1.27) coincide.

2.3. ESTIMATES FOR FUNCTIONS CONNECTED WITH $H(x)$

In order to complete the proof of Theorems 2 and 3, we need several additional auxiliary assertions.

Applying the Taylor formula

$$f(x) = \sum_{0 \leqslant i < k} x^i f^{(i)}(0)/i! + x^k f^{(k)}(|\theta|x)/k! \qquad (1)$$

for $f(h) = A(h)$ and using (1.6), (1.7), (1.9) and (1.10), we have

L e m m a 1. If condition (1.3) is satisfied, we have

$$A(h) = B^2h + \theta h^2 L(R)/2, \qquad (2)$$

$$A(h) = B^2h + h^2\Gamma/2 + \theta h^3 K(R)/3. \qquad (3)$$

Note that by relation (1.5),

$$\Gamma = |\Lambda'''(0)| \leqslant BK^{1/2}(0) \leqslant BK^{1/2}(R). \qquad (4)$$

Now using (4) and making a routine estimate of the constants, from (2) and (3) we obtain

L e m m a 2. If condition (1.3) and either condition (1.20) or (1.29) are satisfied, then

$$|A(h) - hB^2| \leqslant |h| B^2/3. \tag{5}$$

The next lemma is the core of subsequent proofs.

L e m m a 3. If condition (1.21) and either condition (1.20) or (1.29) are satisfied, relations (1.22) and (1.16) hold.

Proof. Because of the symmetry, we consider only the case $x \geqq 0$ and $R > 0$. Since, if (1.21) is satisfied, $h = H_- = (3/4)xB^{-2}$ and $h = H_+ = (3/2)xB^{-2}$ satisfy (1.3), we can use the assertion of Lemma 2 and obtain, under the assumption made above, the inequalities

$$A(H_-) \leqslant (4/3)H_-B^2 = x = (2/3)H_+B^2 \leqslant A(H_+). \tag{6}$$

To complete proving the Lemma, we need to convince ourselves that due to the monotonicity of $A(h)$, relation (6) for $x \geqq 0$ coincides with relation (1.22). Relations (1.21) and (1.22) obviously imply (1.16). ∎

Note that by (1.14) and (1.15) we have the identity

$$\lambda'(x) = H(x).$$

Differentiating (1.14) over x, we obtain $A'(H(x))H'(x) = 1$. From this relation and (1.6), (1.15) and (1.7), we have

L e m m a 4. If condition (1.16) is satisfied, then

$$H'(x) = \lambda''(x) = B^{-2}(H(x)), \tag{8}$$

$$\lambda'''(x) = -\Gamma(H(x))B^{-6}(H(x)), \tag{9}$$

$$\beta'(x) = \Gamma(H(x))B^{-4}(H(x))/2. \tag{10}$$

Note that in deriving equalities (9) and (10), we have used the identity

$$(B^{2k}(H))' = kB^{2(k-2)}(H)\Gamma(H), \tag{11}$$

which follows from (1.6), (8) and from the rule of differentiating a composite function (in (11)-(14), the derivatives are taken with respect to x and the short notation $H = H(x)$ is used). From (11), (1.9) and Lemma 3 we obtain also

$$|(B^4(H))'| = 2|\Gamma(H)| \leqslant 2L(R). \tag{12}$$

If we use inequality (1.5) instead of (1.9) in the last argument, we have

$$|(B^3(H))'| = (3/2)B^{-1}(H)|\Gamma(H)| \leqslant (3/2)K^{1/2}(R). \qquad (13)$$

Finally, from (1.5), (1.6), (1.10), (11), and Lemma 3 one obtains the formula

$$(\Gamma(H)B^{-2m}(H))' = \Lambda^{IV}(H)B^{-2m-2}(H)$$
$$- m(\Gamma(H))^2B^{-2m-4}(H) = \theta(|m|+2)K(R)B^{-2m-2}(H). \qquad (14)$$

Integrating (12) and using (7) and Lemma 3, we obtain that

$$|B^4(H(x)) - B^4| \leqslant 2|x|L(R) \leqslant (2/3)B^4.$$

Similarly, integrating (13), we have

$$|B^3(H(x)) - B^3| \leqslant (3/2)|x|K^{1/2}(R) \leqslant (1/2)B^3.$$

From the last two relations and (8) we obtain

L e m m a 5. If conditions (1.20) and (1.21) are satisfied, then

$$1/3 \leqslant B^4(H(x))/B^4 \leqslant 5/3. \qquad (15)$$

However, if (1.21) and (1.29) are satisfied, then

$$B^{-2}(2/3)^{2/3} \leqslant \lambda''(x) = B^{-2}(H(x)) = e^{-2\beta(x)} \leqslant 2^{2/3}B^{-2}. \qquad (16)$$

Relation (1.5) and Lemmas 4 and 5 obviously imply the following lemma.

L e m m a 6. If conditions (1.21) and (1.29) are satisfied, the following inequalities hold:

$$|\lambda'''(x)| \leqslant 2^{5/3}K^{1/2}(R)B^{-5}, \quad |\beta'(x)| \leqslant K^{1/2}(R)B^{-3}.$$

From (9), (10), (14), (16) and the estimate $5\cdot 2^{8/3} \leqq 32$ we immediately obtain

L e m m a 7. If conditions (1.21) and (1.29) are satisfied, then

$$|\lambda^{IV}(x)| \leqslant 32K(R)B^{-8}, \quad |\beta''(x)| \leqslant 8K(R)B^{-6}.$$

Note that by (1.7), (1.14), (7), (9) and (10) the following equalities are obvious:

$$\lambda(0) = \lambda'(0) = H(0) = 0, \quad \lambda''(0) = B^{-2}, \qquad (17)$$
$$\lambda'''(0) = -\Gamma B^{-6}, \quad \beta'(0) = \Gamma B^{-4}/2. \qquad (18)$$

Substituting (17) and (18) and the assertions of Lemma 7 into (1), for $f(x) = \lambda^{(k)}(x)$ and $f(x) = \beta'(x)$ we obtain

L e m m a 8. If conditions (1.21) and (1.29) hold, then

$$\lambda'''(x) = -\Gamma B^{-6} + \theta 32 x K(R) B^{-8}, \tag{19}$$
$$\lambda''(x) = B^{-2} - x\Gamma B^{6} + \theta 16 x^2 K(R) B^{-8}, \tag{20}$$
$$\lambda'(x) = xB^{-2} - x^2\Gamma B^{-6}/2 + \theta 6 x^3 K(R) B^{-8}, \tag{21}$$
$$\beta'(x) = x\Gamma B^{-4}/2 + \theta 8 x K(R) B^{-6}. \tag{22}$$

To complete proving Theorems 2 and 3, we note that we have shown in Lemma 3 that conditions (1.16) and (1.22) are satisfied under conditions (1.21) and (1.20) or (1.29); and (1.23) follows from lemma 5 under the same conditions after a routine estimation of the constants. Next, from (1.16) and (1.5) for $H = H(x)$ we obtain

$$L^*(H) \leqslant L(R)B^{-3}(H), \quad K^*(H) \leqslant K(R)B^{-4}(H), \tag{23}$$
$$L^*(H) \leqslant K^{1/2}(R)B^{-2}(H), \quad T^{-1} \leqslant 4 \min\{L(R)B^{-2}(H),\ K^{1/2}(R)B^{-1}(H)\}.$$

If we now use (15), we obtain from (23) that

$$L^*(H) \leqslant 3^{3/4} L(R) B^{-3}, \quad T^{-1}(H) \leqslant 4 \cdot 3^{1/2} L(R) B^{-2}, \tag{24}$$

which together with (15) and (1.20) imply

$$T(H) \geqslant R/4, \quad U^*(H) \leqslant B(H)U(R) \leqslant (5/2)BU(R). \tag{25}$$

Substituting (24) and (25) into (1.19) and noting that $6 \cdot 3^{3/4} < 15$, we obtain (1.26). Similarly, from (23) and (16) we have

$$K^*(H) \leqslant 2^{4/3} K(R) B^{-4}, \quad L^*(H) \leqslant 2^{2/3} K^{1/2}(R) \tag{26}$$

and $T^{-1}(H) \leqq 4 \cdot 2^{1/3} K^{1/2}(R)$. Since the last inequality, (1.29) and (16) yield (25), we can now substitute (25) and (26) into (1.28) and (1.33) if only we note that $12 \cdot 2^{4/3} < 32$ but $6 \cdot 2^{2/4} < 10$.

To prove (1.30) and (1.31) one needs to use the Taylor formula (1) for $f(x) = \lambda(x)$ and $f(x) = \beta(x)$ and apply relations (17), (18) and Lemma 7. Similarly, from (1) and (17) we obtain (1.24) and (1.25) if we use the inequalities

$$|\lambda'''(x)| \leqslant 6L(R)B^{-6}, \quad |\beta'(x)| \leqslant 2L(R)B^{-4},$$

which follow immediately from (9), (10), (1.9) and (15) after we have estimated the constants.

Theorems 2 and 3 have been proved completely.

2.4. PROOFS OF THEOREMS 4 AND 5

First we prove some auxiliary assertions; note that Lemma 1 is of independent interest.

L e m m a 1. Suppose that the distribution function $F(x)$ has density $p(x)$ which for some $n \geq 1$ and for all z and x such that $|x| \leq n^{\frac{1}{2}}$, satisfies the conditions

$$p(x) = \varphi(0)e^{-\lambda(x)-\beta(x)}\rho(x), \quad \rho(x) = e^{O(1/n)}, \tag{1}$$

$$p(x+z) \leqslant \varphi(0)\exp(-\lambda(x) - \beta(x) - z\lambda'(x))\rho_+, \quad \rho_+ = O(1), \tag{2}$$

and for the x the following relations hold:

$$|\lambda^{IV}(x)| + |\beta''(x)| + (\lambda'''(0))^2 + (\beta'(0))^2 = O(n^{-1}), \tag{3}$$

$$\lambda(0) = \lambda'(0) = \beta(0) = 0, \quad \lambda''(0) = 1, \tag{4}$$

$$\lambda''(x) = e^{O(1)}. \tag{5}$$

Then for any fixed α, $0 < \alpha < 1$, we have for $|x| \leq \alpha n^{\frac{1}{2}}$

$$F(x) = \Phi(g(x) + O(n^{-1}(1+|x|)^{-1})), \tag{6}$$

where

$$\begin{aligned} g(x) &= a(x) + \delta(x),\\ \delta(x) &= a^{-1}(x)[\beta(x) + \ln a'(x)],\\ a(x) &\equiv x(2x^{-2}\lambda(x))^{1/2} = x + x^2\lambda'''(0)/6 + O(x^3/n)\\ &= x + O(x^2n^{-1/2}) = O(x), \end{aligned} \tag{7}$$

$$\delta(x) = \beta'(0) + \lambda'''(0)/3 + O(x/n) = O(n^{-1/2}). \tag{8}$$

As will follow from the proof, one can see that the absolute constants corresponding to O in (6)-(8) depend only on the absolute constants determining O in (1)-(5), as well as on α.

Almost all of this argument (up to Lemma 6) goes over to proof of Lemma 1; due to the symmetry, we consider only the case $x \geq 0$. We shall assume in the sequel, without stating otherwise, that $n \geq 1$, $|x| \leq n^{\frac{1}{2}}$ and also that conditions (1)-(5) of Lemma 1 hold. We need the following elementary lemma.

L e m m a 2. If $1 + \varepsilon = e^{O(1)}$ and $k = O(1)$, then

$$(1+\varepsilon)^k = 1 + k\varepsilon + O(\varepsilon) = 1 + O(\varepsilon) = O(1). \tag{9}$$

Note next that by (5) we have

$$\lambda'(x) = xe^{O(1)}, \quad 2\lambda(x) = x^2e^{O(1)}. \tag{10}$$

On the other hand, it follows from (3), (4) and the Taylor formula that

$$\lambda''(x) = 1 + x\lambda'''(0) + O(x^2 n^{-1}), \quad (11)$$
$$\lambda'(x) = x(1 + x\lambda'''(0)/2 + O(x^2 n^{-1})), \quad (12)$$
$$2\lambda(x) = x^2(1 + x\lambda'''(0)/3 + O(x^2 n^{-1})), \quad (13)$$
$$\beta'(x) = \beta'(0) + O(xn^{-1}), \quad (14)$$
$$\beta(x) = x\beta'(0) + O(x^2 n^{-1}). \quad (15)$$

The estimates (10) permit us to apply (9) for $\varepsilon = \lambda'(x)/x - 1$ and $\varepsilon = 2\lambda(x)/x^2 - 1$. Noting (12) and (13), we have for $k = O(1)$

$$|a(x)|^{2k} = (2\lambda(x))^k = x^{2k}(1 + kx\lambda'''(0)/3 + O(x^2 n^{-1})) = x^{2k}(1 + O(xn^{-1/2})), \quad (16)$$
$$(\lambda'(x))^{-1} = x^{-1}(1 - x\lambda'''(0)/2 + O(x^2 n^{-1})) = x^{-1}(1 + O(xn^{-1/2})). \quad (17)$$

Relation (7) is a special case of (16).

From (11), (12), (16) and (17), and taking into account (3), we obtain

$$(\ln a'(x))' = \lambda''(x)/\lambda'(x) - \lambda'(x)/(2\lambda(x)) = (1 + x\lambda'''(0))x^{-1}(1 - x\lambda'''(0)/2) - x(1 + x\lambda'''(0)/2) \times x^{-2}(1 - x\lambda'''(0)/3) + O(xn^{-1}) = \lambda'''(0)/3 + O(xn^{-1}). \quad (18)$$

Next, the Taylor formula implies that

$$\ln a'(x) = x\lambda'''(0)/3 + O(x^2 n^{-1}). \quad (19)$$

From (15), (16) and (19) we obtain (8).

Now we estimate $\varepsilon(x) \equiv \delta'(x)/a'(x)$. To this end, we use the representation

$$\varepsilon(x) = -a^{-1}(x)\delta(x) + (\lambda'(x))^{-1}(\beta'(x) + (\ln a'(x)))'. \quad (20)$$

Substituting the estimates (8), (14) and (18) into (20) and using (16) and (17), we obtain

$$\varepsilon(x) = x^{-1}(1 + O(xn^{-1/2}))(-\delta(x) + \beta'(x) + (\ln a'(x))') = O(n^{-1}). \quad (21)$$

Let

$$\Psi(x) = \Phi(g(x)), \quad \psi(x) = \Psi'(x). \quad (22)$$

We need now the following lemma.

L e m m a 3.

$$\psi(x) = p(x)(1 + O(n^{-1})), \quad p(x) = \varphi(g(x))e^{O(1)}. \quad (23)$$

Proof. We note that

$$\begin{aligned}\psi(x) &= \varphi(g(x))a'(x)(1+\varepsilon(x)) \\ &= \varphi(0)e^{-\lambda(x)-\beta(x)}(1+\varepsilon(x))\exp(-\delta^2(x)/2).\end{aligned} \tag{24}$$

The first relation in (23) follows from (8), (21) and (24). Using (1) and (8), we can rewrite (24) in the following form:

$$p(x) = a'(x)\varphi(g(x))\exp(O(n^{-1})). \tag{25}$$

From (25) we obtain the second equality in (23) if we use (10) and the relation $a'(x) = \lambda'(x)/(2\lambda(x))^{\frac{1}{2}}$ in order to obtain the estimate $a'(x) = e^{O(1)}$. ■

L e m m a 4. If $|x| \leqq \alpha n^{\frac{1}{2}}$, then

$$F(x) = \Psi(x) + p(x)O(n^{-1}(1+|x|)^{-1}). \tag{26}$$

Proof. We consider first the case $1 \leqq x \leqq \alpha n^{\frac{1}{2}}$ and use the inequality $\lambda'(x) \geqq x/C$, $0 < C = O(1)$ holding by (10). Noting (2), we have

$$p(x+z) = O(p(x)e^{-z/C}). \tag{27}$$

From (27) we obtain for the x and $R = n^{\frac{1}{2}}$ that

$$1 - F(R) = \int_{z>R-x} p(x+z)\,dz = O(p(x)e^{-(R-x)/C}) = O(p(x)/(nx)). \tag{28}$$

In deriving the last relation in (28) we used the inequalities

$$R - x \geqslant (1-\alpha)R, \quad e^{-(1-\alpha)R/C} = O(R^{-3}) = O(x^{-1}n^{-1}).$$

Similarly, the properties of normal distribution (23) and (27) imply

$$\begin{aligned}1 - \Psi(R) &= O(\varphi(g(R))) = O(p(R)) \\ &= O(p(x)e^{-(R-x)/C}) = O(p(x)/(xn)).\end{aligned} \tag{29}$$

It follows from (27)-(29) and the first relation in (23) that

$$\begin{aligned}|F(x) - \Psi(x)| &\leqslant \int_{x<z<R} |p(x+z) - \psi(x+z)|\,dz + (1 - F(R)) \\ &+ (1-\Psi(R)) = \int_{z>0} O(n^{-1}p(x)e^{-zx/C}) + O(p(x)/(nx)) \\ &= O(p(x)/(nx)).\end{aligned} \tag{30}$$

This estimate yields the validity of (26) for $x \geqq 1$. If $|x| \leqq 1$, then (26) follows from (30) and (23), since for these x we have

$$|F(x) - \Psi(x)| \leqslant \int_{-1<z<1} |p(z) - \psi(z)|\,dz + |F(1) - \Psi(1)| = O(1/n).$$

The case $x \leqq -1$ is easily reduced to $x \geqq 1$ by introducing the distribution function $1 - F(x+0)$.

We need the following lemma.

L e m m a 5. If z and $\tau \geqq 0$ satisfy the condition

$$|z|\tau + \tau^2 \leqslant (\ln 2)/2, \tag{31}$$

then

$$\Phi(z-2\tau) \leqslant \Phi(z) - \tau\varphi(z) \leqslant \Phi(z) + \tau\varphi(z) \leqslant \Phi(z+2\tau).$$

Proof. Note that if (31) is satisfied, we have

$$\varphi(z+2\theta\tau) = \varphi(z)\exp(-2\theta z\tau - 2\theta^2\tau^2) \geqslant \varphi(z)/2,$$

which plus the Taylor formula yield the required assertion, since

$$\Phi(z+2\tau) - \Phi(z) = 2\tau\varphi(z+2\theta\tau) \geqslant \tau\varphi(z),$$
$$\Phi(z-2\tau) - \Phi(z) = 2\tau\varphi(z-2\alpha\tau) \leqslant -\tau\varphi(z). \qquad \blacksquare$$

To complete proving Lemma 1, we rewrite (26) as

$$F(x) = \Phi(g(x)) + \varphi(g(x))\theta C_1 n^{-1}(1+|x|)^{-1} \tag{32}$$

and use relations (7) and (8):

$$|g(x)| \leqslant C_2(|x|+1). \tag{33}$$

Letting $\tau = C_1 n^{-1}(1+|x|)^{-1}$, $z = g(x)$ and noting (33), we obtain that these z and τ satisfy (31) for $n \geqq C_3 \equiv (2/\ln 2)C_1 \cdot (C_1+C_2)$. From (32) and Lemma 5 we have thus obtained the required assertion (6), however, under the additional assumption that $n \geqq C_3$.

We wish to get rid of this constraint. If

$$1 \leqslant n = O(1), \tag{34}$$

then (6) can be rewritten in the form

$$F(x) = \Phi(O(1)) \qquad \text{for } |x| \leqslant \alpha n^{1/2}. \tag{35}$$

Next, $p(x) = e^{O(n)}$ for $|x| \leqq n^{1/2}$ by (1). Integrating the density over the intervals $(\alpha n^{1/2}, n^{1/2})$ and $(-n^{-1/2}, -\alpha n^{-1/2})$, we obtain from the last relation that

$$1 - F(\alpha n^{1/2}) = e^{O(n)}, \quad F(-\alpha n^{1/2}) = e^{O(n)}. \tag{36}$$

To complete proving the Lemma, we need only to see that (36) is equivalent to (35) if (34) is satisfied.

We have thus proved completely relation (6). The remaining assertions of Lemma 1 were proved earlier, prior to lemma 3. We used the condition $\alpha < 1$ in deriving (28) and (29), and we can easily get rid of it if we assume that instead of (1) and (2) a condition of the form (2.26) is satisfied.

We note that the last relations in (1), (3) and (5) imply the existence of constants β^*, ρ_-, λ_-, λ_+ for which we have for $|x| \leqq n^{\frac{1}{2}}$

$$\rho(x) \geqslant \rho_- > 0, \quad \beta_- \leqslant e^{-\beta(x)} \leqslant \beta_+, \tag{37}$$

$$0 < \lambda_- \leqslant \lambda''(x) \leqslant \lambda_+ < \infty. \tag{38}$$

In particular, from (38), (7), (4) and the Taylor formula we obtain

$$|a(x)| = (2\lambda(x))^{1/2} = (x^2\lambda''(\theta x))^{1/2} \leqslant (\lambda_+)^{1/2}|x|, \tag{39}$$

$$a'(x) = \lambda'(x)/a(x) \geqslant \lambda_-/(\lambda_+)^{1/2} \equiv \lambda_0. \tag{40}$$

L e m m a 6. Let $F^{(1)}(x)$ and $F^{(2)}(x)$ be two distribution functions with densities $p^{(1)}(x)$ and $p^{(2)}(x)$. Suppose that each density, $p(x) = p^{(1)}(x)$ or $p(x) = p^{(2)}(x)$, for $|x| \leqq n^{\frac{1}{2}}$, satisfies (1)-(5) and (37), (38) for some $\beta(x) = \beta^{(j)}(x)$, $\lambda(x) = \lambda^{(j)}(x)$, $\rho(x) = \rho^{(j)}(x)$, $j = 1,2$, and the same β_+, β_-, λ_+, λ_-, ρ_+, ρ_- and n. Then if

$$|x|n^{-1/2} \leqslant \alpha^2\lambda^* \equiv (1/3)(\lambda_-/\lambda_+)\alpha^2,$$
$$n \geqslant N \equiv (3/2)(\lambda_+/\lambda_-)^{1/2}(\beta_+/\beta_-)(\rho_+/\rho_-)/\lambda_- \ , \tag{41}$$

we have

$$F^{(1)}(x) = F^{(2)}(x + O(g^{(1)}(x) - g^{(2)}(x)) + O(n^{-1})). \tag{42}$$

Proof. By Lemma 1

$$F^{(j)}(x) = \Phi(x + g^{(j)}(x) + \theta C_4/n) \qquad \text{for} \quad |x| \leqslant \alpha n^{1/2}. \tag{43}$$

We note that to prove the Lemma it suffices to choose a $0 \leqq \tau = \tau(x) = O(g^{(1)}(x) + g^{(2)}(x)) + O(n^{-1})$ such that

$$|x \pm \tau(x)| \leqslant \alpha n^{1/2} \qquad \text{for} \quad |x| \leqslant \lambda^*\lambda^2 n^{1/2}, \tag{44}$$

and the inequalities

$$g^{(2)}(x+\tau) - 2C_4/n \geqslant g^{(1)}(x) \geqslant g^{(2)}(x-\tau) + 2C_4/n \tag{45}$$

are satisfied, since in this case by (43)

$$F^{(2)}(x+\tau) \geqslant \Phi(g^{(2)}(x+\tau) - C_4/n) \geqslant \Phi(g^{(1)}(x) + C_4/n) \geqslant F^{(1)}(x)$$
$$\geqslant \Phi(g^{(1)}(x) - C_4/n) \geqslant \Phi(g^{(2)}(x-\tau) + C_4/n) \geqslant F^{(2)}(x-\tau)$$

yielding $F^{(1)}(x) = F^{(2)}(x+\tau\theta)$.

From (21) and (40), we have for $|x| \leqq n^{\frac{1}{2}}$

$$(g^{(2)}(x))' = (a^{(2)}(x))'(1+\varepsilon^2(x)) \geqslant \lambda_0(1 - C_5/n), \tag{46}$$

and from (8) and (39) we have for $|x| \leqq \lambda^*\alpha^2 n^{\frac{1}{2}}$

$$|g^{(j)}(x)| \leqslant (\lambda_+)^{1/2}|x| + C_6 n^{1/2} \leqslant (\lambda_+)^{1/2}\lambda^*\alpha^2 n^{1/2} + C_6 n^{-1/2}. \tag{47}$$

Let

$$\tau(x) = (|g^{(1)}(x) - g^{(2)}(x)| + 2C_4/n)(1 - C_5/n)^{-1}/\lambda_0 \tag{48}$$

for $n \geqq 2C_5$. By (47), Lemma 2 and the equality $2(\lambda_+)^{\frac{1}{2}}\lambda^* = \lambda_0$ we have

$$\tau(x) \leqslant (2(\lambda_+)^{1/2}\lambda^*\alpha^2 n^{1/2} + 2C_4/n^{-1/2})(1 + O(n^{-1}))/\lambda_0$$
$$= (2/3)\alpha^2 n^{1/2}(1 + O(n^{-1})) \leqslant (2/3)\alpha n^{1/2}(\alpha + C_7/n)$$

yielding (44) for $n \geqq C_7/(1-\alpha)$.

Next, it follows from (46) that if (44) is satisfied and $n \geqq 2C_5$, then

$$g^{(2)}(x-\tau) + \tau\lambda_0(1 - C_5/n) \leqslant g^{(2)}(x) \leqslant g^{(2)}(x+\tau) - \tau\lambda_0(1 - C_5/n). \tag{49}$$

Substituting (48) into (49), we obtain (45).

We have thus proved the Lemma, however, under two additional assumptions: $n \geqq 2C_5$ and $n \geqq C_7/(1-\alpha)$. To do away with these constraints, we note that under condition (34) Lemma 6 can be restated in the form

$$F^{(1)}(x) = F^{(2)}(O(1)) \quad \text{for } |x| \leqslant \lambda^*\alpha^2 n^{1/2}, \tag{50}$$

and to prove (50) we need only to show that

$$F^{(1)}(-\lambda^*\alpha^2 n^{1/2}) \geqslant F^{(2)}(-n^{1/2}), \quad F^{(1)}(\lambda^*\alpha^2 n^{1/2}) \leqslant F^{(2)}(n^{1/2}). \tag{51}$$

As an example, we prove the second assertion in (51). It follows from (38) that

$$(\lambda^{(2)}(n^{1/2}))' \geqslant \lambda_- n^{1/2}, \ \lambda^{(2)}(n^{1/2}) \geqslant \lambda_- n/2,$$

and hence by (2) and (37) we have the estimate

$$p^{(2)}(n^{1/2} + z) \leqslant C_+ \exp(-z\lambda_- n^{1/2}),$$

where

$$C_\pm = \rho_\pm \beta_\pm \exp(-\lambda_- n/2)\varphi(0).$$

From this inequality we obtain

$$1 - F^{(2)}(n^{1/2}) = \int_{z>0} p^{(2)}(n^{1/2} + z)\,dz \leqslant C_+ (\lambda_- n^{1/2})^{-1}. \tag{52}$$

On the other hand, (38) implies

$$\lambda^{(1)}(x) \leqslant \lambda_+ x^2/2 \leqslant \lambda_- n/2 \quad \text{for} \quad |x| \leqslant (\lambda_-/\lambda_+)^{1/2} n^{1/2} \equiv \gamma$$

which yields

$$1 - F^{(2)}(\lambda^* \alpha^2 n^{1/2}) \geqslant \int_{\gamma/3 \leqslant x \leqslant \gamma} p^{(1)}(x)\,dx \geqslant (\gamma - \gamma/3)\,C_- \tag{53}$$

since $3\lambda^*\alpha^2 < \gamma/n^{\frac{1}{2}}$. Next, (52) and (53) imply the second assertion in (50). The first assertion is proved in the similar way, if the interval $(\gamma/3, \gamma)$ in (53) is replaced by $(-\gamma, -\gamma/3)$.

L e m m a 7. If $S \in \mathcal{D}_0(R)$ for $B = 1$ and $R \geqq 4$, then conditions (1)-(5), (37) and (38) hold for

$$n^{1/2} \equiv (2/3)R, \quad \lambda^* = 3^{-5/3}. \tag{54}$$

Furthermore, $N \leqq 16 \leqq n$ for $R \geqq 6$.

Proof. From Lemma 3.5 we obtain that inequality (38) is satisfied for $\lambda_- = (2/3)^{2/3}$, $\lambda_+ = 2^{2/3}$, and hence (5) holds as well as the estimate for λ^* in (54). Lemma 3.7 and equalities (3.17) and (3.18) yield (3) and (4). The estimates (1.50) and Theorem 2 imply that (1) and (2) hold for $|x| \leqq (2/3)R \equiv n^{\frac{1}{2}}$. We also obtain that

$$\rho_- \equiv 3/4 \leqslant 1 - 9/R^2 \leqslant \rho(x) \leqslant 1 + 9/R^2 \leqslant 1 + 6/R \leqslant 2 \equiv \rho_+.$$

Relation (3.16) yields the estimate

$$\beta_- \equiv (2/3)^{1/3} \leqslant e^{-\beta(x)} \leqslant 2^{1/3} \equiv \beta_+.$$

Combining all the arguments, we obtain the assertion of the Lemma for the λ^* and n, as well as the inequality

$$N = (3/2)(3/2)^{2/3} \cdot 2 \cdot (4/3) \cdot 3^{1/3} < 16 \leqslant (2/3)^2 R^2 = n.$$

We proceed to prove Theorems 4 and 5. We note that we can restrict ourselves to the case $B = 1$ only since one can always have it by considering S/B instead of S. In this case, by Lemma 7, Theorem 4 follows from Lemma 1 for $\alpha = (1/2)/(2/3) < 1$ if we express $\beta'(0)$ and $\lambda'''(0)$ using formulas (3.18).

Theorem 5 for $B = 1$ is an immediate corollary of Lemmas 6 and 7 for the values of λ^* and n in (54) and

$$\alpha^2 = (1/10)\,/\,((2/3)\lambda^*) = 3^{8/3}/20 < 27/28 < 1.$$

To show this, it suffices to see that in this case

$$g^{(1)}(x) - g^{(2)}(x) = O(|x|^3 + 1)n^{-1} . \tag{55}$$

But the last relation follows from (7), and (8), since, if (1.55) is satisfied, we have by (3.18) that

$$\lambda^{(j)''}(0) = -\Gamma, \quad \beta^{(j)'}(0) = \Gamma/2 \quad \text{for all } j .$$

Substituting (55) into (42), we obtain (1.56), as required.

2.5. ESTIMATES FOR THE QUANTILES OF CONDITIONAL DISTRIBUTIONS

Let S_1 and S_2 be two independent random variables of class $\mathcal{D}_0(R)$ and let $S_0 = S_1 + S_2$. In this case the density $p(x|y)$ of the conditional distribution of S_1 at a point x, under the condition that $S_0 = S_1 + S_2 = y$, is obviously equal to

$$p(x|y) = p_1(x)p_2(y-x)/p_0(y), \tag{1}$$

where $p_i(x)$ is the density of the random variable S_i. (Here and below, if we do not mention otherwise, the i runs through the values 0,1,2.) Let $F(x|y) = \int_{z<x} p(z|y)\,dz$ denote the conditional distribution function of the random variable S_1 under the condition $S_0 = y$. Our objective in this subsection consists in obtaining estimates for the function $Q(x|y)$, which is a solution of the equation

$$F(x|y) = \Phi(Q(x|y)).$$

We retain for the S_i all the notation introduced before for the variable S, but we shall use an additional subscript i (the notation without this subscript will be used for other variables). Let

$$0 < B_i \equiv (\mathbf{D}S_i)^{1/2} < \infty, \quad B \equiv B_1B_2/B_0. \tag{2}$$

From Lemma 4 (below), which is more precise, although more cumbersome, we have the following

T h e o r e m 6. Under the above assumptions and for

$$|u| \leqslant RB^2/6 , \quad |y| \leqslant RB^2/6 , \quad RB \geqslant 4 \tag{4}$$

we have the equalities

$$F(u|y) = \Phi(v + g(v,\ y/B_0)/B_0 + O(|u|^3 + |y|^3 + B^3)B^{-3}(RB)^{-2}), \tag{4}$$

$$F(u|y) = \Phi(v + O(u^2 + y^2 + B^2)B^{-2}(RB)^{-1}), \tag{5}$$

where

$$\begin{gathered} v = (u - \alpha_1 y)/B, \quad \alpha_i = (B_i/B_0)^2, \quad \Gamma_{i,k} = \Gamma_i B_i^k, \\ g(v,\ w) = -w^2 G_2 - vwG_0(B/B_0)/2 - v^2 G_6/6 - G_6/3 + G_4/2, \\ G_0 = (\Gamma_{1,4} + \Gamma_{2,4})B^2, \quad G_2 = (\Gamma_{1,2} - \Gamma_{2,2})(B/B_0)^2, \\ G_4 = (\Gamma_{1,4} - \Gamma_{2,4})B^2, \quad G_6 = (\Gamma_{1,6} - \Gamma_{2,6})B^4. \end{gathered} \tag{6}$$

Now, let $(S_1^{(1)}, S_2^{(1)}, S_0^{(1)})$ and $(S_1^{(2)}, S_2^{(2)}, S_0^{(2)})$ be two triples of random variables, $S_0^{(j)} = S_1^{(j)} + S_2^{(j)}$ $\forall j$ and

$$0 < \mathbf{D}S_i^{(1)} = B_i^2 = \mathbf{D}S_i^{(2)} < \infty, \quad \mathbf{M}\left(S_i^{(1)}\right)^3 = \Gamma_i = \mathbf{M}\left(S_i^{(2)}\right)^3 \quad \forall i. \tag{7}$$

We retain for the $S_i^{(j)}$ all the notation which was and will be introduced for the S_i, but we endow them with the additional superscript (j). We do assume of course that for all j, the $S_1^{(j)}$ and $S_2^{(j)}$ are independent and belong to class $\mathcal{D}_0(R)$.

T h e o r e m 7. Under the assumptions we have made, we have the equality

$$F^{(1)}(u|y^{(1)}) = F^{(2)}(u + \alpha_1(y^{(2)} - y^{(1)}) + G(u)/2 + O(\varepsilon(u))|y^{(2)}) \tag{8}$$

only if

$$|u| \leqslant RB^2/36, \quad |y^{(1)}| \leqslant RB^2/36, \quad |y^{(2)}| \leqslant RB^2/36, \quad RB \geqslant 12, \tag{9}$$

where

$$\begin{gathered} \varepsilon(u) = (|u|^3 + |y^{(1)}|^3 + |y^{(2)}|^3 + B^3)/(B^3(RB)), \qquad G(u) = [(y^{(2)}/B_0)^2 \\ - (y^{(1)}/B_0)^2]G_2 + (u - \alpha_1 y^{(1)})[(y^{(2)}/B_0) - (y^{(1)}/B_0)]G_0/B_0, \end{gathered} \tag{10}$$

also

$$|G(u)| \leqslant |y^{(1)} - y^{(2)}|(|u| + 2|y^{(1)}| + |y^{(2)}|)(B_0)^{-2}(2R)^{-1}. \tag{11}$$

REMARK 5. Let $\xi, \xi_1, \xi_2, \ldots$ be a sequence of independent identically distributed random variables satisfying the conditions: $\mathsf{M}\xi = 0$, $\mathsf{D}\xi = 1$, $\mathsf{M}\xi^3 = \gamma$ and (1.44). Let $S_1 = \sum_{1\leq j\leq m} \xi_j$, $S_2 = \sum_{m+1\leq j\leq 2m} \xi_j$. As follows from Remark 2, in this case we can find $m_0(F) < \infty$ and $R(F) > 0$ such that $S_i = S_{i,m} \in \mathcal{D}(R(F))$ for $m \geq m_0(F)$. Therefore, for these m and R all the conditions of Theorem 6 are satisfied. We can then rewrite (5) as

$$F(u|y) = \Phi((u - y/2)(m/2)^{-1/2} + \theta C_1(\mathcal{F})(u^2 + y^2 + m)m^{-3/2}). \tag{12}$$

Since now $G_k = 0$ for $k \neq 0$ and $B^2 = m/2$, then (4) and (6) become

$$F(u|y) = \Phi(v - \gamma vy/(4m) + \theta C(\mathcal{F})(|u|^3 + |y|^3 - m^{3/2})m^{-5/2}),$$

where $v = (u - y/2)(m/2)^{-1/2}$.

Relation (12) has been derived in [5].

Note that if $\{\xi_k^{(j)}\}$ are two sequences satisfying the conditions of Remark 5, then for $y = y^{(1)} = y^{(2)}$ relation (10) has an especially simple form.

REMARK 6. Let the conditions of Theorem 6 hold. Introduce two independent variables $S_1(h)$ and $S_2(h)$ having distributions which are the Cramér transformations of the distribution of the variables S_1 and S_2 for the same h. Let $p_h(x|y)$ denote the conditional distribution density of the $S_1(h)$ under the condition that $S_0(h) = S_1(h) + S_2(h) = y$. By (1), (1.13) and the equality $\Lambda_0(h) = \Lambda_1(h) + \Lambda_2(h)$ we have

$$p(x|y) = p_h(x|y). \tag{13}$$

Substituting $h = H$ into (13), where H is the solution of

$$\mathbf{M}S_0(H) = A_0(H) = y, \tag{14}$$

we reduce Theorem 6 to the special case when $y = 0$ but for the random variables $S_i(H)$.

It is easier to prove Theorem 6 for $y = 0$. However, the problem then arises as to how to reduce the solution obtained for $S_i(H)$ to a solution for S_i.

This technique does not make the proof of Theorem 5 shorter; it does, nevertheless, make the methodology of the proof more expository. But we shall not take this approach, because it involves more elaborate notation than the proof given below does.

We proceed to prove Theorem 6. We can always examine the random variables RS_i instead of the random variables S_i. Therefore, we shall assume without loss of generality that $R \equiv 1$, without mentioning this fact again. Since the condition $S \in \mathcal{D}(1)$ is more rigid than (1.29) for $R = \pm 1$, one can use the results of Subsection 2.3 for $R = \pm 1$ for $S = S_1$ and $S = S_2$, as we shall be doing frequently in what follows.

Note the relations

$$B \leqslant B_i \leqslant B_0, \quad 2^{1/2} B \leqslant B_0 \leqslant 2^{1/2} \max_{i=1,2} B_i, \tag{15}$$

$$B_1^{-2} + B_2^{-2} = B^{-2}, \quad B_1^2 + B_2^2 = B_0^2 \tag{16}$$

which follow from (2) and the equality $\mathcal{D}S_0 = \mathcal{D}S_1 + \mathcal{D}S_2$. Furthermore, we also have the inequalities

$$|\Gamma_{i,k}| \leqslant B_i^{2-k}/2 \leqslant B^{2-k} \quad \forall k \geqslant 2, \tag{17}$$

$$|K_i(\pm 1)| \leqslant B_i^2/4. \tag{18}$$

Indeed, for $i = 1$ and $i = 2$, these inequalities follow from the condition $S_i \in \mathcal{D}(1)$ and (3.4), and for $i = 0$ from the identities $\Gamma_0 = \Gamma_1 + \Gamma_2$, $K_0(h) = K_1(h) + K_2(h)$ and (16). In particular, from (15)-(17) we obtain the estimate

$$|G_k| \leqslant 1/2 \quad \forall k. \tag{19}$$

Since for the characteristic functions of the variables $S_i(h)$ we have the inequality

$$|f(t,h)| = |f_1(t,h) f_2(t,h)| \leqslant \min_{i=1,2} |f_i(t,h)|,$$

we have from the condition $S_i \in \mathcal{D}(1)$ and the definition of $U(R)$ that

$$U_0(\pm 1) \leqslant \min_{i=1,2} U_i(\pm 1) \leqslant (B_0 2^{-1/2})^3. \tag{20}$$

In obtaining the last estimate we have also used the second relation in (15).

From Theorem 2, formulas (1.50) and (15) for $S_i \in \mathcal{D}(1)$ we obtain

$$p_i(x_i) = \varphi(0) \exp(-\lambda_i(x_i) - \beta_i(x_i))(1 + 90B^{-2}), \tag{21}$$

$$p_i(x_i + z_i) \leqslant \varphi(0) \exp(-\lambda_i(x_i) - \beta_i(x_i) - z_i H_i(x_i))(1 + 6/B) \tag{22}$$

for $|x_i| \leq (2/3) B_i^2$ and all z_i, $i = 1,2$. Since (18) is satisfied for $i = 0$ and guarantees (1.29) in this case, we can use the assertion of Theorem 3 for $S = S_0$ and $R = 1$. We have

$$p_0(y) = \varphi(0) \exp(-\lambda_0(y) - \beta_0(y)) \rho_0(y) \qquad \text{for } |y| \leqslant (2/3) B_0^2, \tag{23}$$

and by (1.27), (1.32), (18) and (20)

$$|\rho_0(y) - 1| \leqslant \delta_{0,K}(y) \leqslant \max\{B_0 U_0(\pm 1) + 32 B_0^{-4} K_0(\pm 1)\}$$
$$\leqslant B_0 2^{3/2} B_0^{-3} + 8B_0^2/B_0^4 \leqslant 12 B_0^{-2} \leqslant 3/4$$

since $B_0 \geqq B \geqq 4$. This chain of inequalities show that

$$p(x) = \sigma p(a + \sigma x | y) = \sigma p_1(a_1 + x\sigma) p_2(a_2 - x\sigma)/p_0(a_0), \tag{24}$$

only if y satisfies the constraint in (23).

For some a and $\sigma > 0$, we introduce the random variable $S = (S_1 - a)/\sigma$ and denote by $p(x)$ the conditional distribution density of S under the condition that $S_0 = y$. In this case, by (1)

$$\rho_0(y) = \exp(O(B^{-2})), \tag{25}$$

where

$$a_1 = a, \quad a_2 = y - a, \quad a_0 = y. \tag{26}$$

We shall denote by $F(x)$ a distribution function corresponding to the density $p(x)$. By (25),

$$F(u) = F(a + \sigma u | y) . \tag{27}$$

Note that if the condition

$$|a_i| + \delta |x| \leqslant (2/3) B_i^2, \quad i = 1, 2, \tag{28}$$

is satisfied, then from (16) and (26) we have $|y| = |a_0| \leqq |a_1| + |a_2| \leqq (2/3) B_0^2$. Thus, when (28) is satisfied, we can use (21), (22) and (23) for $x_1 = a_1 + \sigma x$, $x_2 = a_2 - \sigma x$, $z_1 = \sigma z$, $z_2 = -\sigma z$. Substituting (21)-(23) into (25), we obtain

$$p(x) = \varphi(x) e^{-\lambda(x) - \beta(x)} \rho(x), \tag{29}$$

$$p(x + z) \leqslant \varphi(0) \exp(-\lambda(x) - \beta(x) - z\lambda'(x)) \rho_+ \tag{30}$$

for

$$\lambda(x) = \lambda_1(a_1 + x\sigma) + \lambda_2(a_2 - x\sigma) - \lambda_0(a_0), \tag{31}$$

$$\beta_0(x) = \beta_1(a_1 + x\sigma) + \beta_2(a_2 - x\sigma) - \beta_0(a_0) - \ln \sigma, \tag{32}$$

$$\rho_+ = (1 + 6/B)^2 / \rho_0(y), \tag{33}$$

$$\rho_- = (1 - 9B^{-2})^2 / \rho_0(y) = \rho(x) \leqslant (1 + 9B^{-2})^2 / \rho_0(y). \tag{34}$$

In particular, from (24) and (34) it follows that

$$\rho(x) = \exp(O(B^{-2})) \quad \text{for} \quad B \geqslant 4 . \tag{35}$$

On the other hand, it follows from (32) and (3.16) for $S = S_1$ and $S = S_2$ that

$$(2/3)^{2/3} \leqslant \exp(-\beta(x) - \beta_0(a_0) - \ln \sigma) \leqslant 2^{2/3}. \tag{36}$$

Note that in deriving (30) from (22), we have used the equality $\lambda'(x) = \sigma H_1(a_1 + x\sigma) - \sigma H_2(a_2 - x\sigma)$, which obtains in a simple way from (31) and (3.7). Moreover, by differentiating (31) and (32) k times, we find that

$$\lambda^{(k)}(x) = \sigma^k\left[\lambda_1^{(k)}(a_1 + x\sigma) + (-1)^k \lambda_2^{(k)}(a_2 - x\sigma)\right], \tag{37}$$

$$\beta^{(k)}(x) = \sigma^k\left[\beta_1^{(k)}(a_1 + x\sigma) + (-1)^k \beta_2^{(k)}(a_2 - x\sigma)\right]. \tag{38}$$

Subsituting estimate (3.16) for $S = S_i$ and $S = S_2$ into (37) for $k = 2$ and noting (16) yields

$$\lambda_- \equiv \sigma^2 B^{-2}(2/3)^{2/3} \leqslant \lambda''(x) \leqslant \sigma^2 B^{-2} 2^{2/3} \equiv \lambda_+. \tag{39}$$

Next, applying Lemma 3.7 for $S = S_1$ and $S = S_2$ and noting inequalities (15) and (18), from (37) for $k = 4$ we obtain

$$\lambda^{IV}(x) = \sigma^4 O(B^{-6}), \tag{40}$$

and from (38) for $k = 2$ we obtain

$$\beta''(x) = \sigma^2 O(B^{-4}). \tag{41}$$

Similary, from Lemma 3.6 and from (37) for $k = 2$ we have

$$\lambda'''(x) = \sigma^3 O(B^{-4}), \tag{42}$$

while it follows from (38) for $k = 1$, that

$$\beta'(x) = \sigma O(B^{-2}). \tag{43}$$

We have thus proved the following

L e m m a 1. If condition (28) is satisfied, relations (29)-(43) hold.

Now, let us choose the numbers a and σ and, therefore, also a_1 and a_2 in (26), in a special way. Let $H \equiv H_0(y)$ be the solution of equation (14). Set

$$a_1 = a \equiv A_1(H), \quad a_2 = y - a, \\ \sigma \equiv B(H) \equiv B_1(H)B_2(H)/B_0(H). \tag{44}$$

By (14) and the equality $\Lambda_0(h) = \Lambda_1(h) + \Lambda_2(h)$, we have

$$a_i = A_i(H), \quad \lambda(0) = 0. \tag{45}$$

Since $H_i(\cdot)$ is the inverse of $A_i(\cdot)$, we have

$$H_i(a_i) = H \quad \forall i. \tag{46}$$

In particular, it follows from (46), (3.7) and (37) for $k = 1$ that

$$\lambda'(0) = \sigma(H - H) = 0. \tag{47}$$

From (44), (46) and (3.8), by analogy with equalities (16), we have

$$\sigma^{-2} = B^{-2}(H) = B_1^{-2}(H) + B_2^{-2}(H) = \lambda_1''(a_1) + \lambda_2''(a_2). \tag{48}$$

In particular, from (48), (16) and (3.16) for $S = S_1$ and $S = S_2$ we obtain

$$(2/3)^{2/3}B^{-2} \leqslant \sigma^{-2} \leqslant 2^{2/3}B^{-2}. \tag{49}$$

We shall need the following

L e m m a 2. If $|y| \leqq (2/3)B_0^2$, then

$$|a_i| \leqslant 2|y|(B_i/B_0)^2. \tag{50}$$

Proof. Since for the y and $R = 1$, for $S = S_0$ condition (1.29) is satisfied by (18), Lemma 3.3 and (1.22) imply that

$$|H| \leqslant (3/2)|y|B_0^{-2}. \tag{51}$$

Since $|H| \leqq 1$, Lemma 3.2 for $R = 1$ and $S = S_i$ implies

$$|A_i(H)| \leqslant (4/3)|H|B_i^2.$$

Substituting the first resulting relation into the second one and using (45), we obtain (50).

L e m m a 3. If

$$|y| \leqslant B_0^2/11, \quad n^{1/2} \equiv (6/11)B^2/\sigma, \quad B \geqslant 4, \tag{52}$$

the density $p(x)$ satisfies all the conditions of Lemma 4.1. Moreover, in this case, relations (4.37) and (4.38) hold for

$$\lambda_+/\lambda_- = \beta_+/\beta_- = 3^{2/3}, \quad \rho_+/\rho_- = (1 + 6/B)^2(1 - 9B^{-2})^{-2}. \tag{53}$$

Proof. First of all note that by (50) and (52), the condition $|x| \leqq n^{1/2}$ is sufficient for (28) and that $n \geqq B^2/6 > 1$ for $B \geqq 4$, as follows from (49). In this case, by Lemma 1, the fact that conditions (4.1) and (4.2) are satisfied follows from (29), (30) and (35). Substituting (49) into (40)-(43) yields

$$|\lambda^{IV}(x)| + |\beta''(x)| = O(B^{-2}),$$
$$|\lambda'''(x)| + |\beta'(x)| = O(B^{-1}),$$

and, in turn, (4.3), in our case. Similarly, relations (39) and (49) imply (4.5). In order to verify (4.4), we need to use equalities (45), (47) and note that $\lambda''(0) = 1$ by virtue of (48) and (37) for $k = 2$, and $\beta(0) = 0$, since (45) and (46) yield the equality $\beta_i(a_i) = \ln B_i(H)$.

Thus, all conditions (4.1)-(4.5) are indeed satisfied. Relations (53) follow from (39), (36) and, respectively, (33) and (34). ■

By Lemma 2 and relations (15),

$$|x| \leqslant (|u| + 2|y|)/\sigma \quad \text{for} \quad x = (u-a)/\sigma \qquad (54)$$

only if $|y| \leqq (2/3)B_0^2$. We, obviously, obtain from this inequality and (15) that

$$|y| \leqslant \alpha(2/11)B^2, \quad |u| \leqslant \alpha(2/11)B^2, \quad 0 < \alpha < 1, \qquad (55)$$

is a sufficient condition for the following estimates:

$$|y| \leqslant \alpha B_0^2/11, \quad |x| \leqslant \alpha(6/11)B^2/\sigma, \quad 0 < \alpha < 1. \qquad (56)$$

This fact, (27) and Lemmas 3 and 4.1 imply the following assertion, which surpasses Theorem 6 in accuracy.

L e m m a 4. If (55) or (56) are satisfied for fixed α, then for $x = (u-a)/\sigma$ we have the equality

$$F(x) = F(a + x\sigma|y) = \Phi(x + g(x) + O(B^{-2}(1 + |x|)^{-1})), \qquad (57)$$

where $g(x)$ is expressed in terms of $\lambda(x)$ and $\beta(x)$ (see (31) and (32)) by the formulas given in Lemma 4.1. In particular,

$$g(x) = x^2\lambda'''(0)/\sigma + \lambda'''(0)/3 + \beta'(0) + O(|x|^3 + 1)B^{-2}. \qquad (58)$$

In order to derive Theorem 6, it remains only to express $\lambda'''(0)$, $\beta'(0)$ and x in (57) and (58) in terms of Γ_i, y and $u = a + x\sigma$. For this we need several auxiliary assertions, which we shall prove. In the sequel, we assume without mentioning this fact, that condition (3) is satisfied and furthermore, that $R = 1$. To avoid cumbersome expressions, we put $b = B_0$ and use the notation α_i, $\Gamma_{i,k}$, G_k and v, introduced in the statement of Theorem 6.

L e m m a 5.

$$a = \alpha_1 y + y^2 b^{-2} G_2/2 + O(y^3 b^{-4}), \qquad (59)$$

$$a_i = \alpha_i y + O(y^2 b^{-2}) = O(y), \quad i = 1, 2. \qquad (60)$$

Proof. It follows from inequalities (3) and (51) that $|H| \leqq 1$. Hence we can use Lemma 3.1 for $R = 1$ and $S = S_1$. We find from (3.3) (18) and (15) that

$$a = \alpha_1 b^2 H + H^2\Gamma_1/2 + O(H^3 b^2). \tag{61}$$

By virtue of (3) and (3.19) for $S = S_0$ and $x = y$ we have

$$H \equiv H_0(y) = yb^{-2} - y^2\Gamma_{0.6}/2 + O(y^3 b^{-5}). \tag{62}$$

It follows from formulas (3), (17) and (62) that

$$H = yb^{-2} + O(y^2 b^{-4}) = O(yb^{-2}),$$

and therefore

$$H^2 = y^2 b^{-4} + O(y^3 b^{-6}).$$

Substituting (62) and the last two relations into (61) and noting relations (15) and (17) for $k = 2$, we obtain

$$a = \alpha_1 y - \alpha_1 y^2\Gamma_{0.4}/2 + y^2\Gamma_1 b^{-4}/2 + O(y^3 b^{-4}). \tag{63}$$

To reduce (63) to the form (59), it suffices to use the identities (16) as well as $\Gamma_0 = \Gamma_1 + \Gamma_2$, from which we have $\Gamma_1 - \alpha_1\Gamma_0 = G_2 b^2$. We get (60) from (59) if we use inequalities (3), (15) and (19). ■

L e m m a 6. If $k = O(1)$, then

$$\sigma^k = B^k(1 + (k/2)yb^{-2}G_0 + O(y^2 B^{-4})), \tag{64}$$
$$\sigma^k = B^k(1 + O(yB^{-2})) = O(B^k). \tag{65}$$

Proof. Substituting the expansions (3.20) into (48) for $S = S_1$ and $x = a_i$ and noting (15), (16) and (18), we have

$$\sigma^{-2} = B^{-2} - a_1\Gamma_{1.6} - a_2\Gamma_{2.6} + O(|a_1|^2 + |a_2|^2)B^{-6}. \tag{66}$$

Estimating a_i in (66) by formula (60) and using (15), (17) and (19), we find that

$$B^2\sigma^{-2} = 1 - yb^{-2}G_0 + O(y^2 B^{-4}) = 1 + O(yB^{-2}). \tag{67}$$

By virtue of inequality (49) we can apply Lemma 4.2 for $1 + \varepsilon = B^2\sigma^{-2}$ and $m = -k/2$, which together with (67) yield (64) and (65). ■

L e m m a 7.

$$\lambda'''(0) = -G_6/B + O(yB^{-3}), \tag{68}$$
$$\beta'(0) = G_4/(2B) + O(yB^{-3}). \tag{69}$$

Proof. We use equality (37) for k = 3 and (3.19) for $S = S_i$ and $a = a_i$. Noting (18) and (15), we have

$$\lambda'''(0) = \sigma^3[-\Gamma_{1,6} + \Gamma_{2,6} + O(|a_1| + |a_2|)B^{-6}]. \tag{70}$$

Next, substituting (60) and (65) into (70) and using (17) yields (68).

Similarly, it follows from (38) for k = 1 and (3.22) that

$$\beta'(0) = \sigma[\Gamma_{1,4}/2 - \Gamma_{2,4}/2 + O(|a_1| + |a_2|)B^{-4}], \tag{71}$$

and we have again used (18) and (15). Applying (60) and (65), we derive (69) from (71).

L e m m a 8.

$$x = v - vyb^{-2}G_0/2 - y^2b^{-2}B^{-1}G_2 + O(|u|y^2 + |y|^3)B^{-5}, \tag{72}$$

$$x^k = v^k + O(|u|^k|y| + |y|^{k+1})B^{-2-k}, \quad k = 1,\ 2. \tag{73}$$

Proof. It follows from (59) that

$$x \equiv (u-a)/\sigma = (B/\sigma)(v - y^2b^{-2}B^{-1}G_2/2 + O(y^3B^{-5})).$$

Substituting here (64) and (65), we obtain (72) and (73) for k = 1 provided we use inequalities (15), (19) and

$$B|v| = |u - \alpha_1 y| \leqslant |u| + |y|. \tag{74}$$

By conditions (3), the case k = 2 in (73) reduces to k = 1. ∎

From relations (57), (73) and Lemma 7, we have

$$g(x) = -v^2G_6/(6B) - G_6/(3B) + G_4/(2B) + O(|u|^3 + |y|^3 + B^3)B^{-5}. \tag{75}$$

In estimating the remainder in the last relation, we have also used (19), (74) and the elementary inequality $|c|d^2 \leqq |c|^3 + |d|^3$.

Note that conditions (3) ensure that conditions (55) be satisfied for $\alpha = 11/21 < 1$. Thus (57) holds under the conditions of Theorem 6. Substituting (72) and (75) into (57), we obtain the required relations (4) and (6), which together with (3), (19) and (74) yield the simpler estimate (5).

Theorem 6 is thus completely proved. Observe now that by estimating the constants in (53), we see from Lemmas 3 and 4.6 that the following lemma holds.

L e m m a 9. If $|y| \leqq (2/11)B^2$ and $B \geqq 12$, the density p(x) satisfies (4.1)-(4.5), (4.37) and (4.38) for

$$n = (6/11)^2B^4/\sigma^2, \quad \lambda^* = 3^{-5/3}.$$

Moreover, in this case $N < (1/4)B^4/\sigma^2 < n$.

In deriving this assertion form Lemma 3, we have also used relations (15) and (39).

We proceed to prove Theorem 7. We equip the above-given variables $p(x)$, $F(x)$, $g(x)$ and σ with a superscript to indicate which triple of the random variables they belong to in a given situation. In this case it follows from Lemmas 9, 4.6 that for fixed α, $0 < \alpha < 1$, and

$$\mu(x) = O(g^{(1)}(x) - g^{(2)}(x)) + O(B^{-2})$$

we have the equality

$$F^{(1)}(x) = F^{(2)}(x + \mu(x)) \tag{76}$$

for

$$|y| \leqslant (2/11)B^2, \quad |x| \leqslant \lambda^* \alpha^2 (6/11) B^2/\sigma^{(1)}. \tag{77}$$

We immediately obtain from (54) that condition (9) is sufficient for (77) to be satisfied for $\alpha^2 = (3/36)/(\lambda^*(6/11)) < 23/24 < 1$. Setting $x = (u-a^{(1)})/\sigma^{(1)}$ in (76) and using representation (27), we see that we have proved the following

L e m m a 10. If conditions (9) are satisfied, for $x = (u-a^{(1)})/\sigma^{(1)}$ we have the relation

$$F^{(1)}(u|y^{(1)}) = F^{(2)}(u - a^{(1)} + a^{(2)} + x(\sigma^{(2)} - \sigma^{(1)}) + \sigma^{(2)}\mu(x)|y^{(2)}). \tag{78}$$

Since G_k and x do not depend on (j), by virtue of (7) it follows from (75) that

$$\mu(x) = O(\varepsilon(u)/B). \tag{79}$$

Next we find from Lemmas 5, 6 and 8 respectively,

$$a^{(2)} - a^{(1)} = V_a + O(|y^{(1)}|^3 + |y^{(2)}|^3)/B^4, \tag{80}$$

$$\sigma^{(2)} - \sigma^{(1)} = BV_\sigma + O(|y^{(1)}|^2 + |y^{(2)}|^2)/B^3 = O(|y^{(1)}| + |y^{(2)}|)/B^2, \tag{81}$$

$$x = (u - \alpha_1 y^{(1)})/B + O(|u|^2 + |y^{(1)}|^2)/B^3, \tag{82}$$

where

$$V_a = \alpha_1(y^{(2)} - y^{(1)}) + [(y^{(2)})^2 - (y^{(1)})^2]b^{-2}G_2/2,$$

$$V_\sigma = B(y^{(2)} - y^{(1)})b^{-2}G_0/2.$$

Substituting (79)-(82) into (78) yields

$$F^{(1)}(u|y^{(1)}) = F^{(2)}(u + V_a + (u - \alpha_1 y^{(1)})V_\sigma + O(\varepsilon(u))|y^{(2)}).$$

Writing this relation in detailed form we see that it is equivalent to (8) and (10). Since (11) follows immediately from (10) and (19), Theorem 7 is completely proved.

2.6. SOME PROPERTIES OF CLASSES $\mathcal{D}$ AND $\mathcal{D}_0$

We retain all the notation and assumptions used in Subsection 2.1. Simple criteria for assigning random variables to class $D(R)$ may be derived from the following assertion.

L e m m a 1. Inequalities (1.11) and (1.12) hold. Moreover,

$$K(\pm R) \leqslant 16 \sum \mathbf{M} |\xi_j|^4 \exp(|R\xi_j|), \tag{1}$$

$$K(\pm R) \leqslant 16R^{-1} \sum \mathbf{M} |\xi_j|^3 \exp(2|R\xi_j|). \tag{2}$$

Proof. To prove (1.11) and (1.12), it suffices to apply successively, for $\xi = \xi_j$ and $k = 3,4$, the known relations:

$$\mathbf{M}|\xi(h) - \mathbf{M}\xi(h)|^k \leqslant 2^k \mathbf{M}|\xi(h)|^k,$$

$$\mathbf{M}|\xi(h)|^h = \mathbf{M}|\xi|^k e^{h\xi}/\mathbf{M}e^{h\xi},$$

$$\mathbf{M}e^{h\xi} \geqslant e^{h\mathbf{M}\xi} \geqslant 1,$$

$$\max_{0 \leqslant h/R \leqslant 1} e^{h\xi} = \max\{1, e^{R\xi}\}.$$

Relation (1) follows from (1.12), and (2) from (1) and from the inequality $|R\xi| \leqq e^{|R\xi|}$. ■

A very simple test for belonging to class $\mathcal{D}(R)$ is given by

L e m m a 2. If $M\xi_j = 0$ for all j and

$$64R^2 \mathbf{M}|\xi_j|^4 \exp(|R\xi_j|) \leqslant \mathbf{D}\xi_j \quad \forall j \tag{3}$$

or

$$\mathbf{P}(|\xi_j| \leqslant 1/(9R)) = 1 \quad \forall j, \tag{4}$$

we have $S = \xi_1 + \cdots + \xi_m \in \mathcal{D}(R)$.

Proof. Note that (3) follows from (4), since in this case the inequality

$$64R^2\xi^4 e^{|R\xi|} \leqslant (8/9)^2 \xi^2 e^{1/9} \leqslant \xi^2$$

holds for $\xi = \xi_j$. On the other hand, if (3) is satisfied, the estimate $4K(\pm R) \leqq D\xi_1 + \cdots + D\xi_m$ follows from (1), thus completing the proof. ■

Since $\Lambda(h) = \int_{0 \leqq t \leqq 1} hA(th)\,dt$, Lemma 3.2 implies

L e m m a 3. If $S \in \mathcal{D}(R)$, we have

$$\ln \mathbf{M}e^{hS} = (1 + \theta/3)h^2 \mathbf{D}S/2 \quad \text{for} \quad |h| \leqslant R.$$

In the sequel, we shall write $K(R,S)$ instead of $K(R)$, to indicate the particular random variable the notion $K(R)$ refers to. In a similar way, we introduce the variables $f(t,h,S)$ and $U(R,S)$. The definitions of the $K(R)$ and $U(R)$ obviously imply

L e m m a 4. If S_1 and S_2 are independent random variables belonging to $\mathcal{D}(R)$, then $S_1 + S_2 \in \mathcal{D}(R)$. Moreover,

$$K(R,\ S_1+S_2)=K(R,\ S_1)+K(R,\ S_2), \tag{5}$$

and if $U(R,S_i) < \infty$ for all i, then

$$U(R,\ S_1+S_2)\leqslant \min_{i=1,2} U(R,\ S_i). \tag{6}$$

Introduce classes $\mathcal{D} \equiv \mathcal{D}(1)$ and $\mathcal{D}_0 \equiv \mathcal{D}_0(1)$. Note that if $S \in \mathcal{D}(R)$, then $RS \in \mathcal{D}$.

A simple test for belonging to $\mathcal{D}_0$ is given by

L e m m a 5. If the random variable S belongs to the class $\mathcal{D}$ and does not depend on the variable S_0 having normal distribution with zero mean, then $S + S_0 \in \mathcal{D}$ for

$$\sigma^2 \equiv \mathbf{D}S_0 \geqslant 8+48|\ln \mathbf{D}(S+S_0)|. \tag{7}$$

Proof. By the properties of normal distribution, we have

$$K(h,\ S_0)=0,\quad |f(t,\ h,\ S_0)|=\exp(-t^2\sigma^2/2)\ \ \forall h. \tag{8}$$

It follows from (5), the first relation in (8) and (1.48) that $S + S_0 \in \mathcal{D}(R)$, whereas from (6) and the second equality in (8) that

$$U(\pm 1)\equiv U(\pm 1,\ S+S_0)\leqslant U(\pm 1,\ S_0)\leqslant \int_{|t|\geqslant 1/4} \exp(-t^2\sigma^2/2)\,dt.$$

From this relation, (7) and the inequalities $\sigma^2 \geqq 8$ and $1-\Phi(x) \leqq x^{-1}\phi(x)$ for $x > 0$, we obtain

$$U(\pm 1)\leqslant 8\sigma^{-2}\exp(-\sigma^2/32)\leqslant \exp(-\sigma^2/32).$$

The last two estimates immediately yield the inequality

$$(\mathbf{D}(S+S_0))^{3/2}U(\pm 1)\leqslant (\mathbf{D}(S+S_0))^{3/2}\exp(-\sigma^2/32).$$

Substituting here σ^2 satisfying (7), we see that (1.49) is satisfied and hence $S + S_0 \in \mathcal{D}_0$. ∎

The next lemma is similar.

L e m m a 6. If the random variable S belongs to class $\mathcal{D}$ and does not depend on the sequence $\nu_1, \nu_2, \ldots$ of independent random variables having uniform distributions on the interval $[-1/9, 1/9]$, then $S + Z_m \in \mathcal{D}_0$ for

$$m \geqslant C_0(1 + |\ln \mathbf{D}(S + Z_m)|), \tag{9}$$

where $Z_m = \nu_1 + \cdots + \nu_m$ and C_0 is an absolute constant.

Proof. It is clear that the bounded random variables $\nu(h) \equiv \nu_1(h)$ (see (1.4)) has for any h a bounded (by the constant $(9/2)e^{|h/9|}$), and hence square-integrable density. Thus, by Parseval's equality, its characteristic function is square integrable, and hence condition (1.44) is satisfied for $k = 2$. Therefore, by Remark 2, relation (1.47) holds, which in this case can be rewritten as

$$U(\pm 1, Z_m) \leqslant C_1 e^{-cm}, \tag{10}$$

where $C_1 < \infty$ and $c > 0$ are absolute (since the distribution of ν is fixed) constants. If we now use inequality (6) for $S_1 = S$ and $S_2 = Z_m$, we have

$$U(R) \equiv U(R, S + Z_m) \leqslant U(R, Z_m). \tag{11}$$

From (10) and (11) we obtain

$$(\mathbf{D}(S + Z_m))^{3/2} U(\pm 1) \leqslant (\mathbf{D}(S + Z_m))^{3/2} C_1 e^{-cm} \leqslant 1 \tag{12}$$

for $m \geqq (3/(2c)) \ln \mathcal{D}(S + Z_m) + c^{-1} \ln C_1$. This shows that for $C_0 = \max\{3/(2c), c^{-1}|\ln C_1|\}$, condition (7) guarantees (12) and hence (1.49).

To complete proving the lemma, we need only to verify that $S + Z_m \in \mathcal{D}$. But this follows from Lemma 3 since $\nu_j \in \mathcal{D}$ by (4).

Now let us prove the following assertion converse, in some sense, to Lemma 4.

L e m m a 7. If the variable $S \in \mathcal{D}(2R)$, $\mathcal{D}S \geqq 2n \geqq 2$ and $R \geqq 1$, it can be represented as a sum $S = S_0 + S_1 + \cdots + S_n$ of independent random variables $S_0, S_1, \ldots, S_n$ so that

$$S_i \in \mathscr{D}(R) \quad \forall i, \quad 1 \leqslant \mathbf{D}S_i \leqslant 5/4 \quad \forall i \neq 0. \tag{13}$$

Proof. If $S \in \mathcal{D}(2R)$, then by (1.48), $\max\limits_{|h| \leqq 2} 16R^2 \times K(h,S) \leqq \mathbf{D}S$. Therefore, one can represent S as a sum of independent random variables $\xi_1, \ldots, \xi_m$ satisfying the condition

$$12 \max_{|h| \leqslant 2R} R^2 \sum \mathbf{M} |\xi_{j,h}|^4 \leqslant \sum \mathbf{D}\xi_j = \mathbf{D}S. \tag{14}$$

(We use the notation $\xi_{j,h}$ from Subsection 2.1, omitting again the limits of summation 1 and m.)

Now we decompose the random variables $\{\xi_j\}$ into two groups by setting

$$J = \left\{j\colon 4R^2 \max_{|h|<2} \mathbf{M}\,|\,\xi_{j,h}\,|^4 \leqslant \mathbf{D}\xi_j\right\}.$$

It is easy to see that in this case, by (14), one has

$$\sum_{j \not\equiv J} \mathbf{D}\xi_j \leqslant 4R^2 \sum \mathbf{M}\,|\,\xi_{j,h}\,| \leqslant \sum \mathbf{D}\xi_j/3. \tag{15}$$

Note that the definition of J and the inequality $M|\xi_{j,0}|^4 = M|\xi_j|^4 \geq D^2\xi_j$ imply

$$\xi_j \in \mathscr{D}(R), \quad \mathbf{D}\xi_j \leqslant 1/(4R^2) \leqslant 1/4 \quad \forall j \in J. \tag{16}$$

By (15),

$$\sum_{j \in J} \mathbf{D}\xi_j \geqslant (2/3) \sum \mathbf{D}\xi_j \geqslant (2/3)\, 2n > (5/4)\, n. \tag{17}$$

It follows immediately from (16) and (17) that it is possible to decompose the set J into disjoint subsets $J(0), J(1), \ldots, J(n)$ so that

$$1 \leqslant \sum_{j \in J(i)} \mathbf{D}\xi_j \leqslant 5/4 \quad \forall j \neq 0. \tag{18}$$

We now put

$$S_0 = \sum_{j \not\equiv J} \xi_j + \sum_{j \equiv J(0)} \xi_j, \quad S_i = \sum_{j \in J(i)} \xi_j \quad \forall i \neq 0. \tag{19}$$

In this case, formulas (18) and (19) ensure that the second of the relations in (13) is satisfied, whereas (16) and Lemma 4 yield the first condition in (13) for $i \neq 0$. We need only to show that $S_0 \in \mathcal{D}(R)$. It follows from (17)-(19) that $D(S - S_0) \leq (5/4)n \leq (2/3)DS$ and hence $DS \geq DS/3$. The last inequality and (14) imply

$$4R^2K(\pm 2R, S_0) \leqslant 4R^2K(\pm 2R, S) \leqslant \mathbf{D}S/3 \leqslant \mathbf{D}S_0,$$

and this fully ensures that $S_0 \in \mathcal{D}$. ∎

In the next assertion, we construct smooth and bounded distributions close to certain distributions from class $\mathcal{D}$.

L e m m a 8. If $S \in \mathcal{D}(8)$ and $3/4 \leq DS \leq 5/4$, there exists a random variable W such that

$$W \in \mathscr{D}, \quad \mathbf{D}W = \mathbf{D}S, \quad \mathbf{M}W^3 = \mathbf{M}S^3, \quad |W| = O(1) \quad \text{a.s.} \tag{20}$$

and W can be represented in the form $W = Z + \nu$, where Z and ν

are independent and ν has uniform distribution on $(-1/9, 1/9)$.

Proof. We assume $\mathrm{D}S = B^2$, $\mathrm{M}S^3 = \Gamma$ and note that by (3.4) and the conditions of the Lemma, one has

$$3/4 \leqslant B^2 \leqslant 5/4, \quad |\Gamma| \leqslant B^2/2R = B^2/16. \tag{21}$$

Let

$$n = 5\cdot 2^8, \quad \sigma^2 = (\mathrm{D}S - \mathrm{D}\nu) = (B^2 - 3^{-5}),$$

$$a_\pm = (\Gamma^2\sigma^{-4}/4 + \sigma^2/n)^{\frac{1}{2}} \pm \Gamma\sigma^{-2}/2 .$$

Furthermore, we consider the sequence $\xi, \xi_1, \ldots, \xi_n$ of independent identically distributed random variables taking on only two values:

$$\mathbf{P}(\xi = a_+) = a_-/(a_+ + a_-), \quad \mathbf{P}(\xi = a_-) = a_+/(a_+ + a_-).$$

In this case,

$$\mathbf{M}\xi = 0, \quad \mathbf{D}\xi = \sigma^2/n, \quad \mathbf{M}\xi^3 = \Gamma/n. \tag{22}$$

Elementary calculations using (21) yield

$$\begin{aligned} |a_\pm| &\leqslant |\Gamma|\sigma^{-2} + \sigma n^{-1/2} \\ &\leqslant B^2(B^2 - 3^{-5})^{-1}/16 + Bn^{-1/2} \\ &\leqslant 5/(4\cdot 16) + (5/(4n))^{1/2} < 1/9. \end{aligned} \tag{23}$$

Now, setting $Z = \xi_1 + \cdots + \xi_n$, we obtain from (22) that

$$\mathbf{D}W = \mathbf{D}S, \quad \mathbf{M}W^3 = \Gamma,$$

and from (23) that $W = O(1)$ and $W \in \mathcal{D}$ by Lemma 2. ■

The following assertion is a corollary of Lemmas 6 and 8.

L e m m a 9. Let $S, W_1, \ldots, W_m$ be a sequence of independent random variables with $S \in \mathcal{D}$, and let each of the variables $W = W_j$ satisfy all the conditions of Lemma 8. Then $S_m \equiv S + W_1 + \cdots + W_m \in \mathcal{D}_0$, only if

$$m \geqslant C_0(1 + |\ln \mathbf{D}S_m|),$$

where $C_0 < \infty$ is an absolute constant.

3. Estimates in the invariance principle under the existence of exponential moments

We shall describe the technique for constructing random variables on a common probability space, mentioned in the Introduction, in Subsection 3.5, as well as an important auxiliary result obtained with the aid of this technique in Subsection 3.6. Since this technique is the refined Komlós-Major-Tusnády method, we shall elaborate on the essence of it in Subsection 3.3. Finally, we apply it to deriving an estimate in the invariance principle for smooth distributions in Subsection 3.4. In Subsections 3.3. and 3.4, the notion of a quantile transformation is essentially used -- we briefly review it in Subsection 3.1. In Subsection 3.2 we give estimates for the transformations necessary for proving the results obtained in Subsections 3.4 and 3.6. Theorem 1 of the Introduction is a corollary of a more general assertion, which will be proved in Subsection 3.7 by combining the results of Subsection 3.4 and Subsection 3.6.

3.1. QUANTILE TRANSFORMATIONS

Let U be some random variable with continuous distribution function F_U and let F_V be an arbitrary fixed distribution function. We define the random variable V to be (any) solution of the equation

$$F_V(V) = F_U(U). \tag{1}$$

It is well known (and not hard to verify directly) that the random variable V has the F_V as its distribution function and $F_U(U)$ is uniformly distributed on $[0,1]$. Following the discussion of [5], we call the variable V the quantile projection of U.

Let us consider now a more complex construction. Suppose that it is required to construct two independent random variables V_1 and V_2 with *a priori* given distributions. We know the joint distribution of $V_0 = V_1 + V_2$ and the (conditional) distribution function $F_V(\cdot|y)$ of the random variable V_1 under the condition $V_0 = y$.

Suppose in addition that a random variable V_0 with the required distribution has already been constructed, such that it does not depend on some variable U with continuous distribution function F_U. In this case we define V_1 to be (any) solution of the equation

$$F_V(V_1|V_0) = F_U(U). \tag{2}$$

Since $F_U(U)$ is uniformly distributed on $[0,1]$, for the fixed value of $V_0 = y$, the random variable V_1 has $F_V(\cdot|y)$ as its distribution function. Therefore, by the total probability formula, the V_1 and

V_0 defined above have the desired joint distribution. In particular, V_1 and $V_2 \equiv V_0 - V_1$ are independent random variables with the given distributions.

Next, if U_1 and U_2 are two independent random variables having distribution densities (with respect to Lebesgue measure), the conditional distribution function $F_U(\cdot|y)$ of the random variable U_1 is a continuous distribution function under the condition $U_0 \equiv U_1 + U_2 = y$ and we have

$$F_U(x|y) = \int_{z<x} (p_1(z)\, p_2(y-z)/p_0(y))\, dz,$$

where $p_i(\cdot)$ is the density of U_i, $i = 0,1,2$.

Consider an even more complex construction. Suppose we are given a random vector W not depending on the independent random variables U_1 and U_2 having distribution density with respect to Lebesgue measure. Furthermore, it is required to construct two independent random variables V_1 and V_2 with given distributions. Assume that $V_0 \equiv V_1 + V_2$ has been constructed, such that it is a function only of W and $U_0 \equiv U_1 + U_2$. In this case, for the fixed values W and $U_0 = y$, the U_1 has the above-defined function $F_u(\cdot|y)$ as its (continuous) distribution function, and hence the quantity $U' = F_U(U_1|U_2)$ for any fixed value U_0 has uniform distribution on $[0,1]$. Thus, U' does not depend on W and U_0 and therefore not on V_0 either. Hence, if for some continuous distribution function F_U we define the random variable U as (any) solution of the equation $F_U(U) = U'$ (of course, $U = U'$ is the simplest case), then the assumptions for the simpler construction of the conditional quantile transformation are satisfied.

Summarizing, we immediately obtain from equality (2) that under the assumptions made in the preceding paragraph, the random variable V_1 defined as (any) solution of the equation

$$F_V(V_1|V_0) = F_U(U) = F_U(U_1|U_0), \tag{3}$$

and the random variable $V_2 = V_0 - V_1$ have joint distribution. We say that the random variable V_1 is the conditional quantile transformation of the random variable U_1 with respect to U_0 and V_0.

Note that if V_1 is the conditional quantile transformation of U_1 with respect to U_0 and V_0, then, by construction, $V_2 = V_0 - V_1$ is the conditional quantile transformation of $U_2 = U_0 - U_1$ with respect to the same U_0 and V_0.

Thus, the quantile transformation enables us to construct random variables with given distributions and the conditional quantile transformation makes it possible to determine pairs of independent random variables with the required distributions when their sum has been constructed.

Note that if we wish to have a representation of the form $U = G(V)$ for a random variable V constructed by formula (1), we need only to prove that $F_V(x) = F_U(G(x))$. Similarly, for a random variable V_1 given by formula (3), it is necessary to derive the identity $F_V(x|y_1) = F_U(G(x, y_1y_2)|y_2)$. The required representations for the functions G have been obtained in Theorems 4-7 for particular cases of significance. In Subsection 3.2 we shall give estimates for quantile transformations -- for convenience of making reference.

3.2. ESTIMATES FOR QUANTILE TRANSFORMATIONS

In Subsection 2.1 we introduced the notions of classes of distributions $\mathcal{D}(R)$ and $\mathcal{D}_0(R) < \mathcal{D}(R)$ for $R > 0$. As follows from the results of Subsection 2.6, if $S = \xi_1 + \cdots + \xi_m$ where the $\xi_1,\ldots,\xi_m$ are independent and have zero means, then for the variable S to belong to $\mathcal{D}(R)$ it is sufficient that the following condition hold:

$$0 < 64R^2 \sum_{1\leqslant j\leqslant m} \mathbf{M}|\xi_j|^3 \exp(2R|\xi_j|) \leqslant \sum_{1\leqslant j\leqslant m} \mathbf{D}\xi_j < \infty. \tag{1}$$

(Note that condition (1) is close, in some sense, to the necessary condition for $S \in \mathcal{D}(R)$ to hold.) Concerning the class of smooth distributions $\mathcal{D}_0(R)$, we emphasize the fact that if $S \in \mathcal{D}(R)$, then by adding to S a normally distributed S-independent random variable with variance of order $\ln \mathbf{D}S$, we obtain a variable of class $\mathcal{D}_0(R)$.

Let us point out that if $S \in \mathcal{D}(R)$, then $\mathbf{M}S = 0$, $0 < \mathcal{D}S < \infty$, and that a random variable having nonsingular normal distribution with zero mean belongs to $\mathcal{D}(R)$ for all R. Recall that if $S \in \mathcal{D}(R)$ (or $S \in \mathcal{D}_0(R)$), then $RS \in \mathcal{D}(1) \equiv \mathcal{D}$ (or, respectively, $RS \in \mathcal{D}_0(1) \equiv \mathcal{D}_0$) and that $\mathcal{D}(R) \subset \mathcal{D}$ and $\mathcal{D}_0(R) \subset \mathcal{D}_0$ for $R \geq 1$.

Lemmas 1-4 follow, respectively, from Theorems 4-7 and the remarks at the end of the preceding subsection.

Let U and V be random variables satisfying the conditions:

$$\mathbf{D}U = B^2 = \mathbf{D}V, \quad U, V \in \mathcal{D}_0. \tag{2}$$

L e m m a 1. If the random variable V is the quantile transformation of the variable U having normal distribution, and, furthermore, condition (2) is satisfied, then

$$U - V = O(1 + |V|B)^2 \tag{3}$$

for

$$|V| \leqq B^2/2 \qquad \text{and} \qquad B \geqq 4 . \tag{4}$$

L e m m a 2. If the random variable V is the quantile transformation of the variable U satisfying condition (2) as well as the condition

$$\mathbf{M}V^3 = \mathbf{M}U^3, \tag{5}$$

we have

$$U - V = O(1 + |V/B|^3)/B \tag{6}$$

for

$$|V| \leqslant B^2/10 \qquad \text{and} \qquad B \geqslant 6. \tag{7}$$

Now, suppose we are given two triples $U_1, U_2, U_0 \equiv U_1 + U_2$ and $V_1, V_2, V_0 \equiv V_1 + V_2$ of random variables, each pair (U_1, U_2) and (V_1, V_2) consisting of independent variables. Assume that the conditions

$$\mathbf{D}U_i = \mathbf{D}V_i \quad \forall i, \; U_1, U_2, V_1, V_2 \in \mathcal{D}_0, \tag{8}$$

are satisfied. Also, let

$$\alpha_i = \mathbf{D}V_i/\mathbf{D}V_0, \quad B^2 = \mathbf{D}V_1 \cdot \mathbf{D}V_2/\mathbf{D}V_0.$$

L e m m a 3. Let U_1 and U_2 have normal distributions and let conditions (8) be satisfied. Then, if V_1 is the conditional quantile transformation of the variable U_1 with respect to U_0 and V_0, we have

$$V_1 - U_1 = \alpha_1(V_0 - U_0) + O(1 + |V_1/B|^2 + |V_0/B|^2) \tag{9}$$

for

$$|V_1| \leqslant B^2/6, \quad |V_0| \leqslant B^2/6, \quad B \geqslant 4. \tag{10}$$

L e m m a 4. Assume that condition (8) and the condition

$$\mathbf{M}(V_i)^3 = \mathbf{M}(U_i)^3 \quad \text{for} \quad i = 1, 2 \tag{11}$$

are satisfied. Then, if V_1 is the conditional quantile transformation of the variable U_1 with respect to U_0 and V_0, then we have

$$V_1 - U_1 = \alpha_1(V_0 - U_0) + g(V_1, V_0, U_0) + O(\varepsilon(V_1, V_0, U_0)) \tag{12}$$

for

$$|V_1| \leqslant B^2/36, \quad |V_0| \leqslant B^2/36, \quad |U_0| \leqslant B^2/36, \quad B \geqslant 12, \tag{13}$$

where

$$|g(V_1, V_0, U_0)| \leqslant |V_0 - U_0|(|V_1| + 2|V_0| + |U_0|)/(4B^2), \tag{14}$$

$$\varepsilon(V_1, V_0, U_0) = (1 + |V_1/B|^3 + |V_0/B|^3 + |U_0/B|^3)/B. \tag{15}$$

3.3. THE KOMLÓS-MAJOR-TUSNÁDY METHOD FOR SMOOTH DISTRIBUTIONS

First we adopt a notation. If we are given a sequence of random variables, e.g., $T_1,\ldots,T_n$, denoted by the same letter (perhaps with superscripts) and numbered starting with one, we write

$$T(l) = \sum_{j<l} T_j, \quad l = 0, \ldots, n. \tag{1}$$

If we are given two such sequences of the same length, e.g., $T_1,\ldots,T_n$ and $S_1,\ldots,S_n$, we write

$$\Delta(T, S) = \max_{l<n} |T(l) - S(l)|. \tag{2}$$

However, if the sequence $T_1,\ldots,T_n$ has a number of terms which is a power of 2, e.g., $n = 2^N$, we write

$$T(i, m) = T(i2^{N-m}) - T((i-2)2^{N-m}) \tag{3}$$

and assume that i ranges from 1 to 2^m, and m from 0 to N. Note that in this case the following useful equality holds:

$$T(i, m) = T(2i-1, m+1) + T(2i, m+1). \tag{4}$$

Next, let $Y_1,\ldots,Y_n$, $n = 2^N$, be a sequence of independent random variables having distribution densities (with respect to Lebesgue measure). Furthermore, it is required to construct a sequence $X_1,\ldots,X_n$ of independent random variables with given distributions (hence we also know the distributions of the $X(j,m)$). The Komlós-Major-Tusnády method is the following. For each $m = 0,1,\ldots,N$ we construct a sequence $\mathcal{X}_m = \{X(1,m), \ldots, X(2^m,m)\}$ of independent random variables having the required distributions and being functions of the random variables $\mathcal{Y}_m = \{Y(1,m), \ldots, Y(2^m,m)\}$ only. In this case, for $m = N$ we obtain the required sequence $\{X_j\} = \mathcal{X}_N$.

For $m = 0$, we define the required random variable $V = X(1,0)$, using the quantile transformation of the random variable $U = Y(1,0)$. If for some $m < N$ we have already constructed the required sequence $\mathcal{X}_m$, then we construct $\mathcal{X}_{m+1}$ by defining for each $i = 1,\ldots,2^m$ the random variable $V_1 = X(2i-1, m+1)$ as the quantile transformation of $U_1 = Y(2i-1, m+1)$ conditional with respect to the variables $V_0 = X(i,m)$ and $U_0 = Y(i,m)$ and extending the definition of $V_2 \equiv X(2i, m+1) = V_0 - V_1$ according to (4).

We show next that the sequence $\mathcal{X}_{m+1}$ constructed is the required one. Note that by construction each pair $X_{i,m} = \{X(2i-1, m+1), X(2i, m+1)\}$ has the required joint distribution and is a function of the variables $Y(2i-1, m+1)$, $Y(i,m)$ and $X(i,m)$ only, therefore $\mathcal{X}_{m+1}$ is a function of $\mathcal{Y}_{m+1}$. We use $F(x,i)$ to denote the *a priori* known distribution function of the random variable $X(i, m+1)$ and introduce the events $A_i = \{X(i, m+1) < x_i\}$ for certain x_i. It remains only to prove that for all x_i one has

$$\mathbf{P}\Big(\bigcap_i A_i\Big) = \prod_i \mathbf{P}(A_i) \equiv \prod_i F(x_i, i), \tag{5}$$

i.e., that the $\mathcal{X}_m$ have the required joint distribution.

By properties of the conditional quantile transformation, the following equalities hold:

$$\mathbf{P}(A_{2i-1}A_{2i} \mid \mathcal{Y}_m) = \mathbf{P}(A_{2i-1}A_{2i} \mid X(i, m)), \tag{6}$$

$$\mathbf{P}(A_{2i}A_{2i-1}) = F(x_{2i-1}, 2i-1)F(x_{2i}, 2i). \tag{7}$$

Since for fixed $\mathcal{Y}_m$ the random variables $Y(2i-1, m+1)$ are (conditionally) independent, the pairs $X_{i,m}$ constructed above are also (conditionally) independent, i.e.,

$$\mathbf{P}\Big(\prod_i A_i\Big) = \mathbf{MP}\Big(\prod_i (A_{2i-1}A_{2i}) \mid \mathcal{Y}_m\Big) = \mathbf{M}\prod_i \mathbf{P}(A_{2i-1}A_{2i} \mid \mathcal{Y}_m). \tag{8}$$

Substituting (6) into (8) and using the independence of $X(1,m), \ldots, X(2^m,m)$ yields

$$\mathbf{P}\Big(\prod_i A_i\Big) = \prod_i \mathbf{P}(A_{2i-1}A_{2i}).$$

The required assertion (5) follows from (9) and (7).

We have thus proved the correctness of the construction. Next, we obtain a convenient representation for $|X(\ell) - Y(\ell)|$. Let $\tau_{\ell,m}$ denote the unique integer satisfying the relation

$$(\tau_{l,m} - 1)2^{N-m} < l \leq \tau_{l,m}2^{N-m}, \tag{10}$$

and set $X_{\ell,k} = X(\tau_{\ell,k}, k)$.

L e m m a 1. If $1 \leq \ell \leq n$ is divisible by 2^{N-m}, then

$$|X(l) - Y(l)| \leqslant \sum_{0 \leqslant k \leqslant m} |X_{l,k} - Y_{l,k}|. \tag{11}$$

Proof. Let $\mu_k = \mu_{\ell,k}$ be the multiple of 2^{N-m} closest to ℓ. (If there are two such numbers, we take the smaller). Let δ_k denote the sign of the difference $\mu_k - \mu_{k-1}$. It follows from (10) that for one of the three values $\delta_k = 0,+1,-1$ the following equalities hold:

$$\begin{aligned} \mu_k &= \mu_{k-1} + \delta_k 2^{N-k}, \\ x_k &\equiv X(\mu_k) - X(\mu_{k-1}) = \delta_k X_{l,k}, \\ y_k &\equiv Y(\mu_k) - Y(\mu_{k-1}) = \delta_k Y_{l,k}. \end{aligned} \tag{12}$$

In order to derive the inequality (11), it suffices to substitute the differences of the relations in (12) into the following obvious identity:

$$X(\mu_m) - Y(\mu_m) = X(\mu_0) - Y(\mu_0) + \sum_{1 \leqslant k \leqslant m} (x_k - y_k)$$

and to note that $\mu_m = \ell$, whereas $X(\mu_0) - Y(\mu_0)$ is equal either to $X_{\ell,0} - Y_{\ell,0}$ or to zero.

L e m m a 2. For all ℓ and m, the random variables

$$X_{l,0} - X_{l,1},\ X_{l,1} - X_{l,2},\ \ldots,\ X_{l,m-1} - X_{l,m}$$

and $X_{\ell,m}$ are collectively independent.

Proof. Note that $X_{\ell,k}$ can be represented (see (3) and (10)) as a sum of variables X_j for $j \in J(k) = \{j: (\tau_{\ell,k} - 1)2^{N-k} < j \leq \tau_{\ell,k}2^{N-k}\}$. Since $J(0) \supset J(1) \supset \cdots \supset J(m)$, the sets $J(0)\backslash J(1), \ldots, J(m-1)\backslash J(m)$ and $J(m)$ do not intersect. This and equality (4) imply that the random variables in the formulation of the Lemma are indeed independent, since they are sums of variables X_j over j belonging to the nonintersecting intervals indicated.. ■

REMARK 7. We formally impose no restrictions on the distributions of the X_j in this method for constructing the random variables $X_1,\ldots,X_n$ with respect to $Y_1,\ldots,Y_n$. However, in order that this method yield good estimates in the invariance principle, we need to require that these distributions satisfy certain smoothness (or lattice) conditions (see [5,6]). Otherwise, it may happen that on the basis of the variable $X(0,0) = X_1 + \cdots + X_n$ at the first stage of this construction one could define uniquely values of all the variables X_j, for instance, if a X_j takes on only values of the form $\pm(1 + 2^{-j})$. We cannot have good estimates for the conditional quantile transformations, because the corresponding conditional distributions are singular in this case. This does not happen, however, for identically distributed variables -- the sum of these variables (even if their values are determined) does not indicate which particular X_j has taken on a certain value. (In fact, this is the essence of the main result obtained in [6].) We shall explain in Subsection 3.5 a technique to overcome this problem in the general case.

3.4. ESTIMATES IN THE INVARIANCE PRINCIPLE FOR SMOOTH DISTRIBUTIONS

Let $\xi_1, \ldots, \xi_n$ be a sequence of independent random variables and let $\eta_1, \ldots, \eta_n$ be a sequence of independent normally distributed variables; in this case

$$\mathbf{M}\xi_j = \mathbf{M}\eta_j = 0, \quad \mathbf{D}\xi_j = \mathbf{D}\eta_j > 0 \quad \forall j. \tag{1}$$

In addition, we impose the following smoothness conditions on the distributions of the ξ_j:

$$\xi(l) - \xi(k) \in \mathcal{D}_0(\lambda) \quad \text{if} \quad \mathbf{D}(\xi(l) - \xi(k)) \geqslant \varepsilon, \tag{2}$$

assuming for convenience that

$$\sigma^2 \max_j \mathbf{D}\xi_j \leqslant \varepsilon \leqslant \mathbf{D}\xi(n) = B^2. \tag{3}$$

(The notation $\xi(\ell)$ above and $\eta(\ell)$ and $\Delta(\varepsilon,\eta)$ below correspond to equalities (3.1) and (3.2).)

T h e o r e m 8. If conditions (1), (2) and (3) are satisfied, the random variables $\xi_1, \ldots, \xi_n$ can be given on a common probability space with the random variables $\eta_1, \ldots, \eta_n$, to obtain

$$\mathbf{M}e^{c\lambda\Delta(\xi,\eta)} \leqslant 1 + \lambda B \exp(\lambda^2\varepsilon), \tag{4}$$

where $c > 0$ is an absolute constant.

As will be shown in Lemma 6, the constants before B and ε on the right side of inequality (4) can be arbitrary if we vary the number $c > 0$ on the left side.

Note that condition (2) is used only in proving Lemma 8 and Lemmas 1-7 are valid if (2) is replaced by the weaker condition:

$$\xi(l) - \xi(k) \in \mathcal{D}(\lambda) \quad \text{if} \quad \mathbf{D}(\xi(l) - \xi(k)) \geqslant \varepsilon. \tag{5}$$

In Subsection 3.7 we shall show that condition (2) can be replaced by (5) in the formulation of Theorem 8 as well (this is in fact the objective of this article).

Since one can always consider $\lambda\xi_j$ instead of ξ_j we may assume without loss of generality that $\lambda = 1$. In the proof we use the obvious inequality

$$\mathbf{P}\left(\max_{i\in I} \zeta_i > x\right) \leqslant \sum_{i\in I} \mathbf{P}(\zeta_i > x) \tag{6}$$

and the following elementary lemma.

L e m m a 1. If

$$\mathbf{P}(|\zeta| > x) \leqslant K^2 e^{-2x},$$

then

$$\mathbf{M}e^{\alpha|\zeta|} \leqslant (2K)^{\alpha} \quad \text{for} \quad 0 \leqslant \alpha \leqslant 1.$$

Proof. Integrating by parts, we have

$$\mathbf{M}e^{|\zeta|} = 1 + \int_{x>0} e^x \mathbf{P}(|\zeta| > x)\,dx \leqslant 1$$

$$+ \int_{0<x\leqslant \ln K} e^x dx + \int_{x>\ln K} e^x K^2 e^{-2x} dx = 2K.$$

The case $\alpha < 1$ follows from Hölder's inequality $\mathbf{M}e^{\alpha|\zeta|} \leqq (\mathbf{M}e^{|\zeta|})^{\alpha}$.

Introduce the integer $N \equiv \max\{m: B^2 \geqq 2^m \max\{\varepsilon+\sigma^2, 4\sigma^2, 80\}\}$. Set $b_m = B^2 2^{-m}$ and $b = b_N$ for $0 \leqq m \leqq N$. On the other hand, if $N < 0$, we set $b \equiv b_0 \equiv b_N \equiv B^2$. Also, let

$$n_i = \min\{m : \mathbf{D}\xi(m) \geqslant ib_N\}, \quad X_i = \xi(n_i) - \xi(n_{i-1}) \tag{7}$$

for $i = 1,\dots,2^N$. Similarly, introduce $Y_i = \eta(n_i) - \eta(n_{i-1})$. It follows from the definition of b_m and the choice of n_i in (7) that

$$(3/4)b_m \leqslant \mathbf{D}X(i, m) = \mathbf{D}Y(i, m) \leqslant (5/4)b_m \quad \forall i \forall m, \tag{8}$$

$$\varepsilon \leqslant \mathbf{D}X_j \quad \forall i, \ \varepsilon \leqslant \mathbf{D}X(i, m) \quad \forall i \forall m. \tag{9}$$

Set

$$X^*(j, m) = \max_j \{|\xi(j) - \xi(n_{i-1})| : n_{i-1} \leqslant j \leqslant \eta_i\} \tag{10}$$

and analogously, replacing ξ by η, define the variables $Y^*(j,m)$.

L e m m a 2. If conditions (1) and (5) are satisfied, then

$$Q_m(x) \leqslant 2 \exp(b_m - x) \quad \forall x, \tag{11}$$

$$Q_m(x) \leqslant 2 \exp(-x^2/(4b_m) \text{ for } |x| \leqslant 2b_m, \tag{12}$$

where

$$Q_m(x) = \max_i \{\mathbf{P}(X^*(i, m) > x), \quad \mathbf{P}(Y^*(i, m)) > x\}.$$

Proof. It follows from (5) and (9) that $X \equiv X(i,m) \in \mathcal{D}$. Hence by Lemma 2.6.3 we have

$$\mathbf{M}e^{hX} \leqslant \exp((2/3)h^2\mathbf{D}X) \leqslant \exp(h^2 b_m) \text{ for } |h| \leqslant 1. \tag{13}$$

In deriving the last relation, we have also used the right-hand inequality in (8).

Next, by (10), $X^* = X^*(i,m)$ is the maximum of the modulus of successive sums of random variables equal to X in the sum. Noting (1), we can use the Kolmogorov inequality with exponential moment, which yields the estimate

$$\mathbf{P}(X^* > x) \leqslant e^{-hx}\mathbf{M}e^{hX} + e^{-hx}\mathbf{M}e^{-hX}. \tag{14}$$

Substituting (13) into (14) yields

$$\mathbf{P}(X^* > x) \leqslant 2 \exp(-hx + h^2 b_m) \quad \text{for} \quad |h| \leqslant 1. \tag{15}$$

Since by the normality of Y_j inequality (15) *a fortiori* holds if X^* is replaced by $Y^* = Y^*(j,m)$, we have that for $h > 1$, (11) follows from (15) and for $h = x/(2b_m)$ we obtain (12) if we note only that $|h| \leqq 1$ for $|x| \leqq 2b_m$.

L e m m a 3. Let $Z(j,m) = \min \{X^2(j,m)/b_m, b_m\}$. Then

$$\max_{i,m} \mathbf{M}e^{Z(i,m)/8} \leqslant 2^{3/2}.$$

Proof. It suffices to verify that $\zeta = Z(i,m)/8$ satisfies the hypothesis of Lemma 1 for $K^2 = 2$. Indeed, for $|x| \leqq b_m/2$, we have from (12) that

$$\mathbf{P}(Z(i, m)/8 > x) \leqslant Q_m((8xb_m)^{1/2}) \leqslant 2e^{-2x},$$

since in this case $(8xb_m)^{\frac{1}{2}} \leqq 2b_m$. For $x > b_m/8$, the inequality $P(\zeta > x) \leqq 2e^{-2x}$ holds automatically since $Z(i,m) \leqq b_m$. ■

Set

$$X^*(m)=\max_i X^*(i,m), \quad Y^*(m)=\max_i Y^*(i,m).$$

By (6) we have

$$\begin{aligned}\mathbf{P}(X^*(m)>x)\leqslant 2^m Q_m(x),\\ \mathbf{P}(Y^*(m)>x)\leqslant 2^m Q_m(x).\end{aligned} \tag{16}$$

Note that by (11) and (16) for $\zeta = X^*(m)/2$, $\zeta = Y^*(m)/2$ one has $P(\zeta > x) \leqq 2^{m+1}\exp(b_m - 2x)$, from which and from Lemma 1 we have the following lemma.

L e m m a 4. If conditions (1) and (5) are satisfied and $0 \leqq \alpha \leqq \frac{1}{2}$, then

$$\begin{aligned}\mathbf{M}\exp(\alpha X^*(m))\leqslant(2^{m+3}\exp(b_m))^\alpha,\\ \mathbf{M}\exp(\alpha Y^*(m))\leqslant(2^{m+3}\exp(b_m))^\alpha.\end{aligned}$$

We need

L e m m a 5. If conditions (1) and (5) are satisfied, for any assignment of the variables $\xi_1,\dots,\xi_n$ and $\eta_1,\dots,\eta_n$ on a common probability space, one has the inequality

$$\mathbf{M}e^{\Delta(\xi,\eta)/12}\leqslant 1+B\exp(B^2/6). \tag{17}$$

Proof. Note that for $\zeta = X^*(0)$ and $\zeta = Y^*(0)$, the Kolmogorov inequality yields the estimate $P(\zeta > x) \leqq \min\{B^2x^{-2}, 1\}$, implying in turn

$$\mathbf{M}|\zeta|^{3/2}=\int_{x>0}\mathbf{P}(\zeta>x)\,dx^{3/2}\leqslant\int_{0<x\leqslant B}dx^{3/2}+\int_{x>B}x^{-2}B^2(3/2)\,x^{1/2}dx=4B^{3/2}.$$

On the other hand, from Hölder's inequality for $p = 3/2$ and $q = 3$ we have

$$\mathbf{M}e^{\zeta/6}\leqslant 1+\mathbf{M}\xi e^{\zeta/6}/6\leqslant 1+(\mathbf{M}\zeta^{3/2})^{2/3}(\mathbf{M}e^{3\zeta/6})/6.$$

The last two relations and Lemma 4 for $m = 0$ yield

$$\mathbf{M}e^{\zeta/6}\leqslant 1+(4B^{3/2})^{2/3}(3\exp(B^2/2))^{1/3}\leqslant 1+B\exp(B^2/6)\,. \tag{18}$$

Now let us use the following coarse estimate:

$$\Delta\equiv\Delta(\xi,\eta)\leqslant X^*(0)+Y^*(0) \tag{19}$$

and Hölder's inequality

$$\mathbf{M}e^{\Delta/12}\leqslant(\mathbf{M}\exp(X^*(0)/6)\mathbf{M}\exp(Y^*(0)/6))^{1/2}.$$

Estimating the right side in the last expression by means of (18), we obtain (17).

We need next the following trivial lemma.

L e m m a 6. If for some absolute constants $c_0 > 0$ and $K < \infty$, one has the inequality

$$\mathbf{M}\exp(c_0\Delta) \leqslant 1 + O(B^K)e^{O(b)}, \tag{20}$$

then for any positive absolute constants $C \leqq K$, $C_1 < \infty$ and $C_2 < \infty$ we can find an absolute constant $c > 0$ such that

$$\mathbf{M}\exp(c\Delta) \leqslant 1 + C_1 B^C \exp(C_2\varepsilon). \tag{21}$$

If, in addition, $B \geqq 1$, (21) holds for any $C \geqq K$.

Proof. We take advantage of the inequality

$$(1+x)^{\alpha\beta} \leqslant 1 + \alpha x^\beta \quad \text{for} \quad 0 < \alpha,\ \beta \leqslant 1,\ x \geqslant 0, \tag{22}$$

which is easily proved by equating the derivatives of both sides with respect to x. Writing the right side in (20) in the form $1 + C_3B^k\exp(C_4b)$ and using Hölder's inequality, we have from (20) and (22) that

$$\mathbf{M}\exp(\alpha\beta c_0\Delta) \leqslant (\mathbf{M}\exp(c_0\Delta))^{\alpha\beta} \leqslant 1 + \alpha C_3 B^{\beta K}\exp(\beta C_4 b). \tag{23}$$

To get (21) from (23), it suffices to consider the following two cases. If $\varepsilon \geqq 20$, then $b < 8\varepsilon$, and (21) follows from (23) for $\beta = \min\{C/K, 8C_4/C_2\}$ since $B \geqq \varepsilon^{\frac{1}{2}} > 1$. However, if $\varepsilon < 20$, then $b < 160$, and (21) follows from (23) for $\beta = C/K$ and $\alpha = (C_1/C_3)\exp(-160\beta C_4)$.

The last assertion of Lemma 6 holds due to the inequality $B^K \leqq B^C$ for $0 \leqq K \leqq C$ and $B \geqq 1$.

In particular, we immediately get from Lemmas 5 and 6 the next

L e m m a 7. If conditions (1), (3), (5) and $N < 0$ are satisfied, then the assertion of Theorem 8 is valid.

Note that if the random variables $\xi_1,\ldots,\xi_n$ and $\eta_1,\ldots,\eta_n$ are given (in any way) on a common probability space, then one has the inequality

$$\Delta \equiv \Delta(\xi,\eta) \leqslant X^*(m) + Y^* + \Delta_m \quad \forall m, \tag{24}$$

where

$$\Delta_m = \max_i \Delta_{i,m},\quad \Delta_{i,m} = |X(i2^{N-m}) - Y(i2^{N-m})|. \tag{25}$$

We proceed to prove Theorem 8. We give the random variables $X_1,\dots,X_{2^N}$ on a common probability space with the variables $Y_1,\dots,Y_{2^N}$, using the method of the preceding subsection, and complete constructing the random variables $\xi_1,\dots,\xi_n$ in an arbitrary way (see Remark 8 at the end of this subsection). To make use of relation (24), we need, by virtue of (25), to obtain "good" estimates for $\Delta_{i,m}$ for all i and m. We shall do this next.

We fix ℓ of the form $\ell = i2^{N-m}$ and for brevity we set

$$X^{(k)} = X_{l,k},\quad Y^{(k)} = Y_{l,k},$$
$$Z^{(m)} = \min\{|X^{(m)}|^2/b_m,\quad b_m\},$$
$$Z^{(k)} = \min\{(X^{(k)} - X^{(k+1)})^2/b_{k+1},\quad b_{k+1}\}\quad \text{for}\quad 0 \leqslant k < m.$$

In this notation we can rewrite Lemma 3.1 in the form

$$\Delta_{i,m} \leqslant \sum_{0 \leqslant k \leqslant m} |X^{(k)} - Y^{(k)}| \equiv \delta_{i,m}. \tag{26}$$

The next lemma is crucial in proving Theorem 8.

L e m m a 8. If all the conditions of Theorem 8 as well as the condition

$$X^*(m) \leqslant b_m/8,\quad m \geqslant 0, \tag{27}$$

are satisfied, then

$$\Delta_{i,m} \leqslant C\Big(m + \sum_{0 \leqslant k \leqslant m} Z^{(k)}\Big). \tag{28}$$

Proof. Since each of the $X^{(k)}$ is the sum of 2^{k-m} summands of the form $X(j,m)$, it follows from (27) that

$$|X^{(k)}| \leqslant b_k/8 \leqslant \mathbf{D}X^{(k)}/6,\ k \leqslant m. \tag{29}$$

Note that the right-hand inequality in (29) follows from relation (8).

Since $X^{(0)} = X(0,0)$ is the quantile transformation of $Y^{(0)} = Y(0,0)$, to estimate the difference $X^{(0)} - Y^{(0)}$ we can use Lemma 2.1 for $U = Y^{(0)}$ and $V = X^{(0)}$. In this case, condition (21) follows from (2), (8), (9), and (2.4) follows from (29), since $b \geqq 80 > 4^2$. Thus, (2.3) yields the estimate

$$|X_0 - Y_0| = O(1 + (X^{(0)})^2/b_0). \tag{30}$$

Since, for $k > 0$, $X^{(k)}$ is the conditional quantile transformation of the variable $Y^{(k)}$ with respect to $Y^{(k-1)}$ and $X^{(k-1)}$, to estimate the difference $X^{(k)}-Y^{(k)}$ we can use the assertion of Lemma 2.3 for $V_1 = X^{(k)}$, $V_0 = X^{(k-1)}$, $U_1 = Y^{(k)}$, $U_0 = Y^{(k-1)}$. Note that by (8)

$$\mathbf{D}X^{(k)}/\mathbf{D}X^{(k-1)} \leqslant ((5/4)b_k)/((3/4)b_{k-1}) = 5/6,$$
$$\mathbf{D}X^{(k)}\mathbf{D}(X^{(k-1)} - X^{(k)})/\mathbf{D}X^{(k-1)} \geqslant (3/4)b_k(3/4)b_k/(5/4)b_{k-1} = (9/80)b_{k-1},$$

and hence in the notation of Lemma 2.3,

$$\alpha_1 \leqslant 5/6,\ B^2 \geqslant b_{k-1}/10 = b_k/5 \geqslant b/5 \geqslant 4^2. \tag{31}$$

Now observe that in the case at hand, condition (2.8) follows from (2), (8), (9), and (2.10) follows from (27), with (29) and (31) taken into account. Thus, we can rewrite (2.9) as follows:

$$|X^{(k)} - Y^{(k)}| \leqslant (5/6)|X^{(k-1)} - Y^{(k-1)}|$$
$$+ O(1 + (X^{(k)})^2/b_k + (X^{(k-1)})^2/b_{k-1}) \tag{32}$$

for $1 \leqq k \leqq m$ if we take into account the estimates from (31) as well.

If we now sum up (30) and (32) over all k and note (26), we obtain $\delta_{i,m} \leqq (5/6)\delta_{i,m} + O(m+Z)$ for

$$Z = \sum_{0<k\leqslant m} (X^{(k)})^2/b_k, \tag{33}$$

and therefore

$$\Delta_{i,m} \leqslant \delta_{i,m} = 6O(m+Z). \tag{34}$$

Using the elementary inequality

$$(3/2)a^2 + 3b^2 - (a+b)^2 = (2^{-1/2}a - 2^{1/2}b)^2 \geqslant 0,$$

we have

$$(X^{(k+1)})^2/b_{k-1} \leqslant (3/2)(X^{(k)})^2/b_{k-1} + 3(X^{(k-1)} - X^{(k)})^2/b_{k-1}. \tag{35}$$

Note next that if inequality (29) is satisfied, then $(X^{(k-1)} - X^{(k)})^2 \leqq Z^{(k)}b_k$ for $0 \leqq k \leqq m$ and $(X^{(m)})^2 \leqq Z^{(m)}b_m$. From these relations, (33) and (35), also noting that $b_{k-1} = 2b_k$, we have

$$Z \leqslant (3/4)\,Z + (3/2) \sum_{0<k<m} Z^{(k)} + Z^{(m)},$$

and therefore

$$Z \leqslant 4\cdot(3/2) \sum_{0<k<m} Z^{(k)}.$$

Substituting the last expression into (34) gives (28). ∎

Noting that by Lemma 3.2 the random variables $Z^{(0)},\dots,Z^{(m)}$ are independent, we have from (28) that

$$R_{i,m} = \mathbf{M}\exp(\Delta_{i,m}/(8C);\ \Omega(m)) \leqslant 2^{m/2} \prod_{0\leqslant k\leqslant m} \mathbf{M}\exp\left(Z^{(k)}/8\right), \tag{36}$$

where the event $\Omega(m) = \{X^*(m) \leqq b_m/8\}$. Since each of the $Z^{(k)}$, for some j depending on k and ℓ, is representable in the form $Z^{(k)} = Z(j,k)$, then from condition (3), the Chebyshev inequality and also (36) we obtain

$$\mathbf{P}(\Delta_{i,m} > x;\ \Omega(m)) \leqslant e^{-x/(8C)} \cdot 2^{m/2+3m/2}, \tag{37}$$

provided the conditions of Lemma 3 are satisfied.

L e m m a 9. If the conditions of Theorem 8 hold, and $N \geqq 0$, then there exists an absolute constant $c > 0$ such that

$$\mathbf{P}(\Delta > x) \leqslant 5 \cdot 2^{3N} \exp(-2cx + 2cb). \tag{38}$$

Proof. First consider the fundamental case where

$$2b_{m+1} \leqslant b_m \leqslant x \leqslant 2b_m, \quad 0 \leqslant m \leqslant N. \tag{39}$$

For these x we have from (24) that

$$\mathbf{P}(\Delta > x) \leqslant P_X + P_Y + P_0, \tag{40}$$

where

$$P_X = \mathbf{P}(X^*(m) > b_m/8),\ P_Y = \mathbf{P}(Y^*(m) > b_m/8),\ P_0 = \mathbf{P}(\Delta_m > b_m/2;\ \Omega_m).$$

It follows from (15) and (12) that

$$P_X + P_Y \leqslant 4 \cdot 2^m \exp(-b_m/(4\cdot 8^2)), \tag{41}$$

whereas (37) and (6) and (25) imply

$$P_0 \leqslant 2^m \cdot 2^{2m} \exp(-b_m/(16C)). \tag{42}$$

Summing (40)-(42) and noting that $m \leqq N$, for x satisfying (39) we obtain (38) for $c = (1/2)\min\{8^{-3}, (32C)^{-1}\}$.

If $x \leqq 2b_0$, (38) follows from (19) and (11) since in that case

$$\mathbf{P}(\Delta > x) \leqslant 2Q(x/2) \leqslant 4\exp(b_0/2 - x/2) \leqslant 4\exp(-x/4).$$

On the other hand, if $x \leqq b_N = b$, (38) holds automatically since its right hand side is greater than one. ∎

Thus, by Lemma 9 the random variable $\zeta = 2c\Delta$ satisfies all the conditions of Lemma 1, and therefore

$$\mathbf{M}e^{2c\Delta} \leqslant 5^{1/2}2^{3N/2}e^{cb} \equiv 5^{1/2}B^3e^{cb}b^{-3/2} \tag{43}$$

for $N \geqq 0$. The assertion of Theorem 8 follows for $N \geqq 0$ from (43) and Lemma 6, since in that case $B^2 \geqq b > 1$.

REMARK 8. Since the distributions of the η_j are already assumed to be smooth, we cannot hope to get a stronger result than Theorem 8: namely, if the conditions of Theorem 8 are satisfied, there exist functions $f_1(x), \ldots, f_n(x)$, $x \in R^n$, such that the random variables $\xi_1 = f_1(\eta)$, ..., $\xi_n = f_n(\eta)$, where $\eta = (\eta_1, \ldots, \eta_n)$, are independent, have the required distributions and satisfy (4).

Indeed, if we analyze the proof of Theorem 8, we can see that we have constructed the random variables $X_1, \ldots, X_{2^N}$ as functions of $Y_1, \ldots, Y_{2^N}$ and have not defined precisely how to construct the variables $\xi_1, \ldots, \xi_n$, since by Lemma 4 this is of no consequence (nevertheless, we do not exclude the possibility of expanding the probability space).

Now, we specify a concrete method of subsequent construction of the variables ξ_j, e.g., for $k \equiv n_{i-1} < j < n_i$. Since the variables $\eta_1, \ldots, \eta_n$ have densities, we can continue the construction, using the method of the previous subsection. We define ξ_{k+1} to be the conditional quantile transformation of the variable η_{k+1} with respect to $Y_i = \eta(n_i) - \eta(k)$ and $X_i = \xi(n_i) - \xi(k)$. If we have defined $\xi_{k+1}, \ldots, \xi_{k+j-1}$, then we take for ξ_j the conditional quantile transformation of the variable η_j with respect to $\eta(n_i) - \eta(k+j-1)$ and $\xi(n_i) - \xi(k+j-1)$. That is, the random variables summing up to X_i are constructed one by one from left to right for all i. As a result, all the $\xi_1, \ldots, \xi_n$ are functions of the $\eta_i, \ldots, \eta_n$ only.

Note that the method for constructing the random variables ξ_j one by one does not yield good estimates of the nearness to η_i. However, by Lemma 4 this is of no consequence for $\mathrm{D}X_i = O(\varepsilon+1)$ when $\lambda = 1$.

3.5. THE BASIC CONSTRUCTION

As in Subsection 3.3., let $Y_1, \ldots, Y_n$, $n = 2^k$, be the given sequence of independent random variables. On the basis of this sequence, it is required to construct a sequence $X_1, \ldots, X_n$ of independent random variables in such a way that each random variable has, on the one hand, a specified distribution and, on the other hand, is a function of the $Y_1,\dots,Y_n$ only. The main requirement is that the distance $\Delta(X,Y)$ be, in some sense, as small as possible.

As mentioned in Remark 7, the construction given in Subsection 3.3 without additional assumptions on distributions of the random variables X_j may not yield good estimates for $\Delta(X,Y)$. Hence we shall outline another construction method, which is more complex but makes it possible, sometimes, to obtain estimates in more general situations that the method of Subsection 3.3. However, by the method of Subsection 3.3, we first construct, using the $\{Y_j\}$ a sequence $W_1,\dots,W_n$ of independent random variables with preferential smooth distributions. Furthermore, we construct the required $X_1,\dots,X_n$ on the basis of $W_1,\dots,W_n$.

The construction suggested is best explained by induction for the number of random variables. If $k = O(1)$, then any method of assigning random variables yields sufficiently good estimates for $\Delta(X,W)$ (see, e.g., Lemma 4.5). We assume that we have a "good" construction method for all $k \leqq N-1$ and proceed to describe the target construction for $k = N$.

We fix some integer A, $0 < A \leqq N$, and, using the method of Subsection 3.3, we construct for $W_1,\dots,W_n$ the auxiliary sequence $Z_1,\dots,Z_n$ of independent random variables whose distributions we have chosen as follows. We assume that for i not divisible by 2^A, the Z_i are identically distributed as X_i, while for i divisible by 2^A, the distributions of these variables are specially chosen and have densities.

After the $Z_1,\dots,Z_n$ have been constructed, we set $X_i = Z_i$ for i not divisible by 2^N, and denote $W_j^* = Z_{j2^A}$, $j = 1,\dots,2^{N-A}$. By an inductive assumption, one can construct on the basis of the sequence $\{W_j^*\}$ a sequence $\{X_j^* = X_{j2^A}\}$ of independent variables having the required distributions and being functions of $W_1^*,\dots,W_{2^k}^*$, $k = N-A$, only. In view of the latter, we thus have two independent families: $X_A = \{X_i : i \neq j2^A\}$ and $X^* = \{X_j^* = X_{ij^A}\}$ of random variables, since X_A and $\{W_j^*\}$ are independent by construction. Since each of the families X_A and X^* consists of independent variables having the desired distributions, we have therefore constructed the required set $X_A \cup X^* = \{X_1,\ldots,X_n\}$ of variables.

This then is our construction. Note that in constructing X^* on the basis of $\{W_j^*\}$, we also use, of course, the technique of introducing the auxiliary sequence, which reduces each time the number of X_j still undefined to fifty percent, at least.

The principal advantage of this method over the method of Subsection 3.3 is that on each step we need to construct on the basis of $\{W_j\}$ a sequence $\{Z_j\}$ in which each 2^A*th* variable has smooth distribution. That is, in particular, the variables $Z(j,m)$ for $m \leqq N-A$ have distribution densities, and these variables play in fact the fundamental role in obtaining estimates for $\Delta(W,Z)$.

REMARK 9. Let the independent variables X_i satisfy the condition

$$X_i \in \mathscr{D},\ \mathbf{D}X_i \approx 1.$$

It is not hard to see that if we now apply the construction of this Subsection $W_j = Y_j$ having normal distributions, we can derive from Theorem 8 that

$$\Delta(X,\ \ Z) \approx \ln n \approx N.$$

Repeating this construction recursively yields

$$\Delta(X, Y) \approx N + (N-A) + (N-2A) + \ldots + 1 \approx N^2,$$

which is worse than the estimate

$$\Delta(X, Y) \approx N$$

obtained in [6] for identically distributed variables.

To avoid this deterioration, we shall construct $\{X_j\}$ from a sequence of variables $\{W_j\}$ having the same first three (rather than two) moments as $\{X_j\}$ have. In this case we are bound to get

$$\Delta(X,Z) \approx 1 \quad \text{and} \quad \Delta(X,Y) \approx N.$$

It is precisely for this accuracy of the estimates that we wanted the estimates, which are so hard to prove, for the quantiles in Section 2.

Note next that the construction suggested above is essentially the addition of a rather unusual smoothing to the Komlós-Major-Tusnády method. The standard way of smoothing will never give the desired result since, if, for example, we add to each of the X_i an independent normally distributed variable ν_i with $D\nu_i = \delta^2$, then, in order to take advantage of Theorem 8, it is necessary that the sum of $\ln n$ summands belong to the class $\mathcal{D}_0$, i.e., that

$$\delta^2 \ln n > 1.$$

On the other hand, to reduce these addends in importance, it is necessary that

$$n\delta^2 < \ln n.$$

Hence this standard smoothing can be effective only for certain distributions, for example those having an absolutely continuous component (see [6]).

3.6. THE KEY THEOREM

Let $Y_1,\ldots,Y_n$ and $X_1,\ldots,X_n$, $n = 2^N$, be two sequences of independent random variables satisfying the condition

$$3/4 \leqslant \mathbf{D}X_j = \mathbf{D}Y_j \leqslant 5/4, \quad X_j \in \mathcal{D}(8) \quad \forall j. \tag{1}$$

(More precisely, at the moment, only the distributions of the random variables $X_1,\ldots,X_n$ are given.)

T h e o r e m 9. If condition (1) is satisfied, then on the basis of sequence of independent normally distributed random variables $Y_1,\ldots,Y_n$ one can construct a sequence $X_1,\ldots,X_n$ of independent random variables with prescribed distributions so that the following inequality is true:

$$\mathbf{M}e^{c\Delta(X,Y)} \leqslant 2n, \tag{2}$$

where $c > 0$ is an absolute constant.

We emphasize the fact that eventually variables of the form $X(\ell)$, $X(j,m)$ and $\Delta(X,Y)$ will be introduced using the same notation as the one adopted in Subsection 3.3. Moreover, the numbers C_1, C_2, C_3 correspond to the same absolute constants each time.

We choose first the distributions of the random variables W_i. Let us assume at once that the variables W_i^* and W_{i2^A} are identically distributed.

L e m m a 1. There exists a sequence of independent random variables $W_1,\ldots,W_n$ satisfying the condition

$$W_i \in \mathscr{D}, \quad \mathbf{D}W_i = \mathbf{D}X_i, \quad \mathbf{M}W_i^3 = \mathbf{M}X_i^3, \quad |W_i| \leqslant C_1 < \infty \quad \forall i. \tag{3}$$

Moreover, the distributions of the W_i have densities, and if the random variable S does not depend on the sequence $\{W_i\}$, then

$$S_m \equiv S + W_{i_1} + \ldots + W_{i_m} \in \mathscr{D}_0 \tag{4}$$

only if

$$m \geqslant C_0(1 + \ln \mathbf{D}S_m), \quad m > 0, \tag{5}$$

where $C_0 < \infty$ is an absolute constant.

The existence of the $\{W_i\}$ with the specified properties follows from Lemmas 2.6.8 and 2.6.9.

L e m m a 2. There exists an absolute constant $0 \leqq L < \infty$ such that for all j,

$$W(j,\ k) \in \mathscr{D}_0, \quad Z(j,\ k) \in \mathscr{D}_0 \quad \text{for } 0 \leqslant k \leqslant N - 2A - L. \tag{6}$$

In other words, the $Z(j,k)$ have sufficiently smooth distributions if they contain at least 2^{A+L} summands of the form W_i.

Proof. Since each of the $W(j,k)$ and $Z(j,k)$ has the form (4), we need only to verify (5) for $m = 2^{N-k-A}$ when $S_m \equiv Z(j,k)$ and for $m = 2^{N-k}$ when $S_m \equiv W(j,k)$. Using the relation $\mathrm{D}Z(j,k) = \mathrm{D}W(j,k) \leqq (5/4)2^{N-k}$ that follows from condition (1), we can rewrite (5) in the form

$$2^{N-k-A} \geqslant C_0(1 + \ln 2^{N-k+1}). \tag{7}$$

However,

$$\ln 2^{N-k+1} = \ln (2^{N-k-A}/2C_0) + \ln (2C_0 2^A) \leqslant 2^{N-k-A}/(2C_0) + 2C_0 2^A - 1.$$

Substituting this into (7), we obtain that for (6) to hold it suffices that the inequality

$$2^{N-k-A} \geqslant 2^{N-k-A}/2 + C_0(2C_0 2^A),$$

i.e., (6) holds for $2^L \geqq (2C_0)^2$. ■

Next, let the $\{W_i\}$ be constructed on the basis of $\{Y_i\}$ by the method of Subsection 3.3, and the $\{Z_i\}$ on the basis of $\{W_i\}$ by the method of Subsection 3.5. By Lemma 2, for $\{W_i\}$ all the conditions of Theorem 8 are satisfied for $\xi_i = W_i$, $\eta_i = Y_i$, $\lambda = 1$, $\sigma^2 \leqq 5/4$, $\varepsilon = 2^{2A+L} \geqq 4 > \sigma^2$, and hence one has

L e m m a 3. There exists an absolute constant $c > 0$ such that

$$\mathbf{M}e^{c\Delta(W,Y)} = O(n^{1/2}\exp(2^{2A+L})). \tag{8}$$

Let us fix some integer ℓ, $0 < \ell \leqq 2^N$ and estimate $\Delta(\ell) = |W(\ell) - Z(\ell)|$. As in Subsection 3.3, let $\tau_k 2^{N-k} \geqq \ell$ be the smallest of the numbers not less than ℓ and divisible by 2^{N-k}. Set

$$W^{(k)} = W(\tau_k,\ k), \quad W^{\{k\}} = W(\tau_k 2^{N-k}) - W(l) \tag{9}$$

and similarly introduce $Z^{(k)}$, $Z^{\{k\}}$, $W^{*(k)}$ and $W^{*\{k\}}$. Since

$$(\tau_k - 1)2^{N-k} < l \leqslant \tau_m 2^{N-m} \leqslant \tau_k 2^{N-k} \quad \text{for} \quad 0 \leqslant k \leqslant m \leqslant N,$$

by Lemma 3.1,

$$\left|W\left(\tau_m 2^{N-m}\right) - Z\left(\tau_m 2^{N-m}\right)\right| \leqslant \sum_{0 \leqslant k \leqslant m} \left|Z^{(k)} - W^{(k)}\right|. \tag{10}$$

Summing (9) and (10), we see that the following lemma holds true.

L e m m a 4. If $0 \leqq m \leqq N$, then

$$|W(l) - Z(l)| \leqslant \sum_{0 \leqslant k \leqslant m} \left|Z^{(k)} - W^{(k)}\right| + \left|W^{\{m\}}\right| + \left|Z^{\{m\}}\right|. \tag{11}$$

Note that by (1),

$$(3/4)2^{N-k} \leqslant \mathbf{D}Z^{(k)} = \mathbf{D}W^{(k)} \leqslant (5/4)2^{N-k}. \tag{12}$$

Furthermore,

$$(3/4)2^{N-k} \leqslant \mathbf{D}(Z^{(k-1)} - Z^{(k)}) \leqslant (5/4)2^{N-k}$$

and hence

$$\alpha^{(k)} \equiv \mathbf{D}Z^{(k)}/\mathbf{D}Z^{(k-1)} \leqslant 5/8, \tag{13}$$

$$(B^{(k)})^2 \equiv \mathbf{D}Z^{(k)}\mathbf{D}(Z^{(k-1)} - Z^{(k)})/\mathbf{D}Z^{(k-1)} \geqslant (3/8)2^{N-k}. \tag{14}$$

In deriving (13) and (14), we have used the inequalities $x/(x+y) \leqq b(a+b)$ and $xy/(x+y) \geqq a/2$ for $0 < a \leqq x$, $y \leqq b < \infty$.

We shall assume that $L \geqq 10$, therefore by (14) the following relations hold:

$$\mathbf{D}Z^{(k)} \geqslant 6^2, \quad B^{(k)} \geqslant 12 \quad \text{for} \quad N - k \geqslant L. \tag{15}$$

Lemma 5. If

$$|Z^{(0)}| \leqslant 2^N/14, \quad N \geqslant 2A + L, \tag{16}$$

then

$$|Z^{(0)} - W^{(0)}| \leqslant C_2(2^{-N/2} + |Z^{(0)}|^3 2^{-2N}). \tag{17}$$

Proof. Since, by construction, $Z^{(0)}$ is a quantile transformation of $W^{(0)}$, we use Lemma 2.2 for $U = W^{(0)}$, $V = Z^{(0)}$ and $B^2 = DZ^{(0)}$. In this case, conditions (2.2) follow from (3) and (6) and (2.5) follows from (3). Noting now that $B^2 \geqq (3/4)2^N$ by (12), we obtain that (17) follows from (2.6) and (2.7) from (15), (16) and the estimate for B^2.

Lemma 6. If $N-k \geqq 2A+L$ and

$$\max\{|Z^{(k)}|,\ |Z^{(k-1)}|,\ |W^{(k-1)}|\} \leqslant 2^{N-k}/96, \tag{18}$$

then

$$|Z^{(k)} - W^{(k)}| \leqslant (5/8 + g_k)|W^{(k-1)} - Z^{(k-1)}| + C_2\varepsilon_k, \tag{19}$$

where

$$g_k \equiv (|Z^{(k)}| + 2|Z^{(k-1)}| + |W^{(k-1)}|)2^{-(N-k)}, \tag{20}$$

$$\varepsilon_k \equiv 2^{-(N-k)/2} + (|Z^{(k)}|^3 + |Z^{(k-1)}|^3 + |W^{(k-1)}|^3)2^{-2(N-k)}. \tag{21}$$

Proof. Since, by construction, $Z^{(k)}$ is a conditional quantile transformation of $W^{(k)}$ with respect to $W^{(k-1)}$ and $Z^{(k-1)}$, we use the assertion of Lemma 2.4 for $U_1 = W^{(k)}$, $U_0 = W^{(k-1)}$, $V_1 = Z^{(k)}$, $V_0 = Z^{(k-1)}$ and $B = B^{(k)}$. In this case, conditions (2.8) follow from (3) and (6), (2.11) follows from (3), and (2.13) from (14), (15) and (18). Comparing (2.14) with (20) and (2.15) with (21), we obtain

$$|g(V_1,\ V_0,\ U_0)| \leqslant |Z^{(k-1)} - W^{(k-1)}|g_k, \quad \varepsilon(V_1,\ V_0,\ U_0) = O(\varepsilon_k).$$

These relations and (2.12) yield (19) only if we note that in this case $\alpha_1 \equiv \alpha^{(k)} \leqq 5/8$. ■

Let

$$\gamma = \min\{1/192,\ (216C_2)^{-1/2}\}, \tag{22}$$

and introduce the variable ω, letting

$$\omega = \max\{m \leqslant N - 2A - L : |Z^{(k)}| \leqslant 2\gamma a(k,\ m) \quad \forall k \leqslant m,\ |Z^{(m)}| \leqslant 2\gamma 2^{N-m}\},$$

where

$$a(k,\ m) = 2^{2(N-k)/3}h^{1/3}(k,\ m), \quad h(k,\ m) = 2^{N+k/8-(9/8)m}.$$

We assume that $\omega = -1$ if either $N < 2A+L$, or $|Z^{(0)}| > 2\gamma 2^N$, or $|Z^{\{0\}}| > 2\gamma 2^N$. We also assume without loss of generality that the L satisfies $2^L \geqq 2C_2/\gamma$, so that

$$C_2 2^{-(N-k)/2} \leqslant (\gamma/6)h(k, \omega) \quad \text{for} \quad 0 \leqslant k \leqslant \omega \leqslant N-L. \tag{23}$$

Note that the definition of ω implies

$$\max\{|Z^{(k)}|, |Z^{\{k\}}|\} \leqslant 2\gamma a(k, \omega) \quad \text{for} \quad 0 \leqslant k \leqslant \omega. \tag{24}$$

Lemma 7. If $0 \leqq k \leqq \omega$, then

$$|Z^{(k)} - W^{(k)}| \leqslant 2\gamma h(k, \omega), \tag{25}$$

$$|W^{(k)}| \leqslant 4\gamma a(k, \omega). \tag{26}$$

Proof. We use induction for k. If $k = 0$, then substituting (24) into (17) yields

$$|Z^{(0)} - W^{(0)}| \leqslant C_2 2^{-N/2} + C_2(2\gamma)^3 h(0, \omega). \tag{27}$$

Noting (22) and (23), it is not hard to get (25) from (27).

We assume next that (25) is satisfied for all $0 \leqq k < m \leqq \omega$. In this case, by (24) and the triangle inequality, relation (26) is also true for the k. If we prove (25) for $k = m$, we automatically get (26), too.

To derive (25) for $k = m$, we substitute (26) for $k = m-1$ and (24) for $k = m-1$ and $k = m$ into (19)-(21), since condition (18) holds automatically for $k = m \leqq \omega$. We have thus obtained

$$g_m \leqslant (1 + 2\cdot 2^{5/8} + 2^{5/8})2\gamma a(m, \omega)2^{-N+m} < 16\gamma \leqslant 1/8, \tag{28}$$

$$\begin{aligned}\varepsilon_m &\leqslant 2^{-(N-m)/2} + (1 + 2(2^{5/8})^3)(2\gamma)^3 a^3(m, \omega)2^{-2N+2m} \\ &\leqslant (1/6 + 1/3)(\gamma/C_2)h(m, \omega).\end{aligned} \tag{29}$$

Note that in deriving (28) and (29), we have used (22) and (23). From (28), (29) and (19) we now have

$$|Z^{(m)} - W^{(m)}| \leqslant (3/4)|Z^{(m-1)} - W^{(m-1)}| + (\gamma/2)h(m, \omega).$$

Substituting here the estimate from (25) for $k = m-1$, we obtain the required inequality (25) for $k = m$.

Note next that by (3)

$$|W^{\{k\}}| \leqslant 2^{N-k}C_1, \quad |W^{*\{k\}}| \leqslant 2^{N-k-A}C_1. \tag{30}$$

Since the right side of (25) is a geometric progression with ratio $2^{1/8}$, from (11), (25), (30) and the definition of ω we have

L e m m a 8. If $\omega \geqq 0$, then

$$|W(l) - Z(l)| \leqslant (2\gamma(1 - 2^{-1/8})^{-1} + 2\gamma + C_1)2^{N-\omega} \equiv C_3 2^{N-\omega}. \tag{31}$$

Let

$$V^{(k)} = Z^{(k)} - W^{*(k)}, \quad V^{\{k\}} = Z^{\{k\}} - W^{*\{k\}}$$

and define ν, replacing in the definition of ω the number 2γ by γ and the random variables Z by V. Now introduce $\beta(N) \equiv \max\{m: |W^{(k)}| \leqq 4\gamma a(k,m)$ for all $k \leqq m\}$ and similarly construct $\beta(N-A)$, replacing in the definition of $\beta(N)$ the number 4γ by $4\gamma 2^{-A}$ and W by W^*. We assume $\beta(N) = -1$ if $|W^0| > 4\gamma a(0,0)$, and make a similar assumption concerning $\beta(N-A)$.

Next, we choose A to be the smallest integer satisfying the inequalities:

$$A \geqslant 1, \quad 2^{(5/2)A}\gamma \geqslant 2^{3/2}C_1. \tag{32}$$

L e m m a 9. If $\nu \geqq 0$, then $\omega \geqq 0$ and

$$2^{N-\beta(N)} + 2^{N-\omega} \leqslant 2^{N-A-\beta(N-A)} + 2 \cdot 2^{N-\nu}. \tag{33}$$

Proof. By Lemma 7,

$$\beta(N) \geqslant \omega, \quad 2^{N-\beta(N)} \leqslant 2^{N-\omega}, \tag{34}$$

and next, it follows from (30) and (32) that

$$|W^{*\{k\}}| \leqslant \gamma 2^{N-k}. \tag{35}$$

Substituting (34) into the left side of (33), we obtain that to prove (33) it suffices to verify the inequality

$$\omega \geqslant \min\{\nu, \beta(N-A) + A + 1\} \equiv \mu. \tag{36}$$

Indeed, if $\omega \geqq \nu$, then in (33) the last summand on the right side of this expression plays the decisive role; however, if $\omega \geqq \beta(N-A) + A + 1$, then the first summand does.

Substituting (35) and the inequality $|Z^{\{k\}}| \leqq |V^{\{k\}}| + |W^{*\{k\}}|$ into the definitions of ω and ν, we see that (36) will be proved if the following relations hold:

$$|V^{(k)}| \leqslant \gamma a(k, \mu) \quad \forall 0 \leqslant k \leqslant \mu, \tag{37}$$

$$|W^{*(k)}| \leqslant \gamma a(k, \mu) \quad \forall 0 \leqslant k \leqslant \mu, \tag{38}$$

since $|Z^{(k)}| \leqq 2\gamma a(k,\mu)$ in this case. But (37) follows from the definition of ν and (38) follows from the definition of the random variable $\beta(N-A)$, however, just for $k \leqq \beta(N-A)$.

It thus remains only to verify (38) for $\beta(N-A) \leqq k \leqq \beta(N-A) + A + 1$. In that case we can use (30) and convince ourselves that

$$|W^{*(k)}| \leqslant 2^{N-k-A}C_1 \leqslant \gamma a(k, \mu),$$

since the last inequality is equivalent to

$$2^{(3/8)\mu} \geqslant 2^{(3/8)k} \geqslant 2^{(3/8)(\mu-A-1)} \geqslant 2^{(3/8)\mu}2^{-A}(C_1/\gamma).$$

This relation holds by the choice of A. ■

Set

$$\delta(N) = |X(l) - W(l)|, \quad \delta(N-A) = |X^*([l2^{-A}]) - W^*([l2^{-A}])|$$

and note that

$$\delta(N) \leqslant \Delta(l) + \delta(N-A). \tag{39}$$

Further, let $S(N) = 2C_3 2^{N-\nu}$ for $\nu \geqq 0$ and $S(N) = V + (3C_1 + 2C_3)2^N$ otherwise, where $V \equiv |V^{(0)}| + |V^{\{0\}}|$. The following coarse relation follows from (11) for $m = 0$ and from (30):

$$\Delta(l) \leqslant V + 3C_1 2^N. \tag{40}$$

Set $\rho(N) = C_3 2^{N-\beta(N)}$ and $\rho(N-A) = C_3 2^{N-A-\beta(N-A)}$. Summing estimates (31), (33), (39) and (40) we see that the following lemma is true.

L e m m a 10. Let

$$\delta(N) + \rho(N) \leqslant \delta(N-A) + \rho(N-A) + S(N).$$

L e m m a 11. Let

$$Q_k(x) = \max\{\mathbf{P}(|V^{(k)}| > x), \quad \mathbf{P}(|V^{\{k\}}| > x)\}.$$

Then

$$Q_k(x) \leqslant 2\exp(2^{N-k} - x) \quad \forall x, \tag{41}$$

$$Q_k(x) \leqslant 2\exp(-x^2 2^{-(N-k)-2}) \qquad \text{for} \quad |x| \leqslant 2^{N-k+1}. \tag{42}$$

Proof. It follows from (1) that $\zeta \in \mathcal{D}$ if $\zeta = V^{(k)}$ or $\zeta = V^{\{k\}}$, since ζ is the sum of independent variables from $\mathcal{D}$. In this case, from Lemma (6.3) of Section 2 one has

$$\mathbf{M}e^{h\zeta} \leqslant \exp((2/3)h^2\mathbf{D}\zeta) \leqslant \exp(h^2 2^{N-k}) \qquad \text{for} \quad |h| \leqslant 1.$$

In proving the last relation, we have used the inequality $\mathbf{D}\zeta \leqq (5/4)2^{N-k}$ which holds by (1). The rest of the proof is like that of Lemma 4.2. ■

Lemma 12. There exists an absolute constant $c > 0$ such that

$$E_1 \equiv \mathbf{M}\{\exp(2c(S(N)+\rho(N)));\ \nu < 0\} = O(1).$$

Proof. It follows from (40) that

$$\mathbf{P}(V > x) \leqslant 2Q_0(x/2), \tag{43}$$

$$E_2 \equiv \mathbf{M}\{e^{2cV};\ \nu < 0\} \leqslant \mathbf{M}\{e^{2cV};\ V > \gamma 2^N\} \leqslant (\mathbf{M}e^{4cV}\mathbf{P}(V > \gamma 2^N))^{1/2}. \tag{44}$$

We have the last step in (44) from Hölder's inequality. Since, by (41) and (43), $V/4$ satisfies the condition of Lemma 4.1, we have

$$\mathbf{M}e^{4cV} = O(\exp(16c2^N)) \text{ for } 0 < 16c \leqslant 1. \tag{45}$$

It follows from (43) and (42) that $P(V > \gamma 2^N) \leqq 4\exp(-\gamma^2 2^N/16)$. Substituting this inequality and (45) into (44), we obtain $E_2 \leqq \exp(8c2^N - \gamma^2 2^N/32)$. Since $E_1 \leqq E_2 \exp(c2^N(3C_1+4C_3))$, the last two relations yield the required assertion for $0 < c < (8 + 3C_1 + 4C_3)^{-1}\gamma^2/32$. ■

Lemma 13. There exists an absolute constant $c > 0$ such that

$$E_3 \equiv \mathbf{M}\{e^{2cS(N)};\ \nu \geqslant 0\} = O(1).$$

Proof. By (42) and the definition of ν we have

$$\mathbf{P}(\nu = m-1) \leqslant \sum_{0 \leqslant k \leqslant m} \left(\mathbf{P}(|V^{(k)}| > \gamma a(k, m)) + \mathbf{P}(|V^{(k)}| > \gamma 2^{N-m})\right)$$

$$\leqslant 4 \sum_{0 \leqslant k \leqslant m} \exp(-\gamma^2 a^2(k, m)/2^{N-k+2}) = 4 \sum_{0 \leqslant k \leqslant m} \exp(-\gamma^2 2^{N-m} 2^{(m-k)/4}/4).$$

Thus, for $2c2C_3 \leqq \gamma^2/16$, $\quad E_3 \leqslant \sum_{0 \leqslant m \leqslant N} \exp(\gamma^2 2^{N-m}/16)\,\mathbf{P}(\nu = m)$

$$\leqslant 4 \sum_{0 \leqslant m \leqslant N} \sum_{0 \leqslant k \leqslant m} \exp(-\gamma^2 2^{N-m} 2^{(m-k)/4}/16) \leqslant 4 \sum_{i,j \geqslant 0} \exp(-\gamma^2 2^{i+j/4}/16).$$

It is seen that this series converges.

Thus, we get from Lemmas 12 and 13 that

$$\mathbf{M}e^{2cS(N)} \leqslant C. \tag{46}$$

The next assertion along with Lemmas 7 and 9 is fundamental in the proof of Theorem 9.

Lemma 14. The independent random variables $X_1, \ldots, X_n$, $n = 2^k$, with the required distributions can be constructed on the basis of $W_1, \ldots, W_n$ so that

$$\mathbf{M}e^{2c(\delta(k)+\rho(k))} = O(C^k) \quad \forall l. \tag{47}$$

Proof. For $k < A$, (47) follows from Lemma 12. Now assume by induction that (47) has been proved for $k < N$. Then, since the sequence

$\{X_j^*\}$ contains 2^{N-A} variables, the following estimate holds:

$$\mathbf{M}e^{2c(\delta(N-A)+\rho(N-A))} = O(C^{N-A}). \tag{48}$$

To prove (47) for $k = N$, we use Lemma 10 and note (and this is most essential) that the variables $\delta(N) + \rho(N)$ and $S(N)$ are independent by construction, since the first variable is a function of $\{W_j^*\}$ and the second variable is a function of $\{V_j\}$, i.e., of the variables $\{X_j\}$ constructed with indices not divisible by 2^A. Thus,

$$\mathbf{M}e^{2c(\delta(N)+\rho(N))} \leqslant \mathbf{M}e^{2c(\delta(N-A)+\rho(N-A))}\mathbf{M}e^{2cS(N)}$$

which together with (46) and (48) yield (47).

Let us complete proving the Theorem. By Lemma 14 we have

$$\mathbf{P}(|X(l) - W(l)| > x) = e^{O(N)-2cx}$$

and therefore

$$\mathbf{P}(\Delta \equiv \Delta(X, W) > x) \leqslant 2^N e^{O(N)-2cx}.$$

Using the latter relation and Lemma 2.1 for $\zeta = c\Delta$, we obtain

$$\mathbf{M}e^{c\Delta} = e^{O(N)}. \tag{49}$$

We assume without loss of generality that the constant $c > 0$ in (8) coincides with the $c > 0$ in (49). Then from (8), (49) and Hölder's inequality one has

$$\mathbf{M}e^{c\Delta(X, Y)/2} = e^{O(N)}. \tag{50}$$

In deriving (50), we have also used the triangle inequality $\Delta(X,Y) \leqq \Delta(X,W) + \Delta(W,Y)$.

Lemma 4.6 and (50) now yield the required assertion of Theorem 9.

3.7. THE MAIN THEOREM

First we prove Theorem 1. Since we can always consider the variables $\lambda\xi_j/64$ instead of ξ_j, we shall be considering only the case when $\lambda = 64$. Under this assumption, the conditions imposed on the ξ_j of the Theorem (making them slightly coarser) can be rewritten in the following form:

$$\mathbf{M}\xi_j = 0, \quad 64\mathbf{M}|\xi_j|^3 \exp(2|\xi_j|) \leqslant \mathbf{D}\xi_j \quad \forall j. \tag{1}$$

Comparing (1) with (2.1) yields

$$\xi(l) - \xi(k) \in \mathscr{D} \quad \forall l > k. \tag{2}$$

Moreover, using (1) and Hölder's inequality, we have

$$\mathbf{D}\xi_j \geqslant 64\mathbf{M}|\xi_j|^3 \geqslant 64(\mathbf{D}\xi_j)^{3/2}. \tag{3}$$

It follows from (2) and (3) that conditions (4.1), (4.3) and (4.5) are satisfied for $\varepsilon = \sigma^2 \leqq (64)^{-2}$. Whence, in particular, the assertion of Theorem 1 follows from $N < 0$, since Lemma 4.7 yields the required estimate:

$$\mathbf{M}e^{c\Delta} \leqslant 1 + B\exp((64)^{-2}) \leqslant 1 + 64B. \tag{4}$$

Now we consider the case $N \geqq 0$. By (2) and (3) we can use the notation of Subsection 3.4 and Lemmas 4.1 - 4.7 for

$$\varepsilon = \sigma^2 \leqslant (64)^{-2}, \quad 80 \leqslant b \leqslant 160, \quad B^2 = b2^N \geqslant b. \tag{5}$$

Since $b^{\frac{1}{2}} > 8$ and $X_j \in \mathcal{D}$ by (2), we have

$$X_j b^{-1/2} \in \mathcal{D}(8) \quad \forall j ,$$

which, by Theorem 9, implies that we can construct the required $\{X_j\}$ on the basis of the given $\{Y_j\}$, to have the inequality

$$\mathbf{M}e^{c\delta} \leqslant 2^{N+1} = 2B^2/b, \tag{6}$$

where $\delta = \Delta(X,Y)$, and by (5) $c = c_1 b^{-\frac{1}{2}} > c_1(160)^{-\frac{1}{2}} > 0$ if we denote by c_1 the absolute constant in Theorem 9.

We assume without loss of generality that $0 < c \leqq 1/6$ in (6) and rewrite (4.24) in the form $\Delta = X + Y + \delta$, where $X = X^*(N)$, $Y = Y(N^*)$. From Hölder's inequality we have

$$\mathbf{M}e^{c\Delta/3} \leqslant (\mathbf{M}e^{cX})^{1/3}(\mathbf{M}e^{cY})^{1/3}(\mathbf{M}e^{c\delta})^{1/3}.$$

Substituting the estimates from (6) and Lemma 4.4 into the last relation yields

$$\mathbf{M}e^{c\Delta/3} = O((2^{cN})^{1/3}(2^{cN})^{1/3}(2^{cN})^{1/3}) = O(B^{(2+2c)/3}). \tag{7}$$

In deriving (7) we have also used inequalities for $b = b_N$ from (5). Comparing (7) with (4.20), we get (4) from Lemma 4.6. Theorem 1 is thus proved for $N \geqq 0$, too.

Let us prove next the following natural generalization of Theorems 1 and 8.

T h e o r e m 10. The assertion of Theorem 8 remains valid if condition (4.2) is replaced by (4.5).

Note that in Theorem 10 it is not claimed that the variables $\xi_1,\dots,\xi_n$ constructed are necessarily functions only of $\eta_1,\dots,\eta_n$, i.e., we do not exclude the possibility of extending the original probability space on which $\eta_1,\dots,\eta_n$ are defined. The fact that the extensions used below exist has been noted, for example, in [13].

Now we proceed to prove the Theorem. Since we can construct the variables $\lambda\xi_j/16$ instead of the ξ_j we consider without loss of generality only the case when $\lambda = 16$. Since $\mathcal{D}(16) \subset \mathcal{D}$, one can retain all the notation introduced in Subsection 3.4 and use Lemmas 4.2 - 4.5 in our case as well -- however, in that case the X_j also satisfy the condition

$$X_j \in \mathcal{D}(16) \quad \forall j \quad \text{for} \quad N \geqslant 0. \tag{8}$$

Next, we assume that $N \geqq 0$, since otherwise Theorem 10 is a corollary of Lemma 4.6. Let L and K be unique integers satisfying the inequalities

$$32 \leqslant L \leqslant b/2 < 2L, \quad L = 2^K. \tag{9}$$

By (8) and Lemma 2.6.7, there exist two mutually independent sequences $S_1,\dots,S_n$, $n = 2^N$, and $V_1,\dots,V_{nL}$, each of which consists of independent variables possessing the properties

$$S_j \in \mathcal{D}(8), \quad V_j \in \mathcal{D}(8), \quad 1 \leqslant \mathbf{D}V_j \leqslant 5/4 \quad \forall j, \tag{10}$$

and what is most essential, for all i the variables X_i and $S_i + V_{(i-1)L+1} + \dots + V_{iL}$ are identically distributed.

Let $U_1,\dots,U_{nL}$ be a sequence of independent random variables not depending on the sequence $S_1,\dots,S_n$, each of the U_j having normal distribution, and $\mathbf{D}V_j = \mathbf{D}U_j$. Set

$$Z_i = S_i + U_{(i-1)L+1} + \dots + U_{iL}. \tag{11}$$

Now assume that we are given only the sequence $\{Y_j\}$ defined in Subsection 3.4, of independent normal random variables, this sequence obviously satisfying the condition $\mathbf{D}Y_j = \mathbf{D}X_j + \mathbf{D}Z_j$ for all j. Concerning the variables $\{S_j, U_j, V_j, Z_j, X_j\}$ we suppose that they have not been defined yet and all the information given above is a necessary requirement for the distributions of these variables which we have yet to construct.

Let us now describe the construction. First, using the sequence $Y_1,\ldots,Y_n$, $n = 2^N$, by the method of Subsection 3.3 we define independent random variables $Z_1,\ldots,Z_n$ with the distributions determined above. Then, extending the probability space, we construct collectively independent random variables $S_1,\ldots,S_n$ and $V_1,\ldots,V_{nL}$ having the required distributions and satisfying equality (11) for the variables $\{Z_i\}$ just defined.

From the sequence $V_1,\ldots,V_{nL}$ already obtained by the method of Subsection 3.5 we construct independent variables $U_1,\ldots,U_n$ with the required distributions. Since the sequences $\{V_i\}$ and $\{S_i\}$ are independent and the independent variables $\{U_i\}$ are functions of $\{V_i\}$ only, the variables

$$X_i = S_i + V_{(i-1)L+1} + \ldots + V_{iL} \tag{12}$$

are independent and have the required distributions.

To complete the construction, it remains only to expand the probability space one more time, in order to complete the definition of the $\xi_1,\ldots,\xi_n$ and $\eta_1,\ldots,\eta_n$ with the required distributions, without changing the variables $\{Y_i,S_i,V_i,U_i,X_i\}$.

We turn now to the estimation of $\Delta \equiv \Delta(\xi,\eta)$. We have

$$\Delta(X,\ Y) \leqslant \Delta(X,\ Z) + \Delta(Z,\ Y), \quad \Delta(X,\ Z) \leqslant \Delta(V,\ U). \tag{13}$$

In deriving the second inequality in (13), we have used the fact that S_i is the same in (11) and (12). Thus, from (13) and (4.18) one has

$$\Delta \leqslant \mu + \delta + X + Y, \tag{14}$$

where

$$\mu = \Delta(Z,\ Y), \quad \delta = \Delta(V,\ U), \quad X = X^*(N), \quad Y = Y^*(N).$$

It follows from Lemma 4.4 that

$$(\mathbf{M}e^{\alpha X})(\mathbf{M}e^{\alpha Y}) \leqslant (2^3 \cdot 2^N e^b)^{2\alpha} = (8B^2 e^b/b)^{2\alpha} \tag{15}$$

for $0 < \alpha \leqq \frac{1}{2}$. Next, by (10) for $X_i = V_i$ and $Y_i = U_i$ the assertion of Theorem 9 holds true and therefore

$$\mathbf{M}e^{c\delta} \leqslant 2nL \leqslant 2(B^2/b)(b/2) = B^2. \tag{16}$$

L e m m a 1. If the conditions of Theorem 10 hold and $N \geqq 0$, then

$$Z_i \in \mathscr{D}_0(8) \qquad \forall i. \tag{17}$$

Proof. We need to show that $8Z_i \in \mathcal{D}_0$. Since $Z_i - S_i$ has normal distribution, then by Lemma 2.6.5 we just have to check

$$8^2\mathbf{D}(Z_i - S_i) \geqslant 8 + 48\ln(5^2\mathbf{D}Z_i) \quad \forall i. \tag{18}$$

Note that by (9),

$$\mathbf{D}(Z_i - S_i) \geqslant L \geqslant b/4 \geqslant (5/4)b/5 \geqslant \mathbf{D}Z_i/5 .$$

Therefore, to prove (18) we just have to be sure that

$$f(x) \equiv x/5 - 8 - 48\ln x \geqslant 0 \qquad \forall x \geqslant 3600, \tag{19}$$

since $x \equiv 8^2 DZ_i \geqq 8^2(3/4)80 \geqq 3600$. But $f'(x) = 1/5 - 48/x \geqq 0$ for $x \geqq 5\cdot 48$, and hence (19) is really true, since $f(3600) > 720 - 8 - 48\cdot 8 > 0$. ∎

Thus by (17), to find an estimate for $\mu = \Delta(Z,Y)$, we can use the assertion of Theorem 8 for $\xi_j = Z_j$, $\eta_j = Y_j$, $\sigma^2 \leqq \varepsilon = (5/4)b$, $\lambda = 1$, and get in the end that

$$\mathbf{M}e^{c\mu} \leqslant 1 + Be^b < 2Be^b, \tag{20}$$

since $B > 1$ for $N \geqq 0$.

Let us conclude the proof of Theorem 10. By Lemma 4.6, we assume that the c in (16) is equal to the c in (20) and they do not exceed 1/2. We have from (14) and Hölder's inequality that

$$\mathbf{M}e^{c\Delta/4} \leqslant (\mathbf{M}e^{c\mu}\mathbf{M}e^{c\delta}\mathbf{M}e^{cX}\mathbf{M}e^{cY})^{1/4}. \tag{21}$$

Substituting into (21) the estimates from (16), (20) and (15) for $\alpha = c$, we obtain

$$\mathbf{M}e^{c\Delta/4} = O(B^{(4c+1+2)/4}e^{(2cb+b)/4}). \tag{22}$$

Comparing (22) with (4.20) and noting that $N \geqq 0$, since $B \geqq 1$, we get the desired assertion from Lemma 4.6.

REFERENCES

[1] Prokhorov, Yu.V. "Convergence of Random Processes and Limit Theorems in Probability Theory." *Theory Probab. Applications*, vol.1, no.2 (1956): 157-214. (English transl.)

[2] Borovkov, A.A. "On the Rate of Convergence for the Invariance Principle." *Theory Probab. Applications*, vol.18, no.2 (1973): 207-225. (English transl.)

[3] Komlós, J., Major, P., and Tusnády, G. "Weak Convergence and Embedding." *Limit Theorems of Probability Theory*. Edited by P. Révész. (Keszthely, Hungary, 1974.) Colloquia Mathematica Societatis János Bolyai, 11, 149-165. Budapest: János Bolyai Mathematical Society and Amsterdam London: North-Holland Publishing Company, 1975.
[4] Strassen, V. "Almost Sure Behavior of Sums of Independent Random Variables and Martingales." *Proceedings of the Fifth Berkeley Sympos. Math. Statist. Probab.*, II. (Part 1), 315-343. Berkeley and Los Angeles: University of California Press, 1967.
[5] Komlós, J., Major, P., and Tusnády, G. "An Approximation of Partial Sums of Independent RV'-s, and the Sample DF. I." *Z. Wahrscheinlichkeitstheorie verw. Gebiete*, B.32, H.2 (1975): 111-131.
[6] ______. "An Approximation of Partial Sums of Independent RV'-s, and the Sample DF. II." *Z. Wahrscheinlichkeitstheorie verw. Gebiete*, B.34, H.1 (1976): 33-58.
[7] Sakhanenko, A.I. "Estimates of the Rate of Convergence in the Invariance Principle." *Soviet Math. Doklady*, vol.15, no.6 (1974): 1752-55. (English transl.)
[8] Arak, T.V. "On an Estimate of A.A. Borovkov." *Theory Probab. Applications*, vol.20, no.2 (1975): 372-373. (English transl.)
[9] Skorokhod, A.V. *Studies in the Theory of Random Processes*. Reading, Mass.: Addison-Wesley, 1965. (English transl.)
[10] Petrov, V.V. *Sums of Independent Random Variables*. New York Berlin Heidelberg: Springer-Verlag, 1975. (English transl.)
[11] ______. "On Large Deviations of Sums of Random Variables" (in Russian). *Vestnik Leningradskogo universiteta* 1 (1961): 25-37.
[12] Zolotarev, V.M. "An Absolute Estimate of the Remainder Term in the Central Limit Theorem." *Theory Probab. Applications*, vol.11, no.1 (1966): 95-105. (English transl.)
[13] Berkes, I., and Philipp, W. "Approximation Theorems for Independent and Weakly Dependent Random Vectors." *Ann. Probab.*, vol.7, no.1 (1979): 29-54.

S.S. Utev

INEQUALITIES FOR SUMS OF WEAKLY DEPENDENT RANDOM VARIABLES AND ESTIMATES OF THE CONVERGENCE RATE IN THE INVARIANCE PRINCIPLE

1. Introduction

This work deals with the approximation of distributions of partial sums of mixing sequences by the normal law.

We consider two kinds of mixing:

(a) strong mixing condition [1], and

(b) uniformly strong mixing condition [2].

In Sections 2-6, we prove assertions which generalize inequalities, known for sums of independent random variables and martingales, to sums of weakly dependent random variables. We prove here results we have announced in [3]-[5]. In Section 6 we give some applications of the results obtained to the strong law of large numbers and convergence of series.

In Sections 7, 8, using these inequalities, we obtain unimprovable estimates in the invariance principle for weakly stationary mixing sequences of random variables. We exploit the approximation theorems of Berkes and Philipp [6] and the unimprovable estimates in the invariance principle for independent random variables, obtained by Sakhanenko [7], [8].

The basic notation is introduced in Section 2.

2. Generalization of Rosenthal's moment inequality

Let $\{\xi_i\}_{i=1}^{\infty}$ be a sequence of random variables with values in a separable Hilbert space H and with zero expectations and let (x,y) and $|x|$ denote the scalar product and norm in H, respectively.

For any σ-algebras F and Q on a common probability space, set

$$\alpha(\mathcal{F}, Q) = \sup_{A\in\mathcal{F}, B\in Q} |\mathbf{P}(AB) - \mathbf{P}(A)\mathbf{P}(B)|,$$

$$\varphi(\mathcal{F}, Q) = \sup_{A\in\mathcal{F}, B\in Q} |\mathbf{P}(B\mid A) - \mathbf{P}(B)|.$$

We denote by M_a^b the σ-algebra generated by the random variables ξ_i, $a \leqq i \leqq b$. Set

$$\alpha(n) = \sup_{1\leqslant k<\infty} \alpha\left(M_1^k, M_{k+n}^{\infty}\right),$$

$$\varphi(n) = \sup_{1\leqslant k<\infty} \varphi\left(M_1^k, M_{k+n}^{\infty}\right),$$

$$L_t(n) = \sum_{i=1}^{n} \mathbf{E}|\xi_i|^t, \quad D_n^2 = L_2(n),$$

$$A_{r,x} = \sum_{i=1}^{n} \left(\mathbf{E}|\xi_i|^x\right)^{r/x},$$

$$S_n = \sum_{i=1}^{n} \xi_i, \quad \|\xi\|_p = \left(\mathbf{E}|\xi|^p\right)^{1/p},$$

$$L_t(n, \delta) = \sum_{i=1}^{n} \| \xi_i \|_{t+\delta}^t, \quad D_n^2(\delta) = L_2(n, \delta),$$

$$\delta > 0, \ j(t) = 2 \min \{k \in N: \ 2k \geqslant t\},$$

$$a(\varphi, t) = \sum_{k=0}^{\infty} \varphi^{1/j(t)}(k)(k+1)^{j(t)-2}, \quad \varphi(0) = 1,$$

$$b(\alpha, t, \delta) = \sum_{k=0}^{\infty} \alpha^{\delta/(j(t)+\delta)}(k)(k+1)^{j(t)-2}, \quad \alpha(0) = 1,$$

$$\varepsilon(\alpha, t, \delta) = \sum_{k=0}^{\infty} \alpha^{\delta/t(t+\delta)}(k),$$

$$c(t) = j(t)(j(t)-1)2^{j(t)-2} j(t)!, \ n = 1, \ldots,$$

$$Q_t(n) = L_t(n), \quad Q_t(n, \delta) = L_t(n, \delta) \qquad \text{if} \qquad t \leqslant 2,$$

$$Q_t(n) = \max\left(L_t(n), D_n^t\right),$$

$$Q_t(n, \delta) = \max\left(L_t(n, \delta), D_n^t(\delta)\right) \qquad \text{if} \qquad t \geqslant 2.$$

T h e o r e m 2.1. For $1 \leqq t < \infty$ we have the inequality

$$|\mathbf{E}|S_n|^t \leqslant c_1(t) a(\varphi, t) Q_t(n), \tag{2.1}$$

where $c'(t) = 16^{j(t)} c(t)$.

T h e o r e m 2.2. If $H = R^p$, then for $t \geqq 1$ we have the inequality

$$\mathbf{E}|S_n|^t \leqslant c_2(t) p^{j(t)/2} b(\alpha, t, \delta) Q_t(n, \delta) Q_t(n, \delta), \tag{2.2}$$

where

$$c_2(t) = 16^{j(t)} (j(t))^2 c(t) e^{\delta} .$$

COROLLARY 2.1. Let $M = \max_{1 \leqq i \leqq n} |\xi_i|^t$. For $t \geqq 1$ we have the inequalities

$$\mathbf{E}\left|\sum_{i=1}^{n} a_i \xi_i\right|^t \leqslant c_1(t) a(\varphi, t) M \left(\sum_{i=1}^{n} a_i^2\right)^{t/2}, \tag{2.3}$$

$$\mathbf{E}|S_n|^t \leqslant c_1(t) a(\varphi, t) M n^{t/2}. \tag{2.4}$$

COROLLARY 2.2. Let

$$M_j = \max_{1 \leqq i \leqq n} \| \xi_i \|_{t+\delta}^t , \qquad H = R^p .$$

For $t \geqq 1$ we have the inequalities

$$\mathbf{E}\left|\sum_{i=1}^{n} a_i \xi_i\right|^t \leqslant p^{j(t)/2} c_2(t) b(\alpha, t, \delta) M_\delta \left(\sum_{i=1}^{n} a_i^2\right)^{t/2}, \tag{2.5}$$

$$\mathbf{E}|S_n|^t \leqslant p^{j(t)/2} c_2(t) b(\alpha, t, \delta) \ M_\delta n^{t/2}. \tag{2.6}$$

In proving (2.1), (2.2), we first prove these inequalities for even-order moments (Section 3) and, next, give some results enabling us, using Rosenthal's moment inequality, to go from even-order moments to arbitrary moments (Section 4). Furthermore, it is not hard to see that in order to obtain estimates in a separable Hilbert space, it suffices to obtain estimates in $H = R^p$, not depending on dimension. Finally, for even-order moments, one can considerably relax the dependence on the mixing coefficients in inequalities (2.3) - (2.6).

For these reasons, it is convenient to state the even-order case and $H = R^p$ as a separate theorem.

Theorem 2.3. Let $x \geqq r \geqq 2$, where r is an even integer, $H = R^p$. We have the inequalities:

$$\mathbf{E}|S_n|^t \leqslant c_3(r)a_1(\varphi, r, x)A_{r,x}, \tag{2.7}$$

$$\mathbf{E}|S_n|^t \leqslant p^{r/2}c_4(r)a_2(\alpha, r, x)A_{r,x}. \tag{2.8}$$

For even r, (2.1) and (2.2) hold, where one can substitute $c(t)$ and $e^{\delta}t^2c(t)$ for the constants $c_1(t)$ and $c_2(t)$. In (2.7), (2.8)

$$a_1(\varphi, r, x) = \sum_{k=0}^{\infty} \varphi^{1-\frac{(r-1)}{x}}(k)(k+1)^{r/2-1}, \qquad c_3(r) = 4r(r-1)3^{r-2}r!,$$

$$a_2(\alpha, r, x) = \sum_{k=0}^{\infty} \alpha^{1-r/x}(k)(k+1)^{r/2-1}, \quad c_4(r) = 3c_3(r).$$

SUMS OF INDEPENDENT RANDOM VARIABLES AND MARTINGALES

Inequality (2.1) (as well as the inversity inequality) for $t \geqq 2$ for sums of independent real-valued random variables has been proved in [9], [10], and for $t \leqq 2$ for real-valued martingales in [11]. For generalizations to more general spaces and to martingales and also for bibliography, see [12], [13], [14], [15]. We note that we use the same order of dependence on $t(\sim(at)^t)$ as in [16].

MIXING

For strictly stationary sequences of real-valued random variables satisfying additional mixing conditions, results similar to inequalities (2.4) and (2.6) are obtained in [17], [18], [19], [20] (for generalizations, see [21], [22, Lemma 3.1], [23]). Analogs of inequalities (2.7) and (2.8) for a sequence of real-valued random variables for even-order moments are announced in [24] (proved only for $t = 4$). In all of these earlier inequalities, one observes no explicit dependence on the order of the moment nor on the mixing coefficients, although some of them are more precise in their own domains. S.N. Bernshtein was the first who pondered the proof of even-order moments [25], [26, pp. 154-155]. His ideas were elaborated in [20, Lemma 2.1]. Inequality (2.1) has been announced in [4], [5].

3. Proof of Theorem 2.3 for even-order moments

First we prove some auxiliary lemmas. By $\vec{i}_s$ we denote the vector with s-components $(i_1,\ldots,i_s)$. Set

$$I(n, s, r) = \{\vec{i}_s : 1 \leqslant i_k \leqslant n,\ k = 1, \ldots, s;\ 0 \leqslant i_k - i_{k-1} \leqslant r,\ k = 2, \ldots, s\},$$

$$I_0(n, s, r) = \{\vec{i}_s \in I(n, s, r):\ k \text{ is such that } i_k - i_{k-1} = r\}, \tag{3.1}$$

$$I(n, s, r, h) = \{\vec{i}_s \in I(n, s, r):\ i_h - i_{h-1} = r\},\quad h = 2, \ldots, s.$$

L e m m a 3.1. Let $s,n,r \in N$, $s \geqq 2$, $a_i \geqq 0$, $i = 1,\ldots,n$. We have the inequalities

$$\sum_{\vec{i}_s \in I_0(n,s,r)} \prod_{j=1}^{s} a_{i_j} \leqslant 4(s-1)\left(\sum_{i=1}^{n} a_i^2\right)^{s/2} (r+1)^{s/2-1}, \tag{3.2}$$

$$\sum_{\vec{i}_s \in I(n,s,r)} \prod_{j=1}^{s} a_{i_j} \leqslant 2\left(\sum_{i=1}^{n} a_i^2\right)^{s/2} (r+1)^{s/2}. \tag{3.3}$$

Proof. Set $a_i = 0$, $i \leqq 0$, $i > n$. The following inequalities are verified by straightforward computations and hence their proofs are omitted.

$$\sum_i \sum_{j=i+r} a_i a_j \leqslant \sum_{i=1}^{n} a_i^2, \tag{3.4}$$

$$\sum_i \sum_{i<j<i+r} a_i a_j \leqslant (r+1) \sum_{i=1}^{n} a_i^2, \tag{3.5}$$

$$\sum_i \sum_{i+r<j<i+2r} a_i a_{i+r} a_j \leqslant (r+1)^{1/2} \left(\sum_{i=1}^{n} a_i^2\right)^{3/2}, \tag{3.6}$$

$$\sum_i \sum_{i-r<j<i} a_j a_i a_{i+r} \leqslant (r+1)^{1/2} \left(\sum_{i=1}^{n} a_i^2\right)^{3/2}, \tag{3.7}$$

$$\sum_{\substack{i<j<k,\\ \max(j-i,k-j)=r}} a_i a_j a_k \leqslant 2(r+1)^{1/2}\left(\sum_{i=1}^{n} a_i^2\right)^{3/2}, \tag{3.8}$$

$$\sum_{\substack{i<j<k\\ \max(j-i,k-j)<r}} a_i a_j a_k \leqslant 2(r+1)^{3/2}\left(\sum_{i=1}^{n} a_i^2\right)^{3/2}, \tag{3.9}$$

$$\sum_{\substack{0<j_k-i_k<b_k,\\ k=1,\ldots,p}} (a_{i_1}a_{j_1})\ldots(a_{i_p}a_{j_p}) \leqslant \prod_{k=1}^{p}(b_k+1)\left(\sum_{i=1}^{n} a_i^2\right)^{p}. \tag{3.10}$$

We proceed to prove the required inequalities. First we prove (3.2). By (3.4) and (3.8), one can assume $s \geqq 4$. Set

$$I_k = \sum_{\vec{i}_s \in I(n,s,r,k)} \prod_{j=1}^{s} a_{i_j}, \quad k = 2, \ldots, s.$$

We estimate I_k. It s is even, then

$$I_k = \sum_{\vec{i}_s \in I(n,s,r,k)} (a_{i_1}\ldots a_{i_{k-2}})(a_{i_{k-1}}a_{i_k})(a_{i_{k+1}}\ldots a_{i_s})$$

$$\leqslant \left(\sum_i a_i a_{i+r}\right) \sum_{\substack{i_1<\ldots<i_{k-2}<i_{k+1}<\ldots<i_s,\\ i_{k+1}-i_{k-2}<3r,\\ i_q-i_{q-1}<r,\ q=2,\ldots,k-2,k+2,\ldots,s.}} a_{i_1}\ldots a_{i_{k-2}}a_{i_{k+1}}\ldots a_{i_s}.$$

By (3.4) and (3.10) we find that

$$I_k \leqslant 4(r+1)^{s/2-1}\left(\sum_{i=1}^{n} a_i^2\right)^{s/2},$$

which yields (3.2) for even s. The case of odd s is distinctive in that in estimating I_k it is necessary to distinguish the triple $a_{i_{k-2}}a_{i_{k-1}}a_{i_k}$ (or $a_{i_{k-1}}a_{i_k}a_{i_{k+1}}$ if $k = 2$) rather than the pair $a_{i_{k-1}}a_{i_k}$, and to use inequality (3.7) (or (3.6)) instead of (3.4).

Next we prove (3.3). If s is even, then the inequality follows from (3.10). If s is odd (by (3.9) one may assume $s \geqq 5$), then

$$\sum_{\vec{i}_s \in I(n,s,r)} \prod_{j=1}^{s} a_{i_j} \leqslant \left(\sum_{\vec{i}_3 \in I(n,3,r)} \prod_{j=1}^{3} a_{i_j}\right)\left(\sum_{\vec{i}_{s-3} \in I(n,s-3,r)} \prod_{j=1} a_{i_j}\right),$$

which together with (3.9) and (3.10) prove the Lemma.

L e m m a 3.2. Let

$$h,\ s,\ r\in N,\ 2\leqslant h\leqslant s,\ \delta\geqslant 0,\ su\leqslant\delta,\ x\leqslant\delta.$$

We have the inequalities

$$\sum_{\vec{i}_s\in I(n,s,r)}\left\|\prod_{j=1}^{s}|\xi_{i_j}|\right\|_{1+u}\leqslant s^{u/(1+u)}(r+1)^{s-1}L_s(n,\delta),$$

$$\sum_{\vec{i}_s\in I(n,s,r,h)}\left\|\prod_{j=1}^{h-1}|\xi_{i_j}|\right\|_{p_h}\left\|\prod_{j=h}^{s}|\xi_{i_j}|\right\|_{q_h}\leqslant cL_s(n,\delta),$$

where

$$p_h=(s+x)/(h-1),\quad q_h=(s+x)/(s-h+1),\quad c=(s^2/4)^{\delta/(s+\delta)}(r+1)^{s-2}.$$

Proof. We prove the first inequality. Since

$$|a_1\cdots a_s|\ \leqq\ (1/s)\sum_{i=1}^{s}|a_i|^s\ ,$$

then

$$\left\|\prod_{j=1}^{s}|\xi_{i_j}|\right\|_{1+u}\leqslant s^{-1/(1+u)}\sum_{k=1}^{s}\left\||\xi_{i_k}|^s\right\|_{1+u}.$$

We have

$$\sum_{\vec{i}_s\in I(n,s,r)}\left\|\prod_{j=1}^{s}|\xi_{i_j}|\right\|_{1+u}\leqslant s^{-1/(1+u)}\sum_{k=1}^{s}\sum_{p=1}^{n}\left\||\xi_p|^s\right\|_{1+u}\left(\sum_{\substack{\vec{i}_s\in I(n,s,r),\\ i_k=p}}1\right)$$

$$\leqslant s^{-1/(1+u)}\sum_{k=1}^{s}\sum_{p=1}^{n}\left\||\xi_p|^s\right\|_{1+u}(r+1)^{s-1}=s^{u/(1+u)}(r+1)^{s-1}L_s(n,\delta),$$

since

$$\left\||\xi_p|^s\right\|_{1+u}\leqslant\left\||\xi_p|\right\|_{s+\delta}^{s}.$$

Next we prove the second inequality. Set $a = (h-1)/s$, $b = (s-h+1)/s$. Applying Hölder's inequality, we obtain

$$\sum_{\vec{i}_s\in I(n,s,r,h)}\left\|\prod_{j=1}^{h-1}|\xi_{i_j}|\right\|_{p_h}\left\|\prod_{j=h}^{s}|\xi_{i_j}|\right\|_{q_h}\leqslant\left(\sum_{\vec{i}_s\in I(n,s,r,h)}\left\|\prod_{j=1}^{h-1}|\xi_{i_j}|\right\|_{p_h}^{1/a}\right)^{a}$$
$$\times\left(\sum_{\vec{i}_s\in I(n,s,r,h)}\left\|\prod_{j=h}^{s}|\xi_{i_j}|\right\|_{q_h}^{1/b}\right)^{b}. \qquad (3.11)$$

Note that by the conditions of the Lemma,

$$\left\| \prod_{j=1}^{h-1} |\xi_{i_j}| \right\|_{p_h}^{1/a} \leqslant \left\| \prod_{j=1}^{h-1} |\xi_{i_j}|^{1/a} \right\|_{1+y}, \tag{3.12}$$

$$\left\| \prod_{j=h}^{s} |\xi_{i_j}| \right\|_{q_h}^{1/b} \leqslant \left\| \prod_{j=h}^{s} |\xi_{i_j}|^{1/b} \right\|_{1+y}, \tag{3.13}$$

where $y = ap_h - 1 = bq_h - 1 = x/s \leqq \delta/s$. Noting (3.12), (3.13) and estimating the factors on the right side of inequality (3.11) as in proving the preceding inequality, we obtain the required result. The Lemma is proved.

L e m m a 3.3. For any $\delta \geqq 0$, $2 < t < \omega$ we have the inequality

$$Q_t(n, \delta) \leqslant (Q_\omega(n, \delta))^{t/\omega}.$$

Proof. It suffices to show that

$$L_t(n, \delta) \leqslant (Q_\omega(n, \delta))^{t/\omega}.$$

since we have

$$\mathbf{E}|XY| \leqslant (\mathbf{E}|X|^r)^{1/r}(\mathbf{E}|Y|^s)^{1/s}$$

only if $r,s > 0$, $1/r + 1/s \leqq 1$, for

$$x + v = t + \delta, \quad x, v > 0, \qquad 1/p + 1/q \leqslant 1, \quad p, q > 0, \tag{3.14}$$

we have

$$L_t(n, \delta) \leqslant \sum_{i=1}^{n} (\mathbf{E}|\xi_i|^{xp})^{t/p(t+\delta)} (\mathbf{E}|\xi_i|^{vq})^{t/q(t+\delta)}.$$

Applying Hölder's inequality, we find that

$$L_t(n, \delta) \leqslant \left(\sum_{i=1}^{n} (\mathbf{E}|\xi_i|^{xp})^{ta/p(t+\delta)} \right)^{1/a} \left(\sum_{i=1}^{n} (\mathbf{E}|\xi_i|^{vq})^{tb/q(t+\delta)} \right)^{1/b}, \qquad 1/a + 1/b = 1, \quad a, b > 0. \tag{3.15}$$

Let

$$xp \leqslant \omega + \delta, \ vq \leqslant 2 + \delta. \tag{3.16}$$

Then

$$L_t(n, \delta) \leqslant \left(\sum_{i=1}^{n} (\mathbf{E}|\xi_i|^{\omega+\delta})^{xta/(t+\delta)(\omega+\delta)} \right)^{1/a} \left(\sum_{i=1}^{n} (\mathbf{E}|\xi_i|^{2+\delta})^{vbt/(t+\delta)(2+\delta)} \right)^{1/b}.$$

To prove the Lemma it is necessary that the relations

$$\frac{xta}{(t+\delta)(\omega+\delta)} = \frac{\omega}{\omega+\delta},$$
$$\frac{vtb}{(t+\delta)(2+\delta)} = \frac{2}{2+\delta}, \qquad (3.17)$$
$$1/a + 2/\omega b = t/\omega$$

hold.

In sum, to prove the Lemma it suffices to show that there exist positive x, v, p, q, a, b such that relations (3.14)-(3.17) hold. This is routinely checked. The Lemma is proved.

Now we can proceed to prove the Theorem.

Let us introduce the notation:

$$\|x\| = \left(\sum_{i=1}^{p} x_i^2\right)^{\frac{1}{2}}, \qquad x = (x_1,\dots,x_p), \qquad \xi_j = (\xi_j^{(1)},\dots,\xi_j^{(p)}), \qquad j = 1,\dots,n$$

$|A|$ is the cardinality of the set A. For any finite subsets of natural numbers A and B, we say that $A < B$ is $x < y$ for any $x \in A$ and $y \in B$. By $\vec{N}_t = (N_1,\dots,N_t)$ we denote the set of nonempty subsets of $\{1,\dots,r\}$ such that $N_i < N_{i+1}$, $i = 1,\dots,t-1$, and $\bigcup_{n=1}^{t} N_i = \{1,\dots,r\}$. Also, Q_t is the union of all such sets $\vec{N}_t$ for fixed t, $1 \leqq t \leqq r$. Let $\vec{j}_r(\vec{q}_r)$ denote a vector $(j_1,\dots,j_r)$ (respectively, $(q_1,\dots,q_r)$) such that $j_k \in \{1,\dots,n\}(q_k \in \{1,\dots,p\})$, $k = 1,\dots,r$. Let ε denote a permutation of the set $\{1,\dots,r\}$. We shall use the notation (3.1) in the sequel. Set

$$I = \{\vec{j}_r\colon\ j_k \leqslant j_{k+1},\ k = 1, \dots, r-1\}\ ',$$

$$I_\varepsilon = \{\vec{j}_k\colon \text{ it follows from } \varepsilon(i) \text{ that } j_k \leqslant j_i\ ,\ 1 \leqslant k\ ,\ i \leqslant r\}\ ,$$

$$M_\varepsilon = \{\vec{q}_r\colon\ q_{\varepsilon(2k)} = q_{\varepsilon(2k-1)},\ k = 1, \dots, r/2\}.$$

Let $\mathbf{A}_t$ denote the class of all functions $\tilde{N}_t$ from I into Q_t. Let $N_k(\vec{j}_r)$ denote the *kth* set from the left in the union $\tilde{N}_t(\vec{j}_r) \in Q_t$. For every permutation we set

$$B_\varepsilon(\widetilde{N}_t) = \sum_{\vec{j}_r\in I}\ \sum_{\vec{q}_r\in M_\varepsilon}\ \prod_{h=1}^{t}\left|\mathbf{E}\left(\prod_{i\in N_h(\vec{j}_r)} \xi_{j_i}^{(q_i)}\right)\right|,$$

$$\beta(\vec{N}_t) = \min_{1\leqslant i\leqslant t} |N_i|, \quad \widetilde{\beta}(\widetilde{N}_t) = \max_{\vec{j}_r\in I} \beta(\widetilde{N}_t(\vec{j}_r)).$$

A few remarks are in order.

1. $\mathbb{A}_1$ consists of one function $\tilde{N}_1$ which carries each vector $\vec{j}_r \in I$ to the unique collection $\vec{N}_1 = (\{1,\dots,r\})$. Hence, for every ε, one has

$$B_\varepsilon(\tilde{N}_1) = \sum_{\vec{j}_r \in I} \sum_{\vec{q}_r \in M_\varepsilon} \left| \mathbf{E} \prod_{i=1}^{r} \xi_{j_i}^{(q_i)} \right|.$$

2. It is not hard to see that by the construction, one has

$$\mathbf{E}\|S_n\|^r = \sum_{j_1,\dots,j_r=1} \sum_{q_1,\dots,q_{r/2}=1} \left(\xi_{j_1}^{(q_1)} \xi_{j_2}^{(q_1)} \dots \right.$$

$$\left. \dots \xi_{j_{r-1}}^{(q_{r/2})} \xi_{j_r}^{(q_{r/2})} \right) \leqslant \sum_{\varepsilon} \sum_{\vec{j}_r \in I_\varepsilon} \sum_{q_1,\dots,q_{r/2}=1} \left| \mathbf{E} \left(\xi_{j_1}^{(q_1)} \xi_{j_2}^{(q_1)} \dots \xi_{j_{r-1}}^{(q_{r/2})} \xi_{j_{r-1}}^{(q_{r/2})} \xi_{j_r}^{(q_{r/2})} \right| \leqslant \right.$$

(the transformations $\vec{j}_r$ corresponding to the permutation ε)

$$\leqslant \sum_{\varepsilon} \sum_{\vec{j}_r \in I} \sum_{\vec{q}_r \in M_\varepsilon} \left| \mathbf{E} \prod_{i=1}^{r} \xi_{j_i}^{(q_i)} \right| = \sum_{\varepsilon} B_\varepsilon(\tilde{N}_1).$$

As a result, we have

$$\mathbf{E}\|S_n\|^r \leqslant r! \ \max_{\varepsilon} B_\varepsilon(\tilde{N}_1). \tag{3.18}$$

3. If $\beta(\tilde{N}_t(\vec{j}_r)) = 1$, then for such a $\vec{j}_r$,

$$\vec{j}_r \sum_{\vec{q}_r \in M_\varepsilon} \prod_{k=1}^{t} \left| \mathbf{E} \prod_{i \in N_k(\vec{j}_r)} \xi_{j_i}^{(q_i)} \right| = 0$$

since $\mathrm{E}\xi_j^{(i)} = 0$ for $1 \leqq j \leqq n$, $1 \leqq i \leqq p$.

4. If $\tilde{\beta}(\tilde{N}_t) = 1$, then $B_\varepsilon(\tilde{N}_t) = 0$. In particular, since $\tilde{\beta}(\tilde{N}_{r/2+1}) = 1$ for every $\tilde{N}_{r/2+1}$, then

$$B_\varepsilon(\tilde{N}_{r/2+1}) = 0 \tag{3.19}$$

for all functions $\tilde{N}_{r/2+1}$ and all permutations ε.

5. The idea of the proof of the theorem is to construct for each function $\tilde{N}_t$, $t \leqq r/2$, a function $\tilde{N}_{t+1}$ not depending on ε and such that

$$B_\varepsilon(\tilde{N}_t) \leqslant B_\varepsilon(\tilde{N}_{t+1}) + c(t, \cdot), \tag{3.20}$$

where $c(t,\cdot)$ in each of the cases (2.1), (2.2), (2.7), (2.8) differs from the right sides of the corresponding inequalities only by a constant depending on t and r.

First let us construct the function $\tilde{N}_{t+1}$ for the function $\tilde{N}_t$, $t \leqq r/2$. Fix $\vec{j}_r \in I$. Let $(N_1,\ldots,N_t)$ correspond to this vector. For every finite subset of integers (if $|A| \leqq 1$, we set $H(A) = 0$), set

$$H(A) = \max_{x\in A}\min_{\substack{y\in A\\ x\neq y}}|x-y|,\quad h(A) = \max\{x\in A\colon \exists y\in A,\ x-y = H(A)\},$$

$$W(A) = \{x\in A\colon x < h(A)\},\ \varepsilon(A) = \{x\in A\colon x \geqslant h(A)\}.$$

Then we set

$$A_k = A_k(\vec{j}_r) = \{j_i\colon i\in N_k\},\ k = 1,\ldots,t,$$

$$v = v(\vec{j}_r) = \max_{1\leqslant k\leqslant t} H(A_k),\quad k_0 = k_0(\vec{j}_r) = \min\{k\colon H(A_k) = v,\ |N_k| > 1\} \tag{3.21}$$

(since $t \leqq r/2$, there exists a k_1 such that $|N_{k_1}| > 1$). Let N'_{k_0} be the set of numbers in N_{k_0} to which the numbers of $W(A_{k_0})$ correspond, and N''_{k_0} the set of numbers in N_{k_0} to which the numbers from $\varepsilon(A_{k_0})$ correspond. By construction, each N'_{k_0} and N''_{k_0} is nonempty and $N'_{k_0} < N''_{k_0}$ (if $|A_{k_0}| = 1$, the construction is obvious).

It remains to set

$$\tilde{N}_{t+1}(\vec{j}_r) = (N_1,\ldots,N_{k_0-1},N'_{k_0},N''_{k_0},N_{k_0+1},\ldots,N_t).$$

The function $\tilde{N}_{t+1}$ is constructed.

Now, we proceed to estimate the difference $B_\varepsilon(\tilde{N}_t) - B_\varepsilon(\tilde{N}_{t+1})$. Since $||a| - |b|| \leqq |a-b|$, we have

$$\begin{aligned} B_\varepsilon(\tilde{N}_t) - B_\varepsilon(\tilde{N}_{t+1}) \leqslant \sum_{\vec{j}_r\in I}\sum_{\vec{q}_r\in M_\varepsilon}\prod_{\substack{k=1,\\ k\neq k_0}}^{t}\left|\mathbf{E}\prod_{i\in N_k}\xi_{j_i}^{(q_i)}\right|\Bigg|\mathbf{E}\Bigg(\prod_{i\in N_{k_0}}\xi_{j_i}^{(q_i)}\Bigg) \\ -\mathbf{E}\Bigg(\prod_{i\in N'_{k_0}}\xi_{j_i}^{(q_i)}\Bigg)\mathbf{E}\Bigg(\prod_{i\in N''_{k_0}}\xi_{j_i}^{(q_i)}\Bigg)\Bigg|, \end{aligned} \tag{3.22}$$

where N_k, k_0, N'_{k_0}, N''_{k_0} depend on $\vec{j}_r$. Note that in (3.22) the summands are nonzero only if

$$\beta(\tilde{N}_t(\vec{j}_r)) > 1. \tag{3.23}$$

Now we fix the vector $\vec{j}_r$ satisfying (3.23) and estimate the remainder (the index of the dependence of N_k, k_0, N'_{k_0}, N''_{k_0} on $\vec{j}_r$ will be omitted). Let

$$T_\varepsilon = \sum_{\vec{q}_r \in M_\varepsilon} \left(\left| \prod_{\substack{k=1,\\ k\neq k_0}}^{t} \mathbf{E}\left(\prod_{i\in N_k} \xi_{j_i}^{(q_i)} \right) \right| \right) \times \left| \mathbf{E}\left(\prod_{i\in N_{k_0}} \xi_{j_i}^{(q_i)} \right) - \mathbf{E}\left(\prod_{i\in N'_{k_0}} \xi_{j_i}^{(q_i)} \right) \mathbf{E}\left(\prod_{i\in N''_{k_0}} \xi_{j_i}^{(q_i)} \right) \right|.$$

Our goal is to estimate T_ε. First we consider the uniform strong mixing case.

L e m m a 3.4. The following inequality holds:

$$T_\varepsilon \leqslant 2\varphi^{1/p_1}(v) \prod_{\substack{k=1,\\ k\neq k_0}}^{t} \mathbf{E}\left(\prod_{i\in N_k} \|\xi_{j_i}\| \right) \left\| \left(\prod_{i\in N'_{k_0}} \|\xi_{j_i}\| \right) \right\|_{p_1} \left\| \left(\prod_{i\in N''_{k_0}} \|\xi_{j_i}\| \right) \right\|_{q_1},$$

where $1/p_1 + 1/q_1 = 1$, $p_1, q_1 > 0$, v is defined while constructing the $\tilde{N}_{t+1}$ (see (3.21)).

Proof. It is not hard to see that it suffices to consider only the case where the $\xi_j^{(i)}$ are simple random variables, at the same time assuming that

$$\begin{aligned} \xi_{j_i}^{(q_i)} &= \sum_{a\in X} \Psi(j_i, q_i, a)\, I(A_a), \quad i\in N_k, \quad k\neq k_0, \\ \xi_{j_i}^{(q_i)} &= \sum_{b\in Y} \Psi(j_i, q_i, b)\, I(B_b), \quad i\in N'_{k_0}, \\ \xi_{j_i}^{(q_i)} &= \sum_{d\in Z} \Psi(j_i, q_i, d)\, I(C_d), \quad i\in N''_{k_0}, \end{aligned} \tag{3.24}$$

where X, Y, Z are finite subsets of integers

$$A_a \in \sigma(\xi_j,\ 1\leqslant j\leqslant n), \quad B_b \in \sigma\left(\xi_{j_i},\ i\in N'_{k_0}\right), \ C_d \in \sigma\left(\xi_{j_i},\ i\in N''_{k_0}\right),$$
$$A_{a_1}A_{a_2} = B_{b_1}B_{b_2} = C_{d_1}C_{d_2} = \Phi$$

if $a_1 \neq a_2$, $b_1 \neq b_2$, $d_1 \neq d_2$. Substituting (3.24) into the expression for T_ε, after simple, but laborious calculations, we obtain

$$T_\varepsilon \leqslant \sum_{\substack{\vec{a}\in X^{t-1}\\ b\in Y, d\in Z}} \left(\prod_{\substack{k=1\\ k\neq k_0}}^{t} \mathbf{P}(A_k) \right) |\mathbf{P}(B_b C_d) - \mathbf{P}(B_b)\mathbf{P}(C_d)| \sum_{\vec{q}_r\in M_\varepsilon} L_\varepsilon(\vec{a}, b, d, \vec{q}_r),$$

where

$$\vec{a} = (a_1, \ldots, a_{k_0-1}, a_{k_0+1}, \ldots, a_t), L_\varepsilon(\vec{a}, b, d, \vec{q}_r) = \prod_{i \in N'_{k_0}} |\Psi(j_i, q_i, b)|$$

$$\times \prod_{i \in N''_{k_0}} |\Psi(j_i, q_i, d)| \prod_{\substack{k=1 \\ k \neq k_0}}^{t} \prod_{i \in N_k} |(\Psi(j_i, q_i, a_k)|.$$

Let us estimate

$$\sum_{\vec{j}_r \in M_\varepsilon} I_\varepsilon(\vec{a}, b, d, \vec{q}_r) \quad .$$

For the moment we ignore the dependece of the functions Ψ on $\vec{a}$, b, d and recall that $q_{\varepsilon(2k)} = q_{\varepsilon(2k-1)}$, $k = 1, \ldots, r/2$ on the set M_ε. Starting from these considerations, we find that

$$\sum_{\vec{q}_r \in M_\varepsilon} L_\varepsilon(\vec{a}, b, d, \vec{q}_r) = \sum_{\vec{q}_r \in M_\varepsilon} \prod_{i=1}^{r} |\Psi(j_i, q_i, \cdot)| =$$

(recall the Cauchy-Buniakowski inequality and note that the permutation is a one-to-one mapping)

$$= \prod_{k=1}^{r/2} \sum_{q=1}^{p} |\Psi(j_{\varepsilon(2k)}, q, \cdot)\, \Psi(j_{\varepsilon(2k-1)}, q, \cdot)| \leqslant \prod_{k=1}^{r} \left(\sum_{q=1}^{p} \Psi^2(j_k, q, \cdot) \right)^{1/2} .$$

Set

$$g(a, k) \doteq \prod_{i \in N_k} \left(\sum_{q=1}^{p} \Psi^2(j_i, q, a) \right)^{1/2}, \quad k \neq k_0, \quad a \in X,$$

$$g_1(b) = \prod_{i \in N'_{k_0}} \left(\sum_{q=1}^{p} \Psi^2(j_i, q, b) \right)^{1/2}, \quad b \in Y,$$

$$g_2(b) = \prod_{i \in N''_{k_0}} \left(\sum_{q=1}^{p} \Psi^2(j_i, q, d) \right)^{1/2}, \quad d \in Z.$$

Recalling the dependence of Ψ on $\vec{a}$, b, d, and noting that $\sum_{i \in A, j \in B} x_i y_j = \sum_{i \in A} x_i \sum_{j \in B} y_j$ for all finite sets A and B, we obtain

$$T_\varepsilon \leqslant \prod_{\substack{k=1, \\ k \neq k_0}}^{t} \left(\sum_{a \in X} g(a, k)\, \mathbf{P}(A_a) \right) \sum_{b \in Y, d \in Z} g_1(b)\, g_2(d)\, |\mathbf{P}(B_b C_d) - \mathbf{P}(B_b)\, \mathbf{P}(C_d)|.$$

Taking into consideration that by the construction

$$\varphi\left(\sigma\left(\xi_{j_i},\ i\in N'_{k_0}\right),\quad \sigma\left(\xi_{j_i},\ i\in N''_{k_0}\right)\right)\leqslant\varphi(v),$$

and using the device employed in Lemma 1.1 [18], we find that

$$\sum_{a\in Z}\mathbf{P}(A_a)\,g(a,k)=\mathbf{E}\Big(\prod_{i\in N_k}\|\xi_{j_i}\|\Big),$$

$$\sum_{b\in Y,d\in Z}g_1(b)\,g_2(d)\,|\,\mathbf{P}(B_bC_d)$$

$$-\mathbf{P}(B_b)\,\mathbf{P}(C_d)\,|\leqslant 2\varphi^{1/p_1}(v)\Big\|\Big(\prod_{i\in N'_{k_0}}\|\xi_{j_i}\|\Big)\Big\|_{p_1}\times\Big\|\Big(\prod_{i\in N''_{k_0}}\|\xi_{j_i}\|\Big)\Big\|_{q_1},$$

where $p_1, q_1 > 0$, $1/p_1 + 1/q_1 = 1$. The foregoing formulas and the inequality for T_ε imply the required result. The Lemma is proved. For the case (2.1), Lemma 3.4 with $p_1 = |N_{k_0}| / |N'_{k_0}|$ and $q_1 = |N_{k_0}| / |N''_{k_0}|$ will be the starting point for investigating (3.22). For the case (2.7), the next lemma will be the result we need.

L e m m a 3.5. For every $x \geqq r$, the following inequality holds:

$$T_\varepsilon\leqslant 2\varphi^{1-(r-1)/x}(v)\prod_{k=1}^{t}\prod_{i\in N_k}\mathbf{E}^{1/x}\|\xi_{j_i}\|^x.$$

Proof. For $k \neq k_0$ we have (because $x \geqq |N_k|$) by Hölder's inequality that

$$\mathbf{E}\Big(\prod_{i\in N_k}\|\xi_{j_i}\|\Big)\leqslant\prod_{i\in N_k}\mathbf{E}^{1/x}\|\xi_{j_i}\|^x. \tag{3.25}$$

Set $q_1 = x / |N''_{k_0}|$, $p_1 = q_1/(1-q_1)$ and note that

$$1/p_1+1/q_1=1,\quad p_1,q_1>0,\quad -p_1^{-1}\leqslant 1-(r-1)/x,\quad p_1\,|N'_{k_0}|\leqslant x.$$

Again by Hölder's inequality,

$$\Big\|\Big(\prod_{i\in N'_{k_0}}\|\xi_{j_i}\|\Big)\Big\|_{p_1}\leqslant\prod_{i\in N'_{k_0}}\mathbf{E}^{1/x}\|\xi_{j_i}\|^x,$$
$$\Big\|\Big(\prod_{i\in N''_{k_0}}\|\xi_{j_i}\|\Big)\Big\|_{q_1}\leqslant\prod_{i\in N''_{k_0}}\mathbf{E}^{1/x}\|\xi_{j_i}\|^x. \tag{3.26}$$

Inequalities (3.25)-(3.26) and Lemma 3.4 yield the required result. The Lemma is proved.

We proceed to estimate T_ε in the strong mixing case. By Davydov's inequality [27, Corollary], [23, Appendix], one has

$$\left|\mathbf{E}\left(\prod_{i\in N_{k_0}} \xi_{j_i}^{(q_i)}\right) - \mathbf{E}\left(\prod_{i\in N'_{k_0}} \xi_{j_i}^{(q_i)}\right)\mathbf{E}\left(\prod_{i\in N''_{k_0}} \xi_{j_i}^{(q_i)}\right)\right|$$
$$\leqslant 6\alpha^{1-1/p_2-1/q_2}(v)\left\|\prod_{i\in N'_{k_0}} \left|\xi_{j_i}^{(q_i)}\right|\right\|_{p_2}\left\|\prod_{i\in N''_{k_0}} \left|\xi_{j_i}^{(q_i)}\right|\right\|_{q_2},$$

where $p_2, q_2 > 0$, $1/p_2 + 1/q_2 \leqq 1$. Noting that

$$|M_\varepsilon| = p^{r/2} \quad \text{and} \quad |\xi_j^{(i)}| \leqslant \|\xi_j\|, \quad j = 1, \dots, n, \quad i = 1, \dots, p,$$

for the case (2.2) we find that

$$T_\varepsilon \leqslant 6p^{r/2}\alpha^{1-1/p_2-1/q_2}(v)\prod_{\substack{k=1,\\ k\neq k_0}}^{t} \mathbf{E}\left(\prod_{i\in N_k}\|\xi_{j_i}\|\right)$$
$$\times\left\|\left(\prod_{i\in N'_{k_0}}\|\xi_{j_i}\|\right)\right\|_{p_2}\left\|\prod_{i\in N''_{k_0}}\|\xi_{j_i}\|\right)\right\|_{q_2}, \tag{3.27}$$

where

$$p_2 = (|N_{k_0}| + \delta)/|N'_{k_0}|, \quad q_2 = (|N_{k_0}| + \delta)/|N''_{k_0}|.$$

For the case (2.8), in Lemma 3.5), we obtain

$$T_\varepsilon \leqslant 6p^{r/2}\alpha^{1-r/x}(v)\prod_{k=1}^{t}\prod_{i\in N_k}\mathbf{E}^{1/x}\|\xi_{j_i}\|^x. \tag{3.28}$$

Thus, using Lemmas 3.4 and 3.5 and inequalities (3.27), (3.28), (3.22), we have obtained: for (2.7)

$$B_\varepsilon(\widetilde{N}_t) \leqslant 6_\varepsilon(\widetilde{N}_{t+1}) + 2\sum_{\vec{j}_r\in I_0}\varphi^{1-(r-1)/x}(v)A(t, \widetilde{N}_t, \vec{j}_r, \xi), \tag{3.29}$$

for (2.8)

$$B_\varepsilon(\widetilde{N}_t) \leqslant B_\varepsilon(\widetilde{N}_{t+1}) + 6p^{r/2}\sum_{\vec{j}_r\in I_0}\alpha^{1-r/x}(v)A(t, \widetilde{N}_t, \vec{j}_r, \xi), \tag{3.30}$$

for (2.1)

$$B_\varepsilon(\widetilde{N}_t) \leqslant B_\varepsilon(\widetilde{N}_{t+1}) + 2\sum_{\vec{j}_r\in I_0}\varphi^{1/r}(v)\prod_{\substack{k=1,\\ k\neq k_0}}^{t}\mathbf{E}\left(\prod_{i\in N_k}\|\xi_{j_i}\|\right)$$
$$\times\left\|\left(\prod_{i\in N'_{k_0}}\|\xi_{j_i}\|\right)\right\|_{p_1}\left\|\left(\prod_{i\in N''_{k_0}}\|\xi_{j_i}\|\right)\right\|_{q_1}, \tag{3.31}$$

for (2.2)

$$B_\varepsilon(\tilde{N}_t) \leqslant B_\varepsilon(\tilde{N}_{t+1}) + 6p^{r/2} \sum_{\vec{j}_r \in I_0} \alpha^{\delta/(r+\delta)}(v) \prod_{\substack{k=1,\\ k\neq k_0}}^{t} \mathbf{E}\Big(\prod_{i\in N_k} \|\xi_{j_i}\|\Big)$$
$$\times \Big\|\Big(\prod_{i\in N'_{k_0}} \|\xi_{j_i}\|\Big)\Big\|_{p_2} \Big\|\Big(\prod_{i\in N''_{k_0}} \|\xi_{j_i}\|\Big)\Big\|_{q_2}, \tag{3.32}$$

where

$$A(t, \tilde{N}_t, \vec{j}_r, \xi) = \prod_{k=1}^{t} \prod_{i\in N_k} \mathbf{E}^{1/x} \|\xi_{j_i}\|^x,$$

$$I_0 = \{\vec{j}_r \in I: \beta(\tilde{N}_t(\vec{j}_r)) > 1\}, \quad p_1 = |N_{k_0}|/|N'_{k_0}|,$$
$$q_1 = |N_{k_0}|/|N''_{k_0}|, \quad p_2 = p_1 + \delta/|N'_{k_0}|, \quad q_2 = p_2 + \delta/|N''_{k_0}|.$$

Note that p_1, q_1, p_2, q_2, v, N_k, k_0, N'_{k_0}, N''_{k_0} depend on $\vec{j}_r$ via $\tilde{N}_t$, $\tilde{N}_{t+1}$.

We have come to the concluding part of the proof. We need to estimate the sums on the right sides of the above inequalities. Note that the estimates for (2.7) and (2.8), actually contain a sum of the form

$$A(\eta, b) = \sum_{\vec{j}_r \in I_0} \eta(v(\vec{j}_r)) \prod_{k=1}^{t} \prod_{i\in N_k(\vec{j}_r)} b_{j_i},$$

where $\eta(v) \geqq 0$, $b_i \geqq 0$ and by construction (see (3.21)),

$$H(\{j_i: i \in N_k(\vec{j}_r)\}) \leqslant v(\vec{j}_r), \quad 0 \leqslant v(\vec{j}_r) \leqslant n-1,$$
$$H(\{j_i: i \in N_{k_0}(\vec{j}_r)\}) = v(\vec{j}_r) = \min_{x\in N'_{k_0}, y\in N''_{k_0}}, j_x - j_y.$$

An estimate for $A(\eta,b)$ is given in the next lemma.

L e m m a 3.6. Under the conditions stated, the following inequality holds:

$$A(\eta, b) \leqslant \Big(\sum_{a=0}^{n-1} \eta(a)(a+1)^{r/2-1}\Big) \Big(\sum_{i=1}^{n} b_i^2\Big)^{r/2} t2^{t+1}(r-1)C_{r-2}^{t-1}.$$

Proof. Let us fix $\vec{N}_t$ so that $\beta(\vec{N}_t) > 1$. Set $|N_k| = s_k$, $k = 1,\dots,t$, $b_i = 0$, $i \leqq 0$, $i > n$,

$$I(\vec{N}_t, a, b) = \{\vec{j}_r \in I_0: \tilde{N}_t(\vec{j}_r) = \vec{N}_t, \quad k_0(\vec{j}_r) = b, \quad v(\vec{j}_r) = a\}.$$

We have

$$A(\eta, b) \leqslant \sum_{a=0}^{n-1} \eta(a) \sum_{\substack{\vec{N}_t \in Q_t, \\ \beta(\vec{N}_t) > 1}} \sum_{b=1}^{t} \sum_{\vec{j}_r \in I(\vec{N}_t, a, b)} \prod_{h=1}^{t} \prod_{i \in N_k} b_{j_i}. \tag{3.33}$$

By construction, we find that

$$\sum_{\vec{j}_r \in I(\vec{N}_t, a, b)} \prod_{k=1}^{t} \prod_{i \in N_k} b_{j_i} \leqslant \left(\sum_{\vec{i}_{s_b} \in I_0(n, s_b, a)} \prod_{j=1}^{s_b} b_{i_j} \right) \times \prod_{\substack{k=1, \\ k \neq k_0}}^{t} \left(\sum_{\vec{i}_{s_k} \in I(n, s_k, a)} \prod_{j=1}^{s_k} b_{i_j} \right).$$

Applying Lemma 3.1 to the individual sums of the right side of the foregoing inequality, we obtain

$$\sum_{\vec{j}_r \in I(\vec{N}_t, a, b)} \prod_{k=1}^{t} \prod_{i \in N_k} b_{j_i} \leqslant 2^{t+1} \left(\sum_{i=1}^{n} b_i^2 \right)^{r/2} (|N_b| - 1)(a+1)^{r/2-1}.$$

Substituting the preceding inequality into (3.33) and noting that $|Q_t \cap \{\vec{N}_t : \beta(\vec{N}_t) > 1\}| \leq C_{r-2}^{t-1}$ yields the required inequality. The Lemma is proved. Next, we estimate (3.31). We fix $\vec{N}_t$, a, b, h such that $\beta(\vec{N}_t) > 1$, $0 \leqq a \leqq n-1$, $1 \leqq b \leqq t$, $1 \leqq h \leqq |N_b| - 1$ (as earlier, $|N_k| = S_k \geqq 2$). Let R denote the sum on the right side of (3.31). Set

$$I(\vec{N}_t, a, b, h) = \{\vec{j}_r : \widetilde{N}_t(\vec{j}_r) = \vec{N}_t, \quad k_0(\vec{j}_r) = b, \quad v(\vec{j}_r) = a, \quad |N_b'| = h, \\ \widetilde{N}_{t+1}(\vec{j}_r) = (N_1, \ldots, N_b', N_b'', \ldots, N_t)\}.$$

We have

$$R \leqslant 2 \sum_{a=0}^{n-1} \varphi^{1/r}(a) \sum_{\vec{N}_t \in Q_t, \beta(\vec{N}_t) > 1} \sum_{b=1}^{t} \sum_{h=1}^{|N_b|-1} D(t, h, \vec{N}_t, b, a), \tag{3.34}$$

where $p_h = |N_b|/h$, $q_h = p_h/(p_h - 1)$,

$$D(t, h, \vec{N}_t, b, a) = \sum_{\vec{j}_r \in I(\vec{N}_t, a, b, h)} \prod_{\substack{k=1, \\ k \neq b}}^{t} \mathbf{E}\left(\prod_{i \in N_k} \|\xi_{j_i}\| \right)$$

$$\times \left\| \left(\prod_{i \in N_b'} \|\xi_{j_i}\| \right) \right\|_{p_h} \left\| \left(\prod_{i \in N_b''} \|\xi_{j_i}\| \right) \right\|_{q_h}.$$

By construction, we obtain

$$D(t, h, \vec{N}_t, b, a) \leqslant \left(\prod_{\substack{k=1, \\ k \neq b}}^{t} \sum_{\vec{i}_{s_k} \in I(n, s_k, a)} \mathbf{E} \left(\prod_{j=1}^{s_k} \| \xi_{i_j} \| \right) \right)$$

$$\times \sum_{\vec{i}_{s_b} \in I(n, s_b, a, h+1)} \left\| \left(\prod_{j=1}^{h} \| \xi_{i_j} \| \right) \right\|_{p_h} \left\| \left(\prod_{j=h+1}^{s_b} \| \xi_{i_j} \| \right) \right\|_{q_h} .$$

Applying Lemma 3.2 to the individual sums on the right side of the preceding inequality, we have

$$D(t, h, \vec{N}_t, b, a) \leqslant (a+1)^{r-2} (|N_b| - 1) \prod_{k=1}^{t} L_{s_k}(n),$$

where $s_k \geq 2$, $\sum_{k=1}^{t} s_k = r$. It follows from Lemma 3.3 that

$$\prod_{k=1}^{t} L_{s_k}(n) \leqslant \prod_{k=1}^{t} \left(\max (L_r(n), D_n^r) \right)^{s_k/r} = Q_r(n).$$

Therefore

$$D(t, h, \vec{N}_t, b, a) \leqslant (a+1)^{r-2} (|N_b| - 1) Q_T(n).$$

which together with (3.34) yields

$$R \leqslant 2 C_{r-2}^{t-1} t (r-1) \sum_{a=0}^{n-1} \varphi^{1/r}(a) (a+1)^{r-2} Q_r(n). \tag{3.35}$$

Now, let R_1 denote the sum on the right side of (3.32). Similarly, just as in estimating (3.31), we obtain

$$R_1 \leqslant 2 e^{\delta} C_{r-2}^{t-1} r^2 t (r-1) p^{r/2} \left(\sum_{a=0}^{n-1} \alpha^{\delta/(r+\delta)}(a) (a+1)^{r-2} \right) Q_r(n, \delta). \tag{3.36}$$

From (3.29) - (3.32), Lemma 3.6 and inequalities (3.35), (3.36) we have: for (2.7)

$$B_\varepsilon(\widetilde{N}_t) \leqslant B_\varepsilon(\widetilde{N}_{t+1}) + C_{t,r}^{(1)} a_1(\varphi, r, x) A_{r,x},$$

for (2.8)

$$B_\varepsilon(\widetilde{N}_t) \leqslant B_\varepsilon(\widetilde{N}_{t+1}) + 3 p^{r/2} C_{t,r}^{(1)} a_2(\alpha, r, x) A_{r,x},$$

for (2.1)

$$B_\varepsilon(\widetilde{N}_t) \leqslant B_\varepsilon(\widetilde{N}_{t+1}) + C_{t,r}^{(2)} a(\varphi, r) Q_r(n),$$

for (2.2)

$$B_\varepsilon(\widetilde{N}_t) \leqslant B_\varepsilon(\widetilde{N}_{t+1}) + C_{t,r}^{(2)} e^\delta r^2 p^{r/2} b(\alpha, r, \delta) Q_r(n, \delta),$$

where

$$C_{t,s}^{(1)} = t2^{t+2}(r-1) C_{r-2}^{t-1}, \quad C_{t,r}^{(2)} = 2C_{r-2}^{t-1} t(r-1).$$

Summing up the above inequalities in t and noting (3.18) and (3.19), we get the desired result. The Theorem is proved.

4. Going from even order moment to arbitrary order in Rosenthal's moment inequality

Let there be given σ-algebras $F = \{F_i\}_{i=1}^n$ and a separable Banach space B with norm $\|x\|$. We say that the set of random variables $\eta = \{\eta_i\}_{i=1}^n$ is adapted to (F, B) if the η_i take on values in B, are measurable with respect to F_i and $E\eta_i = 0$, $i = 1, \ldots, n$. Set

$$L(v, \delta, \eta) = \sum_{i=1}^n \left(\mathbf{E}\|\eta_i\|^{v+\delta}\right)^{v/(v+\delta)},$$

$$D(\eta, \delta) = L(2, \delta, \eta); \tag{4.1}$$

$$Q(v, \delta, \eta) = \begin{cases} \max\left(L(v, \delta, \eta), (D(\eta, \delta))^{v/2}\right), & v \geqslant 2, \\ L(v, \delta, \eta), & 1 \leqslant v \leqslant 2. \end{cases}$$

L e m m a 4.1. For every set of random variables $\eta = \{\eta_i\}_{i=1}^n$ adapted to (F, B), let the following inequality hold:

$$\mathbf{E}\left\|\sum_{i=1}^n \eta_i\right\|^v \leqslant cQ(v, \delta, \eta),$$

where $c \geqq 1$, $\delta \geqq 0$, $v \geqq 1$. Then for all t such that $1 \leqq t \leqq v$, and every set of random variables $\phi = \{\phi_i\}_{i=1}^n$ adapted to (F, B), we have the inequality

$$\mathbf{E}\left\|\sum_{i=1}^n \varphi_i\right\|^t \leqslant c \cdot 16^v \cdot Q(t, \delta, \varphi).$$

REMARK. The validity of Theorems 2.1 and 2.2 follows from Theorem 2.3 and Lemma 4.1.

Proof of Lemma 4.1. Set

$$Q = Q(t, \delta, \varphi), \quad y = Q^{1/t}, \quad \varphi_i = \Psi_i + \eta_i,$$
$$\Psi_i = T_i - \mathbf{E}T_i, \quad \eta_i = Y_i - \mathbf{E}Y_i, \quad T_i = \varphi_i I(\|\varphi_i\| \leqslant y),$$
$$Y_i = \varphi_i I(\|\varphi_i\| > y), \quad i = 1, \ldots, n.$$

By the convexity inequality, we have

$$\mathbf{E}\Big\|\sum_{i=1}^{n}\varphi_i\Big\|^t \leqslant 2^{t-1}\Big(\mathbf{E}\Big\|\sum_{i=1}^{n}\Psi_i\Big\|^t + \mathbf{E}\Big\|\sum_{i=1}^{n}\eta_i\Big\|^t\Big).$$

Applying successively the concavity inequality and convexity inequality, we obtain

$$\mathbf{E}\Big\|\sum_{i=1}^{n}\eta_i\Big\|^t \leqslant \mathbf{E}\Big(\sum_{i=1}^{n}\|\eta_i\|^{t/v}\Big)^v$$
$$\leqslant 2^{v-1}\Big(\mathbf{E}\Big|\sum_{i=1}^{n}(\|\eta_i\|^{t/v} - \mathbf{E}\|\eta_i\|^{t/v})\Big|^v + \Big(\sum_{i=1}^{n}\mathbf{E}\|\eta_i\|^{t/v}\Big)^v\Big).$$

By Lyapunov's inequality,

$$\mathbf{E}\Big\|\sum_{i=1}^{n}\Psi_i\Big\|^t \leqslant \Big(\mathbf{E}\Big\|\sum_{i=1}^{n}\Psi_i\Big\|^v\Big)^{t/v}.$$

As a result, we have

$$\mathbf{E}\Big\|\sum_{i=1}^{n}\varphi_i\Big\|^t \leqslant 2^{t-1}\Big(\mathbf{E}\Big\|\sum_{i=1}^{n}\Psi_i\Big\|^v\Big)^{t/v}$$
$$+ 2^{t+v-2}\Big(\mathbf{E}\Big|\sum_{i=1}^{n}(\|\eta_i\|^{t/v} - \mathbf{E}\|\eta_i\|^{t/v})\Big|^v + \Big(\sum_{i=1}^{n}\mathbf{E}\|\eta_i\|^{t/v}\Big)^v\Big). \tag{4.2}$$

Let $z_i = e(\|\eta_i\|^{t/v} - \mathbf{E}\|\eta_i\|^{t/v})$, $i = 1,\dots,n$, where e is the unit vector in B, $z = \{z_i\}_{i=1}^n$, $\Psi = \{\Psi\}_{i=1}^n$. It is not hard to see that z and Ψ are adapted to (F,B), and we can take advantage of the inequality in the Theorem.

Let us successively estimate all the terms on the right side of inequality (4.2). By the hypothesis of the Theorem,

$$\mathbf{E}\Big\|\sum_{i=1}^{n}\Psi_i\Big\|^v \leqslant c\cdot Q(v,\delta,\Psi).$$

A. Let us estimate $L(v,\delta,\Psi)$. We have

$$L(v,\delta,\Psi) = \sum_{i=1}^{n}(\mathbf{E}\|\Psi_i\|^{v+\delta})^{v/(v+\delta)} \leqslant 2^v\sum_{i=1}^{n}(\mathbf{E}\|T_i\|^{(v+\delta)u})^{v/(v+\delta)u},$$

where $u = \dfrac{v(t+\delta)}{(v+\delta)t} \geqq 1$. Therefore,

$$\mathbf{E}\|T_i\|^{(v+\delta)u} \leqslant y^{((v+\delta)u-(t+\delta))}\mathbf{E}\|\varphi_i\|^{t+\delta},$$

$$L(v,\delta,\Psi) \leqslant 2^v y^{t((v+\delta)u-(t+\delta))/(t+\delta)}\sum_{i=1}^{n}(\mathbf{E}\|\varphi_i\|^{t+\delta})^{t/(t+\delta)} \leqslant 2^v(Q(t,\delta,\varphi))^{v/t}.$$

B. Let $1 \leqq t \leqq v \leqq 2$. Then $Q(v,\delta,\Psi) = L(v,\delta,\Psi)$ and therefore

$$\mathbf{E}\left\|\sum_{i=1}^{n}\Psi_i\right\|^v \leqslant c\cdot 2^v\cdot(Q(t,\delta,\varphi))^{v/t}. \tag{4.3}$$

C. Let $t \leqq 2 \leqq v$. Noting A, $D(\Psi,\delta) = L(2,\delta,\Psi) \leqq 4(Q(t,\delta,\phi))^{2/t}$, we obtain

$$Q(v,\delta,\Psi) \leqslant 2^v(Q(t,\delta,\varphi))^{v/t},$$

i.e., (4.3) is true.

D. Let $2 \leqq t \leqq v$. It is not hard to see that

$$D(\Psi,\delta) = \sum_{i=1}^{n}\left(\mathbf{E}\|\Psi_i\|^{2+\delta}\right)^{2/(2+\delta)} \leqslant 4\sum_{i=1}^{n}\left(\mathbf{E}\|T_i\|^{2+\delta}\right)^{2/(2+\delta)} \leqslant 4Q^{2/t}.$$

Therefore, taking A into account yields (4.3).

Now, let us estimate the second term on the right side in (4.2). By the conditions of the Theorem,

$$\mathbf{E}\left\|\sum_{i=1}^{n} z_i\right\|^v \leqslant c\cdot Q(v,\delta,z).$$

A. Let us estimate $L(v,\delta,z)$. We have

$$L(v,\delta,z) = \sum_{i=1}^{n}\left(\mathbf{E}\left|\|\eta_i\|^{t/v} - \mathbf{E}\|\eta_i\|^{t/v}\right|^{v+\delta}\right)^{\frac{v}{(v+\delta)}}$$

$$\leqslant 2^v\sum_{i=1}^{n}\left(\mathbf{E}\|\eta_i\|^{t(v+\delta)/v}\right)^{v/(v+\delta)} \leqslant 2^{v+t}\sum_{i=1}^{n}\left(\mathbf{E}\|Y_i\|^{t(v+\delta)/v}\right)^{v/(v+\delta)}.$$

Let $u = \dfrac{v(t+\delta)}{(v+\delta)t}$. Then, since $u \geqq 1$, we have

$$L(v,\delta,z) \leqslant 2^{v+t}\sum_{i=1}^{n}\left(\mathbf{E}\|Y_i\|^{t(v+\delta)u/v}\right)^{v/(v+\delta)u}$$

$$= 2^{v+t}\sum_{i=1}^{n}\left(\mathbf{E}\|Y_i\|^{t+\delta}\right)^{t/(t+\delta)} \leqslant 2^{v+t}Q.$$

B. Let $1 \leqq t \leqq v \leqq 2$. Then $Q(v,\delta,z) = L(v,\delta,z)$. Hence

$$\mathbf{E}\left\|\sum_{i=1}^{n} z_i\right\|^v \leqslant c\cdot 2^{v+t}\cdot Q(t,\delta,\varphi). \tag{4.4}$$

C. Let $v \geqq 2$, $t \leqq 2$. Proceeding just as in estimating the $L(v,\delta,z)$ and noting that

$$\frac{v(t+\delta)}{t(2+\delta)} \geqslant 1, \quad \sum_{i=1}^{n}|a_i|^p \leqslant \left(\sum_{i=1}^{n}|a_i|\right)^p$$

for $p \geqq 1$, we obtain

$$D(z,\delta) \leqslant 4^{(1+t/v)} \sum_{i=1}^{n} \left(\mathbf{E}\| Y_i \|^{t(2+\delta)/v}\right)^{2/(2+\delta)}$$

$$\leqslant 4^{(1+t/v)} y^{(t(2+\delta)/v-(t+\delta))} \times \sum_{i=1}^{n} \left(\mathbf{E}\| \varphi_i \|^{t+\delta}\right)^{2/(2+\delta)} \leqslant 4^{(1+t/v)} Q^{2/v}.$$

Therefore,

$$Q(v,\delta,z) = \max\left(L(v,\delta,z), (D(z,\delta))^{v/2}\right) \leqslant 2^{v+t} Q(t,\delta,\varphi).$$

This proves (4.4) in this case.

D. Let $2 \leqq t \leqq v$. We have already estimated for the $L(v,\delta,z)$. Let us estimate $D(z,\delta)$. We have

$$D(z,\delta) \leqslant 4^{(1+t/v)} \sum_{i=1}^{n} \left(\mathbf{E}\| Y_i \|^{t(2+\delta)/v}\right)^{2/(2+\delta)}$$

$$\leqslant 4^{(1+t/v)} y^{2(t/v-1)} \sum_{i=1}^{n} \left(\mathbf{E}\| \varphi_i \|^{2+\delta}\right)^{2/(2+\delta)} \leqslant 4^{(1+t/v)} Q^{2/v}.$$

Therefore, (4.4), is true in this case, too.

Let us estimate next the last term on the right side in (4.2). We have

$$\sum_{i=1}^{n} \mathbf{E}\| \eta_i \|^{t/v} \leqslant 2 \sum_{i=1}^{n} \mathbf{E}\| Y_i \|^{t/v} \leqslant 2Q^{(1/v-1)} \sum_{i=1}^{n} \left(\mathbf{E}\| \varphi_i \|^{t+\delta}\right)^{t/(t+\delta)} \leqslant 2Q^{1/v}.$$

Therefore

$$\left(\sum_{i=1}^{n} \mathbf{E}\| \eta_i \|^{t/v}\right)^{v} \leqslant 2^{v} \cdot Q(t,\delta,\varphi). \tag{4.5}$$

Substituting (4.3) - (4.5) into (4.2) and noting that $c \geqq 1$, after routine calculations, yield the required result. The Theorem is proved.

5. Kolmogorov inequalities for martingales

Let

$$\beta(i) = \beta(i,p,q) = \sup_{1 < k \leqslant \infty} \sup_{\eta \equiv M_1^k} \frac{|\mathbf{E}(\eta, \xi_{k+i})|}{\|\eta\|_p \|\xi_{k+i}\|_q},$$

where $p,q > 0$, $1/p + 1/q \leqq 1$. Let $t = p/(p-1)$. Then $t > 1$ and $q-t = \delta \geqq 0$. Set $\varepsilon = \varepsilon(p,q) = \sum_{i=1}^{\infty} \beta(i,p,q)$.

T h e o r e m 5.1. The following inequality holds:

$$\mathbf{P}\Big(\max_{1\leqslant k\leqslant n} |S_k| \geqslant x\Big) x^{-t}\mathbf{E}|S_n|^t a_1(t)(1+\varepsilon) + x^{-t}L_t(n,\delta)\,\varepsilon(1+\varepsilon)^t a_2(t), \tag{5.1}$$

where

$$a_1(t) = \max(1, 2/t),\ a_2(t) = (t-1)^{t-1}a_1^{\,t}(t).$$

Let

$$\eta \in M_1^k,\ \xi \in M_{k+i}^n,\quad a,\ b > 0,\ 1/a + 1/b \leqslant 1.$$

We have ([18, Lemma 1.1], [25], [3]; the proof routinely extends from the real case to a separable Hilbert space)

$$|\mathbf{E}(\eta, \xi) - (\mathbf{E}\eta, \mathbf{E}\xi)| \leqslant 2\varphi^{1/a}(i)\|\eta\|_a\|\xi\|_b,$$

i.e., $\beta(i,p,q) \leqq 2\phi^{1/p}(i)$. Under the same constraints and for $H = R^m$ [27, Corollary], [23, Appendix],

$$|\mathbf{E}(\eta, \xi) - (\mathbf{E}\eta, \mathbf{E}\xi)| \leqslant 6m(\alpha(i))^{1-1/a-1/b}|\eta|_a|\xi|_b,$$

i.e., $\beta(i,p,q) \leqq 6m(\alpha(i))^{1-1/p-1/q}$. Therefore, we have obtained the following.

COROLLARY 5.1. For $t > 1$, the following inequality holds:

$$\mathbf{P}\Big(\max_{1\leqslant k\leqslant n} |S_k| \geqslant x\Big) \leqslant x^{-t}\mathbf{E}|S_n|^t a_1(t)(1+\varepsilon_1) + x^{-t}L_t(n)\,\varepsilon_1(1+\varepsilon_1)^t a_2(t), \tag{5.2}$$

where

$$\varepsilon_1 = 2\sum_{i=1}^{\infty} \varphi^{(t-1)/t}(i).$$

COROLLARY 5.2. Let $H = R^m$. For $t > 1$, $\delta \geqq 0$, the following inequality holds:

$$\mathbf{P}\Big(\max_{1\leqslant k\leqslant n} |S_k| \geqslant x\Big) \leqslant x^{-t}\mathbf{E}|S_n|^t a_1(t)(1+\varepsilon_2) + x^{-t}L_t(n,\delta)\,\varepsilon_2(1+\varepsilon_2)^t a_2(t), \tag{5.3}$$

where

$$\varepsilon_2 = 6m(\varepsilon(\alpha, t, \delta) - 1) = 6m\sum_{k=1}^{\infty} \alpha^{\delta/t(t+\delta)}(k).$$

Inequality (5.2) is efficient for $t \geqq 2$ (the passing to sequences actually involves convergence of the series $\sum_{i=1}^{\infty} \phi^{\frac{1}{2}}(i)$). However, in the range $1 < t < 2$, inequality (5.2) is inefficient -- hence, in Section 6 we shall give a more convenient inequality. The proof is based on truncation as well as inequality (5.2) for $t = 2$, which again leads to the series $\sum_{i=1}^{\infty} \phi^{\frac{1}{2}}(i)$. The same is true, to an extent, for inequality (5.3). Observe that sometimes (especially in investigating limit theorems of strong-law-of-large-numbers type) it is advantageous to use results which help pass from a moment inequality for a sum (of the type

$$\mathbf{E}\left|\sum_{i=1}^{n} a_i\xi_i\right|^p \leqslant c\left(\sum_{i=1}^{n} a_i^2\right)^{p/2}$$

to a moment inequality for the maximum of partial sums (respectively, to

$$\mathbf{E}\max_{1\leqslant k\leqslant n}\left|\sum_{i=1}^{k} a_i\xi_i\right|^p \leqslant c_p c\left(\sum_{i=1}^{n} a_i^2\right)^{p/2}, \quad p > 2)$$

(see [28] where one can find a bibliography dealing with the same subject). Inequality (5.1) generalizes Kolmogorov's inequality [29]; (for martingales see [29]; [17, Chapter 7]; [30]; [26]; the term is taken from [31]). Some other approaches, with specific problems, may be found in [6, Lemma 4.1.2]; [32]; [34]; [33, Lemma 3]. In the range $1 \leqq t \leqq 2$, similar results have been obtained in [35].

Proof of Theorem 5.1. For $t \geqq 2$, we use the inequality

$$|S_n|^t \geqslant |S_k|^t + t|S_k|^{t-2}(S_k, S_n - S_k), \tag{5.4}$$

and for $1 \leqq t \leqq 2$, the inequality

$$|S_n|^t \geqslant 2/t(|S_k|^t + 2|S_k|^{t-2}(S_k, S_n - S_k)). \tag{5.5}$$

Proof of (5.4) and (5.4) follows from the elementary inequalities $(1+y)^{\alpha} \geqq 1 + \alpha y$ for $|y| \leqq 1$, $\alpha \geqq 1$ and $(1+y)^{\alpha} \geqq \alpha(1+y)$ for $|y| \leqq 1$, $0 \leqq \alpha \leqq 1$ and the identity

$$S_n^2 = S_k^2 + 2(S_k, S_n - S_k) + (S_n - S_k)^2.$$

We prove (5.1) only for $t \geqq 2$, because for $1 < t \leqq 2$ the proof is similar.

Let $B_k = \{|S_k| \geq x,\ |S_i| < x,\ i = 1,\dots,k-1\}$. Just as in the proof of (5.1) by Kolmogorov, in the case of independent summands and $t = 2$ (see [29]; [17]) from (5.4) we obtain

$$\mathbf{E}|S_n|^t \geqslant \sum_{k=1}^{n} \mathbf{E}(|S_k|^t I(B_k)) - t\sum_{k=1}^{n-1} \mathbf{E}(|S_k|^{t-2} I(B_k)(S_k, S_n - S_k))$$
$$\geqslant \omega - t\sum_{k=1}^{n-1}\sum_{j=1}^{n-k} \mathbf{E}(\varphi_k, \xi_{k+j}),$$

where

$$\omega = \sum_{k=1}^{n} \mathbf{E}(|S_k|^t I(B_k)), \quad \varphi_k = |S_k|^{t-2} S_k I(B_k).$$

Noting that $\phi_k \in M_1^k$, we obtain

$$\mathbf{E}|S_n|^t \geqslant \omega - t\sum_{k=1}^{n-1}\sum_{j=1}^{n-k} \beta(j)(\mathbf{E}|S_k|^t)^{1/p} \|\xi_{k+j}\|_q .$$

Transposing the sums and applying Hölder's inequality, we find

$$\mathbf{E}|S_n|^t \geqslant \omega - t\varepsilon\omega^{1/p}(L_t(n, \delta))^{1/t}.$$

Using the elementary inequality

$$xy \leqslant c^a x^a/a + c^{-b} y^b/b,$$

where $1/a + 1/b = 1$, $x,y,a,b,c > 0$ (since $p,q > 0$, $1/p + 1/t = 1$) we have

$$\mathbf{E}|S_n|^t \geqslant \omega(1 - t\varepsilon c^p/p) - \varepsilon L_t(n, \delta) c^{-t},$$

but

$$\omega \geqslant x^t \mathbf{P}\Big(\max_{1\leqslant k\leqslant n} |S_k| \geqslant x\Big).$$

Hence,

$$\mathbf{P}\Big(\max_{1\leqslant k\leqslant n} |S_k| \geqslant x\Big) \leqslant (\mathbf{E}|S_n|^t + \varepsilon L_t(n, \delta) c^{-t}) x^{-t}(1 - t\varepsilon c^p/p)^{-1}.$$

By choosing $c(c^{-p} = t(1+\varepsilon)/p)$, appropriately, we obtain the required inequality. The Theorem is proved.

6. Analogs of the Nagaev-Fook inequality

The additional notation is taken from Sections 4 ((4.1)) and 5.

L e m m a 6.1. For any set of random variables $\eta = \{\eta_i\}_{i=1}^n$ adapted to (F,B), suppose

$$\mathbf{P}\left(\max_{1\leqslant k\leqslant n}\left\|\sum_{i=1}^{k}\eta_i\right\|\geqslant x\right)\leqslant x^{-t}\cdot c\cdot Q(t,\delta,\eta),$$

where $t \geqq 1$, $c \geqq 1$. Then for any set of $\phi = \{\phi_i\}_{i=1}^n$ adapted to (F,B), for $1 \leqq \omega \leqq t$, we have

$$\mathbf{P}\left(\max_{1\leqslant k\leqslant n}\left\|\sum_{i=1}^{n}\varphi_i\right\|\geqslant x\right)\leqslant 3\cdot 2^{t}\cdot x^{-\omega}\cdot c\cdot L(\omega,\delta,\varphi),$$

for $\omega \leqq 2$, and

$$\mathbf{P}\left(\max_{1\leqslant k\leqslant n}\left\|\sum_{i=1}^{k}\varphi_i\right\|\geqslant x\right)\leqslant 3\cdot 2^{t}\cdot c\left(x^{-\omega}L(\omega,\delta,\varphi)+\left(D(\varphi,\delta)\,x^{-2}\right)^{t/2}\right)$$

for $\omega > 2$.

L e m m a 6.2. Let $v \geqq 2$. If for all t such that $2 \leqq t \leqq v$, for every set $\eta = \{\eta_i\}_{i=1}^n$ adapted to (F,B), we have

$$\mathbf{P}\left(\max_{1\leqslant k\leqslant n}\left\|\sum_{i=1}^{k}\eta_i\right\|\geqslant x\right)\leqslant c_t\cdot Q(t,\delta,\eta)\,x^{-t}, \tag{6.1}$$

where $c_t \geqq 1$, then for every set $\phi = \{\phi_i\}_{i=1}^n$ adapted to (F,B), for all $\delta_1 \geqq t/2\delta$, we have

$$\mathbf{P}\left(\max_{1\leqslant k\leqslant n}\left\|\sum_{i=1}^{k}\varphi_i\right\|\geqslant x\right)\leqslant 4^{t}c_t x^{-t}L(t,\delta_1,\varphi)+4^{v}c_v x^{-v}\left(D(\varphi,\delta)\right)^{v/2}. \tag{6.2}$$

L e m m a 6.3. If (6.1) holds for all $t \geqq 2$ and $c_t \leqq A(bt)^{pt}$, then for all $t \geqq 2$, $\delta_1 \geqq t/2\delta$, we have

$$\mathbf{P}\left(\max_{1\leqslant k\leqslant n}\left\|\sum_{i=1}^{k}\varphi_i\right\|\geqslant x\right)\leqslant 4^{t}c_t x^{-t}L(t,\delta_1,\varphi)$$
$$+A\exp\left(pt-pe^{-1}b^{-1}4^{-1/p}\left(x^2/D(\varphi,\delta)\right)^{1/2p}\right).$$

Theorem 2.1, Lemmas 6.1 - 6.3 and Corollary 5.1 imply in the uniform strong mixing case the following theorem.

Theorem 6.1. The following inequalities hold:

$$1)\ \mathbf{P}(\max_{1\leqslant k\leqslant n}|S_k|\geqslant x)\leqslant c_1 a^3(\varphi, 2)\, L_t(n)\, x^{-t}, \tag{6.3}$$

where $1 \leqq t \leqq 2$; c_1 is an absolute constant;

$$2)\ \mathbf{P}(\max_{1\leqslant k\leqslant n}|S_k|\geqslant x)\leqslant c_2 a(\varphi, t)\, a^2(\varphi, 2)\left(L_v(n)\, x^{-v} + D_n^t x^{-t}\right), \tag{6.4}$$

where $2 \leqq v \leqq t$, c_2 depends only on t;

$$3)\ \mathbf{P}(\max_{1\leqslant k\leqslant n}|S_k|\geqslant x)\leqslant c_3 x^{-t} L_t(n) + c_4 \exp(-c_5 (x/D_n)^{1/3}), \tag{6.5}$$

where $t \geqq 2$, $\phi(i) \leqq Aq^i$, $0 \leqq q < 1$; c_3, c_4, c_5 depend on t, A, q.

Set

$$\varepsilon(\alpha, t, \delta) = \sum_{i=0}^{\infty} \alpha^{\delta/t(t+\delta)}(i).$$

Theorem 2.2., Lemmas 6.1 - 6.3 and Corollary 5.2 imply in the strong mixing case the following theorem.

Theorem 6.2. Let $H = R^p$ and $\delta > 0$. The following inequalities hold:

$$1)\ \mathbf{P}(\max_{1\leqslant k\leqslant n}|S_k|\geqslant x)\leqslant c_1 L_t(n, \delta) x^{-t} \varepsilon^{t+1}(\alpha, t, \delta), \tag{6.6}$$

where $1 \leqq t \leqq 2$; c_1 depends only on p;

$$2)\ \mathbf{P}(\max_{1\leqslant k\leqslant n}|S_k|\geqslant x)\leqslant c_2 \varepsilon(\alpha, t, \delta)\,(b(\alpha, t, \delta) + \varepsilon^t(\alpha, t, \delta))\left(x^{-v} L_v(n, \delta) + D_n^t(\delta)\, x^{-t}\right), \tag{6.7}$$

where $2 \leqq v \leqq t$, c_2 depends on p and t;

$$3)\ \mathbf{P}(\max_{1\leqslant k\leqslant n}|S_k|\geqslant x)\leqslant c_3 x^{-t} L_t(n, \delta) + c_4 \exp(-c_5 (x/D_n(\delta))^{1/3}), \tag{6.8}$$

where $t \geqq 2$, $\alpha(i) \leqq Aq^i$, $0 \leqq q < 1$; c_3, c_4, c_5 depend on p, t, A, q, δ.

It is not hard to obtain from Theorems 5.1, 6.1, 6.2 sufficient conditions for the strong law of large numbers as well as convergence of the series. Let us mention some of them. Set

$$r(t) = \min(t, t/2 + 1),\ 1 \leqslant t.$$

COROLLARY 6.1. Let $t \geqq 1$. If

$$a(\varphi, t) < \infty \quad \text{and} \quad \sum_{n=1}^{\infty} \mathbf{E}|\xi_n|^t n^{-r(t)} < \infty,$$

then $S_n/n \to 0$ a.s.

COROLLARY 6.2. Let $H = R^p$ and $t \geqq 1$. If

$$b(\alpha, t, \delta) < 0, \quad \varepsilon(\alpha, t, \delta) < 0 \quad \text{and} \quad \sum_{n=1}^{\infty} \| \xi_n \|_{t+\delta}^{t} n^{-r(t)} < \infty,$$

then $S_n/n \to 0$ a.s.

COROLLARY 6.3. Here $E\xi_i$ may be not equal to zero. If $a(\phi,2) < \infty$ and for some $c > 0$

$$\sum_{i=1}^{\infty} \mathbf{P}(|\xi_i| > c) < \infty,$$

$$\sum_{i=1}^{\infty} \mathbf{E}(\xi_i I(|\xi_i| \leqslant c)),$$

$$\sum_{i=1}^{\infty} \mathbf{D}(\xi_i I(|\xi_i| \leqslant c)) < \infty,$$

then the series $\sum_{i=1}^{\infty} \xi_i$ converges a.s.

COROLLARY 6.4. If for some $\delta \geqq 0$,

$$\sum_{i=1}^{\infty} \beta(i, 2, 2+\delta) < \infty,$$

$$\sum_{i=1}^{\infty} (\mathbf{E}|\xi_i|^{2+\delta})^{2/(2+\delta)} < \infty,$$

then the series $\sum_{i=1}^{\infty} \xi_i$ converges a.s.

Inequalities (6.4), (6.5), (6.7), (6.8) are similar due to the following inequality, known in the case of independent summands as Nagaev-Fook's inequality (see [36, Corollary 4]; the term is taken from [37]; also see the bibliography in [12]):

$$\mathbf{P}(\max_{1 \leqslant k \leqslant n} |S_k| \geqslant x) \leqslant c_t^{(1)} L_t(n) x^{-t} + 2\exp\left(-c_t^{(2)} \frac{x^2}{D_n^2}\right).$$

As is seen, even in the most favorable situation (inequality (6.5)) our exponential term is worse. This is due to the fact that we are going from moment inequalities to probabilities of large deviations, rather than conversely (see [16], [36]). Similar results for the sequence of real-valued random variables and $1 \leqq t \leqq 2$ have been obtained via the martingale-expansion method in [35]. There is yet another approach having to do with the approximation of Berkes and Philipp [6], [22, Proposition 3.1].

Corollaries 6.1 - 6.4 are a generalization of results of Kolmogorov and Brunk-Prokhorov on sufficient conditions for the strong law of large numbers and convergence of the series (for the bibliography, see [38], [12], [14], [15]). For the case of mixing, similar results are derived in [32], [35] (see also [39, Lemma 4.2.4, Appendix], [22, Proposition 4.1]).

The proof of Lemma 6.1 is obvious. It suffices to truncate at the level x.

The proof of Lemma 6.2 is almost the same as that of Lemma 4.1.

To prove Lemma 6.3, it suffices to minimize the variance component in (6.2), i.e., to find

$$\inf_{v \geqslant t} A 4^v x^{-v} (bv)^v (D(\varphi, \delta))^{v/2}.$$

Proof of Corollaries 6.1-6.4. Let

$$x_r = \max_{2^r < k \leqslant 2^{r+1}} \left| \sum_{n > 2^r}^{k} \xi_n \right|, \quad r = 1, 2, \ldots$$

It suffices to show that in the case of the strong law of large numbers (see [40], [14])

$$\sum_{r=1}^{\infty} \mathbf{P}(x_r \geqslant \varepsilon 2^r) < \infty, \tag{6.9}$$

whereas in the case of convergence of the series

$$\sum_{r=1}^{\infty} \mathbf{P}(x_r \geqslant \varepsilon) < \infty \tag{6.10}$$

for every $\varepsilon > 0$.

To check (6.9), (6.10), it is necessary to use the "appropriate" inequality from Theorems 6.1, 6.2, 5.1.

7. Results on the convergence rate in the invariance principle for weakly stationary sequences with mixing

Let $\{\xi_i\}_{i=1}^{\infty}$ be a weakly stationary sequence of random variables with zero expectations and unit variances. Set

$$S_k = \sum_{i=1}^{k} \xi_i, \quad t_k = k/n, \quad S_0 = t_0 = 0,$$

$$L_t = n^{-t/2} \sum_{i=1}^{n} \mathbf{E} |\xi_i|^t, \quad A_t = \sup_i \mathbf{E} |\xi_i|^t,$$

and assume that

$$0 < \sigma = \mathbf{E}\xi_1^2 + 2 \sum_{i=2}^{\infty} \mathbf{E}\xi_1\xi_i < \infty. \tag{7.1}$$

We denote by η_n the random polygonal line with nodes at $(t_k, S_k(n\sigma)^{-\frac{1}{2}})$, by w the standard Wiener process, by P_n the distribution of η_n in $C[0,1]$, and by W the distribution of w in $C(0,1]$. We denote by $L(P_n,W)$ the Lévy-Prokhorov distance between the measures P_n and W, $\|x\|$ being the norm in $C[0,1]$.

T h e o r e m 7.1. For some t, $2 < t < 5$, let $\phi(k) \leqq Ak^{-g}$, $k = 1,2,\ldots$, where $g > j(u)(j(u)-1)$, $u = \frac{(2+5t)}{2(5-t)}$, $j(u) = 2\min\{k \in N\colon 2k \geqq u\}$. Then

$$L(P_n, W) \leqslant cL_t^{1/(t+1)}, \tag{7.2}$$

where c depends on A, g, t, σ.

T h e o r e m 7.2. For some t, $2 < t < 5$, let $\alpha(k) \leqq Ak^{-g}$, $k = 1,2,\ldots$, where $g > \max(\delta_1^{-1} j(u)(j(u)+\delta_1), \delta_2^{-1} j(2t)(j(2t)+\delta_2))$, $u = \max(2t, \frac{(2+5t)}{2(5-t)})$, $\delta_1 = (t-2)(t(t+1))^{-1}$, and $j(u)$ is defined above. Then

$$L(P_n, W) \leqslant cA_t n^{(2-t)/2(t+1)}, \tag{7.3}$$

where c depends on A, g, t, δ.

Let $\varepsilon > 0$.

COROLLARY 7.1. If $\phi(k) \leqq Aq^k$, $0 \leqq q < 1$, $k = 1,2,\ldots$, then for $2 < t < 5$ we have (7.2).

COROLLARY 7.2. If $\alpha(k) \leqq Aq^k$, $0 \leqq q < 1$, $k = 1,2,\ldots$, then for $2 < t < 5$ we have (7.3).

COROLLARY 7.3. If $\phi(k) \leqq Ak^{-(110+\varepsilon)}$, then for $2 < t \leqq 4$ we have (7.2).

COROLLARY 7.4. If $\alpha(k) \leqq Ak^{-(776+\varepsilon)}$, then for $3 \leqq t \leqq 4$ we have (7.3).

The problem on the rate of convergence of the invariance principle was posed by Yu.V. Prokhorov [41]. The definitive result on the order of smallness for $2 < t \leqq 3$ and independent summands was obtained by A.B. Borovkov [37]. In [42] the problem was solved for independent and identically distributed summands. In [43] the unimprovability of Borovkov's estimate was proved. In [44] the result from [37] was extended to the range $3 < t < 5$. In [7] a solution of the problem for non-identically distributed independent random variables was announced. In [45] the exponential-order estimate (7.3) $(n^{(2-t/2(t+1)}\ln n)$ was obtained for the range $2 < t \leqq 3$ and a strictly stationary process in which the strong mixing coefficients decrease with the geometric rate $(\alpha(k)) \leqq Aq^k$, $0 \leqq q < 1)$. Coarse estimates, although sufficient to be applied to the iterated-logarithm-type laws, can be found in [39], [46].

8. Proof of Theorems 7.1 and 7.2

We shall prove these theorems simultaneously and in two stages: first for the uniform strong mixing condition and next for the strong mixing condition. We note that by the conditions of Theorem 7.1 we *a priori* have $g > 12$, i.e., $\alpha(k) \leqq Ak^{-12}$, similarly $\alpha(k) \leqq Ak^{-100/(t-2)}$. It follows trivially from the last relations that the series (7.1) converges. We now prove two auxiliary lemmas.

L e m m a 8.1. If $\{\eta_i\}_{i=1}^{\infty}$ is weakly stationary, $|E\eta_1\eta_j| \leqq A|j-1|^{-p}$, $p > 2$, then

$$\sup_n \left| \mathbf{E}\left(\sum_{i=1}^n \eta_i\right)^2 - an \right| < \infty, \qquad \text{where} \qquad a = \mathbf{E}\eta_1^2 + 2\sum_{j=2}^{\infty} \mathbf{E}\eta_1\eta_j.$$

Proof. It follows from the identity

$$\mathbf{E}\left(\sum_{i=1}^n \eta_i\right)^2 = n\mathbf{E}\eta_1^2 + 2\sum_{j=2}^{n} \mathbf{E}\eta_1\eta_j(n-j+1)$$

that

$$\left|\mathbf{E}\left(\sum_{i=1}^n \eta_i\right)^2 - na\right| \leqslant n\sum_{j=n+1}^{\infty} |\mathbf{E}\eta_1\eta_j| + 2\sum_{j=1}^{\infty}(j-1)|\mathbf{E}\eta_1\eta_j| ,$$

which together with the conditions of the Lemma prove the required assertion.

L e m m a 8.2. If

$$|E\eta_i\eta_j| \leqq A|j-i|^{-p}, \qquad p > 3,$$

then

$$\sup_{k,l,m} \sum_{i=k}^{k+l} \sum_{j=k+l+1}^{k+l+m} |\mathbf{E}\eta_i\eta_j| < \infty.$$

Proof. Indeed,

$$\sum_{i=k}^{k+l}\sum_{j=k+l+1}^{k+l+m}|\mathbf{E}\eta_i\eta_j|\leqslant A\sum_{i=k}^{k+l}\sum_{j=k+l+1}^{k+l+m}|j-i|^{-p}=A\sum_{i=0}^{l}\sum_{j=l+1}^{l+m}|j-i|^{-p}$$
$$=A\sum_{r=1}^{l+m}r^{-p}\sum_{i=0}^{l}I(l+1-i\leqslant r\leqslant l+m-i)$$
$$=A\sum_{r=1}^{l+m}r^{-p}\sum_{u=0}^{l}I(u+1\leqslant r\leqslant u+m)\leqslant A\sum_{r=1}^{l+m}r^{1-p}.$$

Hence

$$\sup_{k,l,m}\sum_{i=k}^{k+l}\sum_{j=k+l+1}^{k+l+m}|\mathbf{E}\eta_i\eta_j|\leqslant A\sum_{r=1}^{\infty}r^{1-p}<\infty.$$

The lemma is proved.

Set $\varepsilon_n = n^{(2-t)/2(t+1)}$, $x_t = L^{1/(t+1)}$. Note that $A_t \geqq 1$, $x_t \geqq \varepsilon_n$.

I. This is necessary only for the strong mixing case. Let

$$i=1,\ 2,\ \ldots,\ n,\quad \xi_i^{(0)}=\xi_i I(|\xi_i|\leqslant\sqrt{n}\varepsilon_n)-\mathbf{E}\xi_i I(|\xi_i|\leqslant\sqrt{n}\varepsilon_n),\quad S_k^{(0)}=\sum_{i=1}^{k}\xi_i^{(0)}.$$

Denote by $\eta_n^{(0)}$ the random polygonal line with nodes $(t_k, S_k^{(0)}(n,\sigma)^{-\frac{1}{2}})$. Noting that

$$n^{-1/2}\sum_{i=1}^{n}\mathbf{E}|\xi_i I(|\xi_i|>\sqrt{n}\varepsilon_n)|\leqslant A_t\varepsilon_n,$$

we obtain

$$\mathbf{P}(\|\eta_n-\eta_n^{(0)}\|>A_t\varepsilon_n)\leqslant\sum_{i=1}^{n}\mathbf{P}(|\xi_i|>\varepsilon_n\sqrt{n})\leqslant A_t\varepsilon_n.$$

Hence,

$$\mathbf{P}(\|\eta_n-\eta_n^{(0)}\|>A_t\varepsilon_n)\leqslant A_t\varepsilon_n\ . \tag{8.1}$$

II. The second stage. Set

$$S_0^{(1)}=0,\quad k=1,\ 2,\ \ldots,\ n,\quad \xi_k^{(1)}=\xi_k I\times(|\xi_k|\leqslant y\sqrt{n})-\mathbf{E}\xi_k I(|\xi_k|\leqslant y\sqrt{n}),\quad S_k^{(1)}=\sum_{i=1}^{k}\xi_k^{(1)},$$

where $y = L_t^{1/t}$ for the uniform strong mixing case and $y = n^{1/t-1/2}$ for the strong mixing case. Denote by $\eta_n^{(1)}$ the random polygonal line with nodes $(t_k, S_k^{(1)}(\sigma n)^{-\frac{1}{2}})$.

1. It follows from the condition of Theorem 7.1 that

$$a(\varphi, t) = \sum_{i=0}^{\infty} \varphi^{1/j(t)}(i)(i+1)^{j(t)-2} < \infty.$$

Hence, using (6.4) of Theorem 6.1, we obtain

$$\mathbf{P}(\|\eta_n - \eta_n^{(1)}\| \geqslant a) \leqslant ca^{-t}\left(n^{-t/2}\sum_{i=1}^{n}\mathbf{E}|\xi_i - \xi_i^{(1)}|^t + n^{-t/2}\left(\sum_{i=1}^{n}\mathbf{E}(\xi_i - \xi_i^{(1)})^2\right)^{t/2}\right).$$

Here and below, the letter c designates distinct constants not depending on n. From Lemma 1 of [44] we have

$$n^{-t/2}\sum_{i=1}^{n}\mathbf{E}|\xi_i - \xi_i^{(1)}|^t \leqslant 2^t L_t,$$

$$n^{-t/2}\left(\sum_{i=1}^{n}\mathbf{E}(\xi_i - \xi_i^{(1)})^2\right)^{t/2} \leqslant L_t.$$

Therefore

$$\mathbf{P}(\|\eta_n - \eta_n^{(1)}\| \geqslant Rx_t) \leqslant cR^{-t}x_t. \tag{8.2}$$

2. Let $v = 2t$, $\delta = (t-2)(t(t+1))^{-1}$. It follows from the conditions of Theorem 7.2 that

$$\varepsilon(\alpha, v, \delta) < \infty, \quad b(\alpha, v, \delta) < \infty.$$

Hence we can use inequality (6.7) of Theorem 6.2. We have

$$\mathbf{P}(\|\eta_n^{(1)} - \eta_n^{(0)}\| \geqslant a) \leqslant ca^{-v}n^{-v/2} \times\left(\sum_{i=1}^{n}\|\xi_i^{(1)} - \xi_i^{(0)}\|_{v+\delta}^{v} + \left(\sum_{i=1}^{n}\|\xi_i^{(1)} - \xi_i^{(0)}\|_{2+\delta}^{2}\right)^{v/2}\right). \tag{8.3}$$

It is not hard to see that

$$n^{-v/2}\sum_{i=1}^{n}\|\xi_i^{(1)} - \xi_i^{(0)}\|_{v+\delta}^{v} \leqslant cA_t n^{1-v/2}(\varepsilon_n\sqrt{n})^r,$$

$$r = (v+\delta-t)v/(v+\delta).$$

By elementary calculations we convince ourselves that for $s \geqq 2t$, $0 \leqq \delta_1 \leqq t-2$ one has

$$n^{1-s/2}(\varepsilon_n\sqrt{n})^{(s+\delta_1-t)t/(s+\delta_1)} \leqslant \varepsilon_n^{s+1}. \tag{8.4}$$

On the other hand,

$$n^{-v/2}\left(\sum_{i=1}^{n}\|\xi_i^{(1)} - \xi_i^{(0)}\|_{2+\delta}^{2}\right)^{v/2} \leqslant cA_t^v n^{(1-t/(2+\delta))v/2} y^{(2+\delta-t)v/(2+\delta)}.$$

Once again, for $s \geqq 2t$, $\delta_2 \geqq (t-2)(t(t+1))^{-1}$, one has

$$n^{(1-t/(2+\delta_2))s/2} y^{(2+\delta_2-t)s/(2+\delta_2)} \leqslant \varepsilon_n^{s+1} . \tag{8.5}$$

We have thus derived from (8.3) - (8.5) that

$$\mathbf{P}\left(\|\eta_n^{(1)} - \eta_n^{(0)}\| \geqslant R\varepsilon_n\right) \leqslant cR^{-v}\varepsilon_n A_t^v. \tag{8.6}$$

III. Partition into blocks. Let

$$l = b + h, \quad n = ml + r, \quad l = n^{3/(t+1)-\varepsilon},$$
$$\varepsilon = 2(5-t)/3(t+1), \quad h = n^{\tilde{\sigma}}, \quad \tilde{\sigma} < 3/(t+1) - \varepsilon,$$
$$\Psi_k^{(1)} = \sum_{i=1}^{b} \xi_{(k-1)l+i}^{(1)}, \quad \Psi_k = \sum_{i=1}^{b} \xi_{(k-1)l+i},$$
$$\varphi_k^{(1)} = \sum_{i=1}^{h} \xi_{(k-1)l+b+i}^{(1)}, \quad \varphi_k = \sum_{i=1}^{h} \xi_{(k-1)l+b+i},$$
$$\Psi_{m+1}^{(1)} = \sum_{i=1}^{r} \xi_{ml+i}^{(1)}, \quad \Psi_{m+1} = \sum_{i=1}^{r} \xi_{ml+i},$$
$$k = 1, 2, \ldots, m.$$

Let $\eta_n^{(2)}$ denote the random polygonal line with nodes $(t_{k\ell}, S_{k\ell}^{(1)}(\sigma n)^{-\frac{1}{2}})$, $k = 0,1,\ldots,m$, and for $r > 0$ with last point $(1, S_n^{(1)}(\sigma n)^{-\frac{1}{2}})$.

1. Let $u = \frac{(2+5t)}{2(5-t)}$. By the conditions of the Theorem,

$$a(\varphi, u) = \sum_{i=0}^{\infty} \varphi^{1/j(u)}(i)(i+1)^{j(u)-2} < \infty.$$

we have from (6.4) that

$$\mathbf{P}\left(\|\eta_n^{(2)} - \eta_n^{(1)}\| \geqslant a\right) \leqslant \sum_{k=1}^{m+1} \mathbf{P}\left(\max_{1 \leqslant i \leqslant l} |S_{i+kl}^{(1)} - S_{kl}^{(1)}| \geqslant a(n\sigma)^{1/2}\right)$$
$$\leqslant c\left(a^{-t}L_t + a^{-u}(n/l)^{1-u/2}\right).$$

Since $x_t \geqq \varepsilon_n$, then to prove Theorem 7.1 it suffices to show that

$$(n/l)^{1-u/2} \leqslant \varepsilon_n^{u+1}, \tag{8.7}$$

i.e., $u \geqq 2 + \frac{3(t-2)}{(t+1)\varepsilon}$. The latter is routine. Hence,

$$\mathbf{P}\left(\|\eta_n^{(2)} - \eta_n^{(1)}\| \geqslant a\right) \leqslant \left(a^{-t}L_t + a^{-u}\varepsilon_n^{u+1}\right). \tag{8.8}$$

2. Just as in the uniform strong mixing case, but using (6.7), we obtain $\delta = t - 2$, $d = \max\left(2t, \frac{(2+5t)}{2(5-t)}\right)$ then

$$\mathbf{P}\left(\|\eta_n^{(2)} - \eta_n^{(1)}\| \geqslant a\right) \leqslant ca^{-d}\left(\sum_{i=1}^{n} \|\xi_i^{(1)}\|_{d+\delta}^{d} + A_t^d \varepsilon_n^{d+1}\right).$$

The first term of the preceding inequality taking account of (8.4), may be evaluated just as during the second stage. Hence

$$\mathbf{P}\left(\|\eta_n^{(2)} - \eta_n^{(1)}\| \geqslant R\varepsilon_n\right) \leqslant cR^{-d}\varepsilon_n A_t^d. \tag{8.9}$$

IV. Let $\eta_n^{(3)}$ denote the random polygonal line with nodes

$$\left(t_{kl}, \sum_{i=1}^{k} \Psi_i^{(1)} (\sigma n)^{-1/2}\right), \quad k = 0, 1, \ldots, m+1.$$

1. We obtain from (6.4) that

$$\mathbf{P}\left(\|\eta_n^{(3)} - \eta_n^{(2)}\| \geqslant a\right) \leqslant c\left(a^{-t} L_t + a^{-u} (h/l)^{u/2}\right), \tag{8.10}$$

where u was defined in III.

2. Just as in the uniform strong mixing case, we have for $\delta = t - 2$, $d = \max\left(2t, \frac{(2+5t)}{2(5-t)}\right)$ that

$$\mathbf{P}\left(\|\eta_n^{(3)} - \eta_n^{(2)}\| \geqslant a\right) \leqslant c\left(a^{-d} A_t \varepsilon_n^{d+1} + a^{-d} A_t^d (h/l)^{d/2}\right). \tag{8.11}$$

V. Using the approximation theorems of Berkes and Philipp [6, Theorems 1 and 2], we obtain without loss of generality that for the uniform strong mixing

$$\mathbf{P}\left(\|\eta_n^{(4)} - \eta_n^{(3)}\| \geqslant c\varphi(h)\, n/l\right) \leqslant c\varphi(h)\, n/l, \tag{8.12}$$

and for the strong mixing

$$\begin{aligned} \mathbf{P}\left(\|\eta_n^{(4)} - \eta_n^{(3)}\| \geqslant c\,\frac{n}{l}\left(\frac{\ln n}{n} + n(\alpha(h))^{1/2} + l\,n^{-2}\right)\right) \\ \leqslant cn/l\left(\frac{\ln n}{n} + n(\alpha(h))^{1/2} + l\,n^{-2}\right), \end{aligned} \tag{8.13}$$

where $\eta_n^{(4)}$ is the random polygonal line with nodes

$$\left(t_{kl}, \sum_{i=1}^{k} \Psi_i^{(2)} (\sigma n)^{-1/2}\right),$$

$k = 0,1,\ldots,m+1$, the $\Psi_i^{(2)}$ are independent and distributed like the $\Psi_i^{(1)}$.

VI. Using the result of A.I. Sakhanenko [7], [8] and the truncation method of [44], we obtain

$$\mathbf{P}\left(\|\eta_n^{(4)}-\eta_n^{(5)}\|\geqslant q_s^{1/(s+1)}\right)\leqslant q_s^{1/(s+1)},\quad s>2,\tag{8.14}$$

where $\eta_n^{(5)}$ is the random polygonal line with nodes

$$\left(t_{kl},\ \sum_{i=1}^{k} Y_i(\sigma n)^{-1/2}\right),$$

$k = 0,1,\ldots,m+1$, the Y_i are independent normally distributed random variables,

$$\mathbf{E}Y_i=0,\quad \mathbf{D}Y_i=\mathbf{D}\Psi_i^{(1)},\quad q_s=c\left(\sum_{i=1}^{m+1}\mathbf{E}\,|\Psi_i^{(1)}|^s\right)\left(\sum_{i=1}^{m+1}\mathbf{D}\Psi_i^{(1)}\right)^{-s/2},$$

$i = 1,2,\ldots,m+1$.

VII. To begin with, let us estimate

$$\sum_{i=1}^{m+1}\mathbf{E}\,|\Psi_i^{(1)}|^s.$$

1. Let $s = \frac{(2+5t)}{2(5-t)}$. It follows from (2.1) of Theorem 2.1:

$$n^{-s/2}\sum_{i=1}^{m+1}\mathbf{E}\,|\Psi_i^{(1)}|^s\leqslant c\left(L_t^{s/t}+(n/l)^{1-s/2}\right).$$

Noting (8.7), we find that

$$n^{-s/2}\sum_{i=1}^{m+1}\mathbf{E}\,|\Psi_i^{(1)}|^s\leqslant c\quad x_t^{s+1}.\tag{8.15}$$

2. Taking account of (8.4) and (8.7), from inequality (2.2) of Theorem 2.2 we obtain for $s = \max\left(2t, \frac{(2+5t)}{2(5-t)}\right)$, $\delta = t,2$ that

$$n^{-s/2}\sum_{i=1}^{m+1}\mathbf{E}\,|\Psi_i^{(1)}|^s\leqslant c\varepsilon_n^{s+1}A_t^s.\tag{8.16}$$

VIII. Estimation of the variances.

1. First we prove that

$$\max_{1\leqslant k\leqslant m+1}\left|\sum_{i=1}^{k}\left(\mathbf{D}\Psi_i-\mathbf{D}\Psi_i^{(1)}\right)\right|\leqslant cnL_t^{2/t}.\tag{8.17}$$

Denoting the right side of the last inequality by I_1, we have

$$\begin{aligned}I_1\leqslant 2n\sum_{1\leqslant i<j\leqslant n}&\left|\mathbf{E}\left(n^{-1/2}\xi_iI(|\xi_i|\leqslant yn^{1/2})\,n^{-1/2}\xi_jI(|\xi_j|>yn^{1/2})\right)\right|\\&+\left|\mathbf{E}\left(n^{-1/2}\xi_iI(|\xi_i|>yn^{1/2})\,n^{-1/2}\xi_jI(|\xi_j|\leqslant yn^{1/2})\right)\right|\\&+\left|\mathbf{E}\left(n^{-1/2}\xi_iI(|\xi_i|>yn^{1/2})\,n^{-1/2}\xi_jI(|\xi_j|>yn^{1/2})\right)\right|.\end{aligned}$$

Estimate, for instance, the first summand in the preceding inequality. Denoting it by I_2, we have

$$I_2 \leqslant c \sum_{1\leqslant i\leqslant j\leqslant n} \varphi^{1/t}(j-i)\,\mathbf{E}^{1/t}|n^{-1/2}\xi_i|^t\,\mathbf{E}^{(t-1)/t}|n^{-1/2}\xi_j|^t \quad y^{2-t}.$$

Hence, by Hölder's inequality,

$$I_2 \leqslant c_1\left(\sum_{r=0}^{n-1}\varphi^{1/t}(r)\right)L_t^{1/t}L_t^{(t-1)/t}y^{2-t} \leqslant c_2 L_t^{2/t}.$$

Next, by Lemmas 8.1 and 8.2 and the conditions of the Theorem, we have

$$\max_{1\leqslant k\leqslant m+1}\left|\left(\sum_{i=1}^{k}\mathbf{D}\Psi_i\right)-kl\sigma\right| \leqslant c\,(l+nh/l). \tag{8.18}$$

Therefore

$$\max_{1\leqslant k\leqslant m+1}\left|\left(\sum_{i=1}^{k}\mathbf{D}\Psi_i^{(1)}\right)-kl\sigma\right| \leqslant c\left(l+nh/l+nL_t^{2/t}\right). \tag{8.19}$$

2. Just as in the uniform strong mixing case, (8.18) is true. Now let us prove an analog of inequality (8.17). We also estimate I_2. For $p > 0$, $1/p + 1/t < 1$, we have

$$I_2 \leqslant c \sum_{1\leqslant i\leqslant j\leqslant n} \alpha^{1-1/t-1/p}(j-i)\,\mathbf{E}^{1/t}|n^{-1/2}\xi_i|^t\,\mathbf{E}^{1/p}|n^{-1/2}\xi_i|^t y^{1-t/p}.$$

Therefore

$$I_2 \leqslant c\left(\sum_{r=0}^{n-1}\alpha^{1-1/t-1/p}(r)\right)A_t n^{1/2-t/2p}y^{1-t/p},$$

where $y = n^{1/t-1/2}$. By elementary calculations, for p such that $1/p > 1/t + (t-2)/(t+1)$, we obtain

$$n^{1/2-t/2p}y^{1-t/p} \leqslant c\varepsilon_n^2(\ln n)^{-2} \tag{8.20}$$

and, since $\alpha(k) \leqq Ak^{-g}$ and $g > 2t(t+1)(t-2)^{-1}$ (this follows from the conditions of the Theorem), we can find p such that $1/p > 1/t + (t-2)/(t+1)$ and $1 - 1/t - 1/p > 2/g$ as well. Therefore, for the strong mixing case one has

$$\max_{1\leqslant k\leqslant m+1}\left|\left(\sum_{i=1}^{k}\mathbf{D}\Psi_i^{(1)}\right)-kl\sigma\right| \leqslant c\left(l+nh/l+A_t n\varepsilon_n^2(\ln n)^{-2}\right). \tag{8.21}$$

IX. Noting that $L_t = o(1)$ (otherwise Theorem 7.1 is trivial) and that $\ell = o(n)$, $h = o(\ell)$ by construction, we have

$$\sum_{i=1}^{m+1} \mathbf{D}\Psi_i^{(1)} = \mathbf{D}S_n(1+o(1)) = \sigma n(1+o(1)).$$

Therefore, in (8.14), $q_s^{1/(s+1)} \leqq cx_t$ for the uniform strong mixing and $q_s^{1/(s+1)} \leqq c\varepsilon_n A_t$ for the strong mixing.

X. From (8.19) for the uniform strong mixing and from (8.21) for the strong mixing, and from Lemma 2 of [44], we find that

$$\mathbf{P}\left(\|\eta_n^{(5)} - w\| \geqslant c\sqrt{a_1}\right) \leqslant c\sqrt{a_1}, \tag{8.22}$$

where

$$a_1 = \left(l/n + h/l + L_t^{2/t}\right)|\ln L_t| \tag{8.23}$$

for the uniform strong mixing case and

$$a_1 = \left(l/n + h/l + A_t\varepsilon_n^2(\ln n)^{-2}\right)\ln n \tag{8.24}$$

for the strong mixing case.

XI. Noting that the Lévy-Prokhorov distance is no greater than the Ky-Fan distance (see Lemma 1.2 of [41]), we have the following.

1. Using formulas (8.2), (8.8), (8.10), (8.12), (8.14), (8.22), (8.23) and taking into account IX., we see that to prove Theorem 7.1 it remains to find $h = \tilde{n}^{\sigma}$, $\tilde{\sigma} > 0$, such that

$$(h/l)^{u/2} \leqslant \varepsilon_n^{u+1},\ (n/l)\,h^{-g} \leqslant \varepsilon_n,$$
$$\left(l/n + h/l + L_t^{2/t}\right)|\ln L_t| \leqslant cL_t^{2/(t+1)},$$

where

$$u = \frac{(2+5t)}{2(5-t)},\quad l = n^{3/(t+1)-\varepsilon},\quad \varepsilon = \frac{2(5-t)}{3(t+1)},\ \varepsilon_n = n^{(2-t)/2(t+1)}.$$

Since $L_t^{1/(t+1)} \geqq \varepsilon_n$, then the preceding system of inequalities is equivalent to the following one:

$$\begin{gathered}(\tilde{\sigma} - 3/(t+1) + \varepsilon)u/2 \leqslant (u+1)(2-t)/2(t+1),\\ ((t-2)/(t+1)+\varepsilon) - g\tilde{\sigma} \leqslant (2-t)/2(t+1),\\ \max(3/(t+1) - 1 - \varepsilon, \tilde{\sigma} - 3/(t+1) + \varepsilon) < (2-t)/(t+1).\end{gathered} \tag{8.25}$$

The conditions of Theorem 7.1 (i.e., the choice of the rate of decrease of the uniform strong mixing coefficient g) are selected so that (8.25), (8.7) and

$$\sum_{k=0}^{\infty} \varphi^{1/j(u)}(k)(k+1)^{j(u)-2} < \infty$$

be satisfied for $u \geqq t$, $\varepsilon > 0$, $\tilde{\sigma} > 0$, i.e., $g > j(u)(j(u)-1)$. Roughly speaking, in solving the system of inequalities we try to minimize one of the unknowns, namely, g. We omit all the necessary calculations since they are elementary, though somewhat laborious.

In particular, in (8.25) one needs to take $\tilde{\sigma} = (\frac{3(t-2)}{2(t+1)} + \varepsilon)^{-1} g$.

2. Just as in the uniform strong mixing case, but using formulas (8.1), (8.6), (8.9), (8.11), (8.13), (8.14), (8.22), (8.24) and taking IX. into account, we arrive at the system of inequalities:

$$(\tilde{\sigma} - 3/(t+1) + \varepsilon)d/2 \leqslant (d+1)(2-t)/2(t+1),$$
$$-g\tilde{\sigma}/2 + 2 - 3/(t+1) + \varepsilon \leqslant (2-t)/2(t+1),$$
$$\max(3/(t+1) - 1 - \varepsilon, \tilde{\sigma} - 3/(t+1) + \varepsilon) < (2-t)/(t+1),$$

where $d = \max(2t, \frac{(2+5t)}{2(5-t)})$. Here we take $\tilde{\sigma} = (11t+8)(6g(t+1))^{-1}$.

The theorems are proved.

REMARK. The method used is actually Bernshtein's method [25]. In [47] it is noted that the estimate of order $n^{-\frac{1}{4}}$ in the central limit theorem for stationary mixing sequences of random variables is the top of this method. That is why we give estimates only for $t < 5$ -- because for $t > 5$ we have $\varepsilon_n << n^{-\frac{1}{4}}$.

REFERENCES

[1] Rozenblatt, M.A. "Central Limit Theorem and a Strong Mixing Condition." *Proc. Nat. Acad. Sci. USA*, vol.42, no.1 (1956): 43-47.

[2] Ibragimov, I.A. "Some Limit Theorems for Stationary Narrow-Sense Probability Processes." *Soviet Math. Doklady*, vol.125, no.4 (1959): 711-718. (English transl.)

[3] Utev, S.A. "Some Inequalities and Limit Theorems for Weakly Dependent Random Variables." In *XV Vsesoyuznaya shkola-kollokvium po teorii veroyatnostej i matematicheskoj statistike*, Tezisy dokladov (The Fifteenth All-Union School-Colloquium on Probability Theory and Mathematical Statistics, Abstracts), 27-28. Tbilisi, 1981.

[4] ______. "Some Inequalities for Weakly Dependent Random Variables." In *Tret'ya Vilniusskaya konferentsiya po teorii veroyatnostej i matematicheskoj statistike,* Tezisy dokladov (The Third Vilnius Conference on Probability Theory and Mathematical Statistics, Abstracts), 204-205 (vol.2). Vilnius, 1981.

[5] ______. "On Some Limit Theorems for Random Variables with Uniformly Strong Mixing." *Theory Probab. Applications*, vol.27, no.1 (1982): 212. (English transl.)

[6] Berkes, István, and Philipp, Walter. "Approximation Theorems for Independent and Weakly Dependent Random Vectors." *The Annals of Probability*, vol.7, no.1 (1979): 29-54.

[7] Sakhanenko, A.I. "On the Rate of Convergence of the Invariance Principle." In *Tret'ya Vilniusskaya konferentsiya po teorii veroyatnostej i matematicheskoj statistike*, Tezisy dokladov (The Third Vilnius Conference on Probability Theory and Mathematical Statistics, Abstracts), 135-36 (vol.2). Vilnius, 1981.

[8] ______. "A Rate of Convergence of the Invariance Principle for Non-identically Distributed Variables with Exponential Times." In *Predel'nye teoremy dlya summ sluchajnykh velichin* (Limit Theorems for Sums of Random Variables). Novosibirsk. In press.

[9] Rosenthal, H.P. "On the Subspaces of L_p $(p > 2)$ Spanned by Sequences of Independent Random Variables." *Israel J. Math.*, vol.8, no.3 (1970): 273-303.

[10] ______. "On the Span in L^p of Sequences of Independent Random Variables." In *Proceedings of the Sixth Berkeley Sympos. Math. Statist. Probab.*, II, 149-68. Berkeley and Los Angeles: University of California Press, 1972.

[11] Bahr, B. von, and Esseen, C.G. "Inequalities for the rth Absolute Moment of a Sum of Random Variables." *The Annals of Math. Statist.*, vol.36, no.1 (1965): 299-303.

[12] Nagaev, S.V. "Large Deviations of Sums of Independent Random Variables." *The Annals of Probability*, vol.7, no.7 (1979): 745-98.

[13] Burkholder, L.L. "Distribution Function Inequalities for Martingales." *The Annals of Probability*, vol.1, no.1 (1973): 19-42.

[14] Pinelis, I.F. "On the Distribution of Sums of Sums of Independent Random Variables with Values in a Banach Space." *Theory Probab. Applications*, vol.23, no.3 (1978): 608-15. (English transl.)

[15] Acosta, A. de. "Inequalities for B-valued Random Vectors with Applications to the Strong Law of Large Numbers." *The Annals of Probability*, vol.9, no.1 (1981): 157-61.

[16] Nagaev, S.V., and Pinelis, I.F. "Some Inequalities for the Distribution of Sums of Independent Random Variables." *Theory Probab. Applications*, vol.22, no.2 (1977): 248-56. (English transl.)

[17] Doob, J.L. *Stochastic Processes*. New York: John Wiley & Sons, Inc., 1953.

[18] Ibragimov, I.A. "Some Limit Theorems for Stationary Processes." *Theory Probab. Applications*, vol.7, no.4 (1962): 349-82. (English transl.)

[19] Yokoyama, R. "Moment Bounds for Stationary Mixing Sequences." *S. Wahrscheinlichkeitstheorie verw. Gebiete*, B.52, H.1 (1980): 45-57.

[20] Khas'minskij, R.Z. "On the Stochastic Processes Defined by Differential Equations with a Small Parameter." *Theory Probab. Applications*, vol.11, no.2 (1966): 211-28. (English transl.)
[21] Babu, G.J., Chosh, H., and Singh, K. "On Rates of Convergence to Normality for Φ-mixing Processes." *Sankhya, Indian J. of Statist.*, ser. A, vol.40, no.3 (1978): 278-93.
[22] Kuelbs, J., and Philipp, W. "Almost Sure Invariance Principles for Partial Sums of Mixing B-valued Random Variables." *The Annals of Probability*, vol.8, no.6 (1980): 1003-36.
[23] Hipp, C. "Convergence Rates of the Strong Law for Stationary Mixing Sequences." *Z. Wahrscheinlichkeitstheorie verw. Gebiete*, B.52, H.1 (1979): 49-62.
[24] Ken-ichi, Yoshihara. "Moment Inequalities for Mixing Sequences." *Kodai Math. J.*, vol.1, no.2 (1978): 316-28.
[25] Bernstein, S.N. "Sur l'extension du théorème du calcul des probabilités aux sommes de quantités dépendantes." *Math. Ann.*, B.97 (1926): 1-59. (French transl.)
[26] ______. *Sobranie sochinenij* (Collected works). VOl.4. Moscow: Nauka, 1964.
[27] Davydov, Yu.A. "Convergence of Distributions Generated by Stationary Stochastic Processes." *Theory Probab. Applications*, vol.13, no.4 (1968): 691-96. (English transl.)
[28] Moricz, F. "Moment Inequalities and the Strong Laws of Large Numbers." *Z. Wahrscheinlichkeitstheorie verw. Gebiete*, B.35, H.4 (1976): 299-314.
[29] Kolmogorov, A.N. "Über die Summen durch den Zufall bestimmter unabhängiger Grössen." *Math. Ann.*, B.99, H.2 (1928): 309-19). (German transl.)
[30] Bernstein, S.I. "On Some Modifications of the Chebyshev Inequality" (in Russian). *Doklady Akademii Nauk*, vol.17, no.6 (1937): 275-77.
[31] Gikhman, I.I., and Skorokhod, A.V. *Introduction to the Theory of Random Processes*. Saunders, Philadelphia: Scripta Technica, 1969. (English transl.)
[32] Cohn, H. "On a Class of Dependent Random Variables." *Rev. Roum. Math. Pures et Applications*, vol.10, no.10 (1965): 1593-1606.
[33] Reznik, M.Kh. "The Law of the Iterated Logarithm for Some Classes of Stationary Processes." *Theory Probab. Applications*, vol.13, no.4 (1968): 606-21. (English transl.)
[34] Borodin, A.I. "Quasimartingales." *Theory Probab. Applications*, vol.23, no.3 (1978): 637-41. (English transl.)
[35] McLeish, D.L. "A Maximal Inequality and Dependent Strong Laws." *The Annals of Probability*, vol.3, no.5 (1975): 829-39.
[36] Fook (Fuk), D.Kh., and Nagaev, S.V. "Probability Inequalities for Sums of Independent Random Variables." *Theory Probab. Applications*, vol.16, no.4 (1971): 643-60. (English transl.)
[37] Borovkov, A.A. "On the Rate of Convergence for the Invariance Principle." *Theory Probab. Applications*, vol.18, no.2 (1973): 207-25. (English transl.)
[38] Petrov, V.V. *Sums of Independent Random Variables*. New York Berlin Heidelberg: Springer-Verlag, 1975. (English transl.)

[39] Philipp, Walter, and Stout, William. "Almost Sure Invariance Principles for Partial Sums of Weakly Dependent Random Variables." *Memoirs of the Amer. Math. Society*, vol.2, issue 2, no.161 (July 1975). Providence, Rhode Island: American Math. Society.

[40] Prokhorov, Yu.V. "On a Strengthened Law of Large Numbers" (in Russian). *Izvestiya Akademii Nauk SSSR*, ser. matem., vol.14, no.6 (1950): 523-36.

[41] ______. "The Convergence of Random Processes and Limit Theorems in Probability Theorems." *Theory Probab. Applications*, vol.1, no.2 (1956): 157-214. (English transl.)

[42] Komlós, J., Major, P., and Tusnády, G. "An Approximation of Partial Sums of Independent RV'-s, and the Sample DF. I." *Z. Wahrscheinlichkeitstheorie verw. Gebiete*, B.34, H.1 (1976): 33-58.

[43] Sakhanenko, A.I. "Estimates of the Rate of Convergence in the Invariance Principle." *Soviet Math. Doklady*, vol.15, no.6 (1974): 1752-55. (English transl.)

[44] Utev, S.A. "A Note on the Rate of Convergence in the Invariance Principle" (in Russian). *Sibirskij matematicheskij zhournal*, vol.22, no.5 (1981): 206-209.

[45] Gorodetskij, V.V. "Estimates of the Rate of Convergence of the Invariance Principle for Strong Mixing Sequences." In *Tret'ya Vil'niusskaya konferentsiya po teorii veroyatnostej i matematicheskoj statistike*, Tezisy dokladov (The Third Vilnius Conference on Probability Theory and Mathematical Statistics, Abstracts), 149-50 (vol.1). Vilnius, 1981.

[46] Ken-ichi, Yoshihara. "Convergence Rates of the Invariance Principles for Absolutely Regular Sequences." *Yokohama Math. J.*, vol.27, no.1 (1979): 49-55.

[47] Stein, Ch. "A Bound for the Error in the Normal Approximation to the Distribution of a Sum of Dependent Random Variables." In *Proc. of the Sixth Berkeley Sympos. Math. Statist. Probab.* II, 583-602. Berkeley and Los Angeles: University of California Press, 1972.

V.R. Khodzhibaev

ASYMPTOTIC ANALYSIS OF DISTRIBUTIONS IN TWO-BOUNDARY PROBLEMS FOR CONTINUOUS-TIME RANDOM WALKS

Let $\xi(t)$, $t \geqq 0$, $\xi(0) = 0$, be a homogeneous process with independent increments and right-continuous sample trajectories, $M\xi(1) = 0$. For $-a < 0 < b$, define the random variables

$$T = \inf\{t\colon \xi(t) \notin (-a, b)\}, \quad Y = \xi(T).$$

We consider the probabilities

$$\mathbf{P}(\xi(t) \in A, T > t), \quad A \subset (-a, b);$$
$$\mathbf{P}(Y \in B, T < t), \quad B \subset (-\infty, -a] \cup [b, \infty), \tag{1}$$

and their integral transformations

$$H(u, A) = \int_0^\infty e^{-ut}\mathbf{P}(\xi(t) \in A, T > t)\, dt,$$
$$\Pi(u, B) = \int_0^\infty e^{-ut}\mathbf{P}(Y \in B, T \in dt). \tag{2}$$

Our objective in this work is to obtain complete asymptotic expansions of the probabilities (1) as $t \to \infty$, as $a = a(t) \to \infty$, $b = b(t) \to \infty$.

Of the extensive literature on the study of boundary value functionals of random processes with independent increments, we cite only those directly related to the subject of our study. For discrete-time random walks with one boundary $(a = \infty)$, the complete asymptotic expansions of distributions of boundary value functionals are given in [1]-[3]. Complete asymptotic expansions of boundary value functionals for the same walks on a segment have been obtained, under the most general assumptions, in [4]-[6]. In several cases they are also derived for continuous-time random walks. In [7], asymptotic formulas for $\Pi(u,B)$ and the corresponding complete asymptotic expansions are found for $a = \infty$, $B = [b,\infty)$ and under specific conditions. For $a = C_1\sqrt{t}$, $b = C_2\sqrt{t}$, $t \to \infty$ (in what follows, the letter C with or without subscripts designates positive constants), complete asymptotic expansions have been obtained under Cramér's condition in [9] -- for the distribution of the exit time of a process without positive jumps through the lower boundary, and in [10] -- for the probabilities $P(T < t)$, $P(Y \geq b, T < t)$ under the condition that positive jumps of a process have a particular distribution. Exact expressions for the distribution of T in the absence of positive (negative) jumps of a process in the case $a = \infty$ are well known from [11], [12].

In Section 1, asymptotic representations of transformations (2) are found for $a \to \infty$, $b \to \infty$. We use some of the results obtained in [7] and, therefore, impose the following conditions (from [7]) on the process:

1. $Me^{\lambda\xi(1)} < \infty$ for $\lambda_- \leqq \lambda \leqq \lambda_+$, $-\infty < \lambda_- < 0 < \lambda_+ < \infty$. Let

$$\psi(\lambda) = \ln Me^{\lambda\xi(1)} = \alpha\lambda + \sigma^2\lambda^2/2 + \int_{-\infty}^{\infty} (e^{\lambda x} - 1 - \lambda x)\, dS(x),$$

where α, σ are real numbers. Also, $S(-\infty) = S(+\infty) = 0$.

2. If $\sigma^2 > 0$, then for $\lambda_- \leqq \operatorname{Re}\lambda \leqq \lambda_+$, $|\operatorname{Im}\lambda| > C_4$ and some $p > 0$

$$|\psi_1(\lambda)| = |\psi(\lambda) - \alpha\lambda - \sigma^2\lambda^2/2| \leqslant C_3|\lambda|^{2-p}.$$

3. If $\sigma^2 = 0$, then $S(x)$ is a function of bounded variation of $(-\infty,\infty)$,

$$(-\infty, \infty), \quad \gamma = \alpha - \int_{-\infty}^{\infty} x\,dS(x) \neq 0$$

and

$$\overline{\lim_{|\lambda|\to\infty}} |\psi_2(\lambda)|/\psi_2(\operatorname{Re}\lambda) < 1 \qquad \text{for} \qquad \lambda_- \leqslant \operatorname{Re}\lambda \leqslant \lambda_+,$$

where

$$\psi_2(\lambda) = \int_{-\infty}^{\infty} e^{\lambda x}\,dS(x) - \mu, \quad \mu = S(+0) - S(-0).$$

Our method for obtaining asymptotic representations of the integral transformations (2) includes, at the first level, finding identities for the double transformations of the distributions (1), which contain the factorization of the function $u/(u - \psi(\lambda))$, with subsequent determination of singularities of the factors. The properties of such factors, known from [7], enable us to invert the double transformations with respect to the spatial variable, isolating thereby the principal part. This makes our method similar to the investigations of discrete-time random walks in [1]-[5]. We have also taken several techniques from [2], [4] and [7]. In Section 2, using the asymptotic representations of (2), we find complete asymptotic expansions of the probabilities (1) in a range of deviations of the boundaries $a = o(t) \to \infty$, $b = o(t) \to \infty$, $a + b \geqq C\sqrt{t}$, $t \to \infty$. As in [2]-[4], we apply a modified saddle-point method to investigate the principal parts of the asymptotic representtations obtained in Section 1.

1. Asymptotic representations of integral transformations

1. Let

$$V(u,\lambda)=\int_0^\infty e^{-ut}\int_{-a}^{b} e^{\lambda x}\mathbf{P}(\xi(t)\in dx,\ T>t)\,dt,\quad \operatorname{Re}u>0,\ \operatorname{Re}\lambda=0;$$

$$V_{u+}(\lambda)=\int_0^\infty e^{-ut}\int_b^\infty e^{\lambda x}\mathbf{P}(Y\in dx,\ T\in dt);$$

$$V_{u-}(\lambda)=\int_0^\infty e^{-ut}\int_{-\infty}^{-a} e^{\lambda x}\mathbf{P}(Y\in dx,\ T\in dt),\quad \operatorname{Re}u\geqslant 0,\ \operatorname{Re}\lambda=0.$$

It is known from [13] and [15] that for $\operatorname{Re}u>0$, $\operatorname{Re}\lambda=0$,

$$(u-\psi(\lambda))V(u,\lambda)=1-V_{u+}(\lambda)-V_{u-}(\lambda)\,. \tag{3}$$

As in [2], let V be the collection of complex-valued functions $v(t)$, $-\infty<t<\infty$, having bounded variation, and let

$$\mathfrak{B}(\mu_-,\mu_+)=\left\{\int_{-\infty}^\infty e^{\lambda t}dv(t):v(\cdot)\in V,\quad \int_{-\infty}^\infty e^{Re\lambda t}\,|dv(t)|<\infty,\ \mu_-\leqslant\operatorname{Re}\lambda\leqslant\mu_+\right\},$$

$$\mathfrak{B}_\pm(\mu_\pm)=\left\{\pm\int_0^{\pm\infty} e^{\lambda t}dv(t):v(\cdot)\in V,\quad \pm\int_0^{\pm\infty} e^{Re\lambda t}|dv(t)|<\infty,\ \pm\operatorname{Re}\lambda\leqslant\pm\mu_\pm\right\}.$$

We norm $\mathfrak{B}(\mu_-,\mu_+)$ as follows (see also [7]):

$$\|f\|=\max_{\mu_-\leqslant c\leqslant\mu_+}\int_{-\infty}^\infty e^{cx}\,|dF(x)|\qquad\text{for}\qquad f(\lambda)=\int_{-\infty}^\infty e^{\lambda x}dF(x).$$

Let $\mathfrak{R}(\mu_-,\mu_+)$ denote the subset of $\mathfrak{B}(\mu_-,\mu_+)$ formed by the elements $g(\lambda)=\int_{-\infty}^\infty e^{\lambda x}dv(x)$ with absolutely continuous $v(\cdot)$. The sets $\mathfrak{R}_\pm(\mu)$ are defined similarly. For

$$f(\lambda)=\int_{-\infty}^\infty e^{\lambda x}dF(x),\qquad F(\cdot)\in V,\ \operatorname{Re}\lambda=0$$

set

$$[f(\lambda)]^A=\int_A e^{\lambda x}dF(x),\quad A\subset(-\infty,\infty).$$

Let $r_u(\lambda)=r_{u+}(\lambda)r_{u-}(\lambda)$ be the infinitely divisible factorization of the function $r_u(\lambda)=u/(u-\psi(\lambda))$ for $\operatorname{Re}u>0$, $\operatorname{Re}\lambda=0$. Then $r_{u+}(\lambda)$

$(r_{u-}(\lambda))$ is the Laplace transformation for $\operatorname{Re}\lambda \leqq 0$ $(\operatorname{Re}\lambda \geqq 0)$ of an infinitely divisible distribution whose support is contained on the non-negative (nonpositive) half-axis, $r_{u\pm}(\lambda) \in \mathfrak{B}_{\pm}(0)$ (see [8]).

Let us mention some properties of the function $\psi(\lambda)$ known from [7]. The function $\psi(\lambda)$ is analytic inside the strip $\lambda_- \leqq \operatorname{Re}\lambda \leqq \lambda_+$. Moreover, it is convex for $\lambda_- \leqq \lambda \leqq \lambda_+$ (see also [14]) and, hence, attains a minimum on this segment. Since $M\xi(1) = 0$, the point $\lambda = 0$ is the minimum point of $\psi(\lambda)$, $\psi(0) = 0$. It follows from the convexity that the equation $u = \psi(\lambda)$ for

$$\lambda_- \leqslant \lambda \leqslant \lambda_+, \quad 0 < u \leqslant \min\{\psi(\lambda_-),\ \psi(\lambda_+)\}$$

has exactly two roots

$$\lambda_-(u),\ \lambda_+(u),\ \lambda_-(u) < 0 < \lambda_+(u),\quad \lambda_-(0) = \lambda_+(0) = 0\ .$$

If $u_0 = \min\{\psi(\lambda_-),\ \psi(\lambda_+)\}$, then $0 < \lambda_+(u) < \lambda_+$, $\lambda_- < \lambda_-(u) < 0$ for $0 < u < u_0$. The functions $\lambda_{\pm}(u)$ can analytically be continued to the region

$$U_{\varepsilon_1,\delta} = \{u\colon -\varepsilon_1 < \operatorname{Re} u < u_0 - \varepsilon_2\ ,$$
$$|\operatorname{Im} u| < \delta\} \setminus \{u\colon -\varepsilon_1 < \operatorname{Re} u = u \leqslant 0\}\ ,$$

where $\delta > 0$ is sufficiently small $\varepsilon_1 = \varepsilon_1(\delta) > 0$, $\varepsilon_2 = \varepsilon_2(\delta) \to 0$ as $\delta \to 0$. Here the extensions obtained continue to be solutions of the equation $u = \psi(\lambda)$. The functions $\lambda_-(u)$ and $\lambda_+(u)$ are branches of a two-valued analytic function with branching point $u = 0$, and may be viewed as analytic in $U_{\varepsilon_1,\delta}$ and schlicht.

Next, consider the function

$$\psi_3(\lambda) = \begin{cases} \alpha\lambda + \sigma^2\lambda^2/2 - u_1 & \text{if} \quad \sigma^2 > 0\ , \\ \gamma\lambda - u_1 & \text{if} \quad \sigma^2 = 0\ , \end{cases}$$

where $u_1 > 0$ is such that $\max\{\psi_3(\lambda_-),\ \Psi_3(\lambda_+)\} > -1$. This function possesses all the above-mentioned properties of the function $\psi(\lambda)$. In particular, for $u > -u_1$, $\delta^2 > 0$ the solutions $\mu_{\pm}(u)$ of the equation $u = \psi_3(\lambda)$ are defined. If $\sigma^2 = 0$, then one of the functions $\mu_{\pm}(u)$ is defined depending on the sign of γ.

Under the assumptions made, the function $\psi_1(\lambda,u) = (u-\psi(\lambda))/(u-\psi_3(\lambda))$ for $\operatorname{Re} u > 0$, $\operatorname{Re}\lambda = 0$ has the canonical factorization

$$\psi_1(\lambda, u) = \psi_{1+}(\lambda, u)\psi_{1-}(\lambda, u),\ \psi_{1+}^{\pm 1}(\lambda, u) \in \mathfrak{B}_+(0),\ \psi_{1-}^{\pm 1}(\lambda, u) \in \mathfrak{B}_-(0).$$

with components having the following properties. There is a $\delta_1 > 0$ such that for

$$u \in K_{\varepsilon_1,\delta} = U_{\varepsilon_1,\delta} \cup \{u: -\varepsilon_1 < \operatorname{Re} u = u < 0\}$$

the functions

$$\begin{aligned}&[\psi_{1+}(\lambda, u)(\lambda - \lambda_+ - 1)/(\lambda - \lambda_+(u))]^{\pm 1} \in \mathfrak{B}_+(\delta_1),\\ &[\psi_{1-}(\lambda, u)(\lambda - \lambda_- + 1)/(\lambda - \lambda_-(u))]^{\pm 1} \in \mathfrak{B}_-(-\delta_1)\end{aligned} \tag{4}$$

are analytic and uniformly bounded in the norm $u \in K_{\varepsilon_1,\delta}$, and for

$$u \in \widetilde{U}_{\varepsilon_1,\delta} = \{u: \operatorname{Re} u > -\varepsilon_1\} \setminus \{u: -\varepsilon_1 < \operatorname{Re} u < u_0, \quad |\operatorname{Im} u| < \delta\}$$

the functions

$$[\psi_{1+}(\lambda, u)]^{\pm 1} \in \mathfrak{B}_+(\delta_1), \; [\psi_{1-}(\lambda, u)]^{\pm 1} \in \mathfrak{B}_-(-\delta_1) \tag{4'}$$

are analytic and uniformly bounded in u. In [7], the relationship is given between the infinitely divisible factorization of the function $r_u(\lambda)$ and the sufficiently small function $\psi_1(\lambda,u)$: For $\operatorname{Re} u > 0$, $\operatorname{Re} \lambda \leqq 0$ one has

$$r_{u+}(\lambda) = \begin{cases} \dfrac{\psi_{1+}(0,u)\,\mu_+(u)}{\psi_{1+}(\lambda,u)\,(\mu_+(u) - \lambda)} & \text{if } \sigma^2 > 0 \text{ or } \sigma^2 = 0,\ \gamma > 0, \\ \dfrac{\psi_{1+}(0,u)}{\psi_{1+}(\lambda,u)} & \text{if } \sigma^2 = 0,\ \gamma < 0; \end{cases} \tag{5}$$

for $\operatorname{Re} u > 0$, $\operatorname{Re} \lambda \geqq 0$ one has

$$r_{u-}(\lambda) = \begin{cases} \dfrac{\psi_{1-}(0,u)\,\mu_-(u)}{\psi_{1-}(\lambda,u)\,(\mu_-(u) - \lambda)} & \text{if } \sigma^2 > 0 \text{ или } \sigma^2 = 0,\ \gamma < 0, \\ \dfrac{\psi_{1-}(0,u)}{\psi_{1-}(\lambda,u)} & \text{if } \sigma^2 = 0,\ \gamma > 0. \end{cases} \tag{6}$$

T h e o r e m 1. There exist $\delta > 0$, $\delta_1 > 0$, $\varepsilon > 0$ such that for sufficiently large a and b, uniformly in $U_\delta = \{u\colon 0 \leqq \operatorname{Re} u < \varepsilon,\ |\operatorname{Im} u| < \delta,\ u \neq 0\}$ for $B \subset [b,\infty)$

$$\Pi(u, B) = \frac{F_1(u,B)\left(1 - \mu^a(u)\,H_1(u)\right)}{e^{\lambda_+(u)b}\left(1 - \mu^{b+a}(u)\,H_2(u)\right)} + O\left(e^{-\delta_1 \min(x, x-b+a)}\right), \quad x = \inf_{z \in B} z;$$

and for $B \subset (-\infty,-a]$

$$\Pi(u, B) = \frac{F_2(u,B)\,e^{\lambda_-(u)a}\left(1 - \mu^b(u)\,H_3(u)\right)}{1 - \mu^{b+a}(u)\,H_2(u)} + O\left(e^{\delta_1 \max(x, x+a-b)}\right),$$

$$x = \sup_{z \in B} z,$$

where

$$\int_b^\infty e^{\lambda x} F_1(u, dx) = e^{\lambda b} v_u(\lambda_+(u))/v_u(\lambda) ,$$

$$\int_{-\infty}^{-a} e^{\lambda x} F_2(u, dx) = w_u(\lambda_-(u))/(w_u(\lambda)\lambda^a) ,$$

$$\mathrm{Re}\,\lambda = 0 ,$$

$$v_u(\lambda) = (\lambda - \lambda_+(u)) r_{u+}(\lambda)/\lambda_+(u) ,$$

$$w_u(\lambda) = (\lambda - \lambda_-(u)) r_{u-}(\lambda)/\lambda_-(u) = \exp(\lambda_-(u) - \lambda_+(u)) ,$$

$$H_1(u) = w_u(\lambda_-(u))/w_u(\lambda_+(u)), \quad H_3(u) = v_u(\lambda_+(u))/v_u(\lambda_-(u)),$$

$$H_2(u) = H_1(u) H_3(u) .$$

We need several lemmas to prove the Theorem. In [13], the following lemma has been proved.

L e m m a 1. For $\mathrm{Re}\,u > 0$, $\mathrm{Re}\,\lambda \leqq 0$ one has

$$V_{u+}(\lambda) = r_{u+}^{-1}(\lambda)\,[r_{u+}(\lambda)\,(1 - V_{u-}(\lambda))]^{[b,\infty)}.$$

The next lemma is proved similarly.

L e m m a 2. For $\mathrm{Re}\,u > 0$, $\mathrm{Re}\,\lambda \geqq 0$ one has

$$V_{u-}(\lambda) = r_{u-}(\lambda)\left[r_{u-}^{-1}(\lambda)\,(1 - V_{u+}(\lambda))\right]^{(-\infty,-a]} .$$

L e m m a 3. (See also [2] If $\{v_u\}$ is a collection of functions from $\mathfrak{B}(0,\mu_+)$, $(\mathfrak{B}(\mu_-,0))$, uniformly bounded in the norm, then in the representation

$$v_u(\lambda) = \int_{-\infty}^{\infty} e^{\lambda x} d\omega_u(x), \ \mathrm{Re}\,\lambda = 0$$

the estimate

$$\int_x^\infty |\, d\omega_u(t)\,| = O\left(e^{-\mu_+ x}\right), \quad x \to \infty, \quad \left(\int_{-\infty}^{x} |\, d\omega_u(t)\,| = O\left(e^{-\mu_- x}\right), \quad x \to -\infty\right)$$

is valid uniformly in u and x.

Proof. Follows from the fact that for $\mathrm{Re}\,\lambda = 0$ one has

$$\int_x^\infty |\, d\omega_u(t)\,| = \int_x^\infty e^{-\mu_+ t} e^{\mu_+ t} |\, d\omega_u(t)\,| \leqslant e^{-\mu_+ x} \int_x^\infty e^{\mu_+ t} |\, d\omega_u(t)\,| .$$

We recall now the following lemma from [2].

Lemma 4. If $\nu \in \mathfrak{B}(\mu_-,\mu_+)$ and $\nu(\mu_0) = 0$, $\mu_- \leqq \mathrm{Re}\,\mu_0 \leqq \mu_+$, then $r(\lambda) = \nu(\lambda)/(\lambda-\mu_0) \in \mathfrak{R}(\mu_-,\mu_+)$ and in the representation

$$r(\lambda) = \int_{-\infty}^{\infty} e^{\lambda x} p(x)\,dx \quad |p(x)| = O\left(e^{-\mu_+ x}\right) \text{ as } x \to \infty$$

one has $|p(x)| = O(e^{-\mu_+ x})$ as $x \to \infty$ and $p(x) = O(e^{-\mu_- x})$ as $x \to -\infty$.

Lemma 5. Let the functions $f_u(\lambda) \in \mathfrak{B}(-\delta_1, 0)$ be uniformly bounded in the norm in U_δ. Then for $\mathrm{Re}\,\lambda = 0$ one has

$$\left[\frac{f_u(\lambda)}{\lambda - \lambda_-(u)}\right]^{(-\infty,-a]} = \frac{f_u(\lambda_-(u))\, e^{\lambda_-(u)a}}{e^{\lambda a}(\lambda - \lambda_-(u))} + \theta(u,\lambda),$$

where

$$\theta(u,\lambda) = \int_{-\infty}^{-a} e^{\lambda x}\theta_u(x)\,dx$$

and uniformly in $x \leqq -a$ and $u \in U_\delta$, $|\theta_u(x)| = O(e^{\delta_1 x})$, $x \to -\infty$. Moreover, the $(\lambda - \lambda_-(u))\theta(u,\lambda) \in \mathfrak{B}(-\delta_1, 0)$ are uniformly bounded in u, $u \in U_\delta$, and in the representation

$$(\lambda - \lambda_-(u))\,\theta(u,\lambda) = \int_{-\infty}^{-a} e^{\lambda x} d\eta_u(x), \quad \int_{-\infty}^{x} |d\eta_u(t)| = O\left(e^{\delta_1 x}\right)$$

uniformly in u.

Proof.

$$\frac{f_u(\lambda)}{\lambda - \lambda_-(u)} = \frac{f_u(\lambda_-(u))}{\lambda - \lambda_-(u)} - \frac{1}{\lambda - \lambda_-(u)}[f_u(\lambda_-(u)) - f_u(\lambda)].$$

The expression in square brackets is zero for $\lambda = \lambda_-(u)$. The second summand, by Lemma 4, belongs to class $\mathfrak{R}(-\delta_1, 0)$. The desired estimate for $\theta_u(x)$ follows from Lemmas 3, 4. Here

$$\left[\frac{1}{\lambda - \lambda_-(u)}\right]^{(-\infty,-a]} = \int_{-\infty}^{-a} e^{(\lambda - \lambda_-(u))x}\,dx = \frac{e^{-(\lambda-\lambda_-(u))a}}{\lambda - \lambda_-(u)}.$$

The proof of the second part of the Lemma follows from the relation

$$\left[\frac{f_u(\lambda) - f_u(\lambda_-(u))}{\lambda - \lambda_-(u)}\right]^{(-\infty,-a]} = \frac{[f_u(\lambda)]^{(-\infty,-a]}}{\lambda - \lambda_-(u)} - \frac{e^{-(\lambda-\lambda_-(u))a}}{\lambda - \lambda_-(u)}\,[f_u(\lambda)]^{(-\infty,-a]}_{\lambda=\lambda_-(u)}.$$

Similarly, we prove

L e m m a 6. Let the functions $g_u(\lambda) \in \mathfrak{B}(0,\delta_1)$ be uniformly bounded in the norm in U_δ. Then for $\operatorname{Re}\lambda = 0$ one has

$$\left[\frac{g_u(\lambda)}{\lambda-\lambda_+(u)}\right]^{[b,\infty)} = \frac{g_u(\lambda_+(u))\, e^{\lambda b}}{e^{\lambda_+(u)b}(\lambda-\lambda_+(u))} + \varphi(u,\lambda),$$

where

$$\varphi(u,\lambda) = \int_b^\infty e^{\lambda x}\varphi_u(x)\,dx, \quad |\varphi_u(x)| = O\left(e^{-\delta_1 x}\right), \quad x\to+\infty,$$

uniformly in u, $u \in U_\delta$, $(\lambda-\lambda_+(u))\phi(u,\lambda) \in \mathfrak{B}(0,\delta_1)$ and in the representation

$$(\lambda-\lambda_+(u))\,\varphi(u,\lambda) = \int_b^\infty e^{\lambda x} d\zeta_u(x) \int_x^\infty |d\zeta_u(t)| = O\left(e^{-\delta_1 x}\right)$$

as $x \to \infty$ uniformly in u.

Proof of Theorem 1. Lemmas 1, 2 for $\operatorname{Re} u > 0$, $\operatorname{Re}\lambda = 0$ give two equations with two unknowns: $V_{u+}(\lambda)$, $V_{u-}(\lambda)$. By (4), (5) and Lemma 4 for $u \in U_\delta$ we have:

for $\sigma^2 > 0$

$$v_u(\lambda)(\lambda-\lambda_+-1)^{-1} \in \mathfrak{R}_+(\delta_1), \; w_u(\lambda)(\lambda-\lambda_-+1)^{-1} \in \mathfrak{R}_-(-\delta_1),$$
$$v_u^{-1}(\lambda) \in \mathfrak{B}_+(\delta_1), \quad w_u^{-1}(\lambda) \in \mathfrak{B}_-(-\delta_1); \tag{7}$$

for $\sigma^2 = 0$, $\gamma > 0$

$$v_u(\lambda)(\lambda-\lambda_+-1)^{-1} \in \mathfrak{R}_+(\delta_1), \; w_u(\lambda)(\lambda-\lambda_-+1)^{-1} \in \mathfrak{B}_-(-\delta_1),$$
$$v_u^{-1}(\lambda) \in \mathfrak{B}_+(\delta_1), \quad w_u^{-1}(\lambda) \in \mathfrak{R}_-(-\delta_1); \tag{8}$$

for $\sigma^2 = 0$, $\gamma < 0$

$$v_u(\lambda)(\lambda-\lambda_+-1)^{-1} \in \mathfrak{B}_+(\delta_1), \; w_u(\lambda)(\lambda-\lambda_-+1)^{-1} \in \mathfrak{R}_-(-\delta_1),$$
$$v_u^{-1}(\lambda) \in \mathfrak{R}_+(\delta_1), \quad w_u^{-1}(\lambda) \in \mathfrak{B}_-(-\delta_1). \tag{9}$$

These functions are analytic inside V_δ, continuous and uniformly bounded in the norm, including the boundary $\operatorname{Re} u = 0$. Since for any $\delta_1 > 0$ the functions $V_{u+}(\lambda) \in \mathfrak{B}(-\delta_1,0)$, $V_{u-}(\lambda) \in \mathfrak{B}(0,\delta_1)$ are analytic in U_δ and continuous, including the boundary, then, using Lemmas 1-6 and relations (7)-(9), we have:

for $0 \leqq \operatorname{Re}\lambda \leqq \delta_1$, $u \in U_\delta$,

$$
\begin{aligned}
V_{u+}(\lambda) &= \frac{\lambda-\lambda_-(u)}{w_u(\lambda)}\left[\frac{w_u(\lambda)(1-V_{u+}(\lambda))}{\lambda-\lambda_-(u)}\right]^{(-\infty,-a]} \\
&= \frac{e^{\lambda_-(u)a}w_u(\lambda_-(u))(1-V_{u+}(\lambda_-(u)))}{e^{\lambda a}w_u(\lambda)} + (\lambda-\lambda_-(u))\,\theta_1(u,\lambda);
\end{aligned}
\tag{10}
$$

for $-\delta_1 \leqq \operatorname{Re}\lambda \leqq 0$, $u \in U_\delta$,

$$
\begin{aligned}
V_{u+}(\lambda) &= \frac{\lambda-\lambda_+(u)}{v_u(\lambda)}\left[\frac{v_u(\lambda)(1-V_{u-}(\lambda))}{\lambda-\lambda_+(u)}\right]^{[b,\infty)} \\
&= \frac{e^{\lambda b}v_u(\lambda_+(u))(1-V_{u-}(\lambda_+(u)))}{e^{\lambda_+(u)b}v_u(\lambda)} + (\lambda-\lambda_+(u))\,\varphi_1(u,\lambda);
\end{aligned}
\tag{11}
$$

the estimates from Lemmas 5 and 6, respectively, hold for the last summands in (10) and (11). Substituting $\lambda = \lambda_+(u)$ and $\lambda = \lambda_-(u)$ into (10) and (11) yields two equations and, in turn,

$$
\begin{aligned}
V_{u+}(\lambda_-(u)) &= \frac{\mu^b(u)H_3(u)-\mu^{b+a}(u)H_2(u)}{1-\mu^{b+a}(u)H_2(u)} \\
&- \frac{(\lambda_-(u)-\lambda_+(u))\bigl(\mu^b(u)H_3(u)\,\theta_1(u,\lambda_+(u))-\varphi_1(u,\lambda_-(u))\bigr)}{1-\mu^{b+a}(u)H_2(u)}.
\end{aligned}
\tag{12}
$$

Note that the functions $H_1(u)$ and $H_3(u)$ are analytic in $K_{\varepsilon_1,\delta}$ cut by the ray $u \leqq 0$, and in some neighborhood of zero have a series expansion in powers of $\sqrt{u}$, $H_1(0) = H_2(0) = 1$. These properties follow from (4)-(9). The function $|\lambda_-(u) - \lambda_+(u))||1-\mu^{a+b}(u)H_2(u)|^{-1}$ is uniformly bounded in $K^1_{\varepsilon_1,\delta} = K_{\varepsilon_1,\delta} \cap \{u: |\arg u - \pi| \geqq \pi/4\}$ for sufficiently small δ, ε_1 and sufficiently large $a+b$. Indeed (see also [4]),

$$
\begin{aligned}
&|\lambda_+(u)-\lambda_-(u)||1-\mu^{b+a}(u)H_2(u)|^{-1} \\
&\leqslant |\lambda_+(u)-\lambda_-(u)|(1-|\mu(u)|)^{-1}(1+|\mu(u)|+|\mu^2(u)+\ldots \\
&+|\mu^{b+a-1}(u)|+|\mu^{b+a}(u)|(1-|H_2(u)|)(1-|\mu(u)|)^{-1}),
\end{aligned}
$$

the function $|\lambda_+(u) - \lambda_-(u)|(1 - |\mu(u)|)^{-1}$ is uniformly bounded in $K^1_{\varepsilon_1,\delta}$ and by the above properties of the functions $H_1(u)$ and $H_3(u)$ one has

$$
1+|\mu(u)|+\ldots+|\mu^{b+a-1}(u)|+|\mu^{b+a}(u)|(1-|H_2(u)|)(1-|\mu(u)|)^{-1} > 1
$$

for sufficiently large $a+b$, $u = 0$. Since the expression on the left side of the inequality is continuous in a neighborhood of $u = 0$, the inequality is also true for $u \in U_{\varepsilon_1,\delta}$ for some $\delta = \delta(a+b) > 0$ and $\varepsilon_1 = \varepsilon_1(a+b) > 0$, and, moreover, as a function of $a+b$ this quantity does not decrease. One may likewise prove the uniform boundedness in $K^1_{\varepsilon_1,\delta}$ of the functions

$$(1-\mu^a(u)H_1(u))(1-\mu^{b+a}(u)H_2(u))^{-1}.$$

Next, $|\mu(u)| \leq 1$ in U_δ, $|H_3(u)|$ is uniformly bounded, and

$$|\theta_1(u, \lambda_+(u))| = O(e^{-\delta_1 a}), \quad |\varphi_1(u, \lambda_-(u))| = O(e^{-\delta_1 b}). \tag{13}$$

The estimates in (13) are uniform in U_δ and follow from Lemmas 5 and 6. Relations (12), (13) yield

$$V_{u+}(\lambda_-(u)) = \frac{\mu^b(u)H_3(u) - \mu^{b+a}(u)H_2(u)}{1-\mu^{b+a}(u)H_2(u)} + O(e^{-\delta_1 a} + e^{-b_1\delta}) \tag{14}$$

for $u \in U_\delta$. Analogously, one has uniformly in U_δ that

$$V_{u-}(\lambda_+(u)) = \frac{\mu^a(u)H_1(u) - \mu^{b+a}(u)H_2(u)}{1-\mu^{b+a}(u)H_2(u)} + O(e^{-\delta_1 a} + e^{-\delta_1 b}). \tag{15}$$

It follows from Lemmas 5, 6 and relations (10), (11), (14), (15) that for $u \in U_\delta$, for sufficiently large a and b and for the above choice of $\delta > 0$, one has

$$V_{u-}(\lambda) = \frac{e^{\lambda_-(u)a}w_u(\lambda_-(u))\left(1-\mu^b(u)H_3(u)\right)}{e^{\lambda a}w_u(\lambda)\left(1-\mu^{b+a}(u)H_2(u)\right)} + \int_{-\infty}^{-a} e^{\lambda x}d\varepsilon_u(x), \tag{16}$$

$$V_{u+}(\lambda) = \frac{e^{\lambda b}v_u(\lambda_+(u))\left(1-\mu^a(u)H_1(u)\right)}{e^{\lambda_+(u)b}v_u(\lambda)\left(1-\mu^{b+a}(u)H_2(u)\right)} + \int_b^{\infty} e^{\lambda x}d\rho_u(x), \tag{17}$$

where uniformly for $u \in U_\delta$ and x,

$$\int_{-\infty}^{x} |d\varepsilon_u(y)| = O(e^{\delta_1 x} + e^{\delta_1(x+a-b)}), \quad x \leqslant -a;$$

$$\int_{x}^{\infty} |d\rho_u(y)| = O(e^{-\delta_1 x} + e^{-\delta_1(x-b+a)}), \quad x \geqslant b.$$

This completes the proof of Theorem 1.

For

$$a=\infty, \ B=[b, \infty), \ \Psi_+(\lambda, u) = \psi_{1+}(0, u)\,\mu_+(u)\,v_u^{-1}(\lambda)$$

we obtain

COROLLARY 1.1. For $u \in U_\delta$,

$$u \int_0^\infty e^{-ut} \mathbf{P}(\bar{\xi}(t) \geqslant b)\, dt = \Psi_+(0, u)\, \Psi_+^{-1}(\lambda_+(u), u)\, e^{-\lambda_+(u)b} + O\left(e^{-\delta_1 b}\right),$$

where

$$\bar{\xi}(t) = \max_{0 < s < t} \xi(s).$$

This assertion is also known from [7].

Let

$$[(H_1(u) - 1)u^{-1/2}]_{u=0} = \gamma_1, \quad [(H_2(u) - 1)u^{-1/2}]_{u=0} = \eta_1,$$
$$[(H_3(u) - 1)u^{-1/2}]_{u=0} = \zeta_1, \quad [\lambda_+(u)u^{-1/2}]_{u=0} = \alpha_1.$$

COROLLARY 1.2. For $B \subset [b, \infty)$,

$$\mathbf{P}(Y \in B) = \frac{a - \gamma_1/(2\alpha_1)}{b + a - \eta_1/(2\alpha_1)} F_1(0, B) + O\left(e^{-\delta_1 \min(x, x-b+a)}\right), \quad x = \inf_{z \in B} z,$$

and for $B \subset (-\infty, -a]$,

$$\mathbf{P}(Y \in B) = \frac{b - \zeta_1/(2\alpha_1)}{b + a - \eta_1/(2\alpha_1)} F_2(0, B) + O\left(e^{\delta_1 \max(x, x+a-b)}\right), \quad x = \sup_{z \in B} z.$$

Using relations (3), (16), (17), one may easily prove

T h e o r e m 2. There exist $\delta_1 > 0$, $\delta > 0$, $\varepsilon > 0$ such that for $u \in U_\delta = U_\delta(\varepsilon)$ (see Theorem 1) and sufficiently large a and b, for any $A \subset (-a, b)$ uniformly in u, one has

$$uH(u, A) - u \int_0^\infty e^{-ut} \mathbf{P}(\xi(t) \in A)\, dt = \mu^a(u)\, \Pi_2(u)\, F_4(u, A)$$
$$- \mu^b(u)\, \Pi_1(u)\, F_3(u, A) + \lambda_+(u)\, O\left(e^{-\delta_1 \min\{\inf(b-A),\, \inf(a+A)\}}\right),$$

where

$$\Pi_1(u) = (1 - \mu^a(u) H_1(u))/(1 - \mu^{b+a}(u) H_2(u));$$
$$\Pi_2(u) = (1 - \mu^b(u) H_3(u))(1 - \mu^{b+a}(u) H_2(u))^{-1},$$
$$z + A = \{z + y\colon\ y \in A\};$$

$$F_3(u, A)$$
$$= \lambda_+(u)\, \lambda_-(u) \int_A e^{-\lambda_-(u)x} dx\, v_u(\lambda_+(u))\, w_u(\lambda_-(u))\, (\lambda_+(u) - \lambda_-(u))^{-1};$$

$$F_4(u, A) = F_3(u, A) \int_A e^{\lambda_+(u)x} dx \Big/ \int_A e^{-\lambda_-(u)x} dx.$$

It is not hard to observe that the principal parts of the asymptotic formulas for $\Pi(U,B)$ and $H(u,A)$ may analytically be continued to $U_{\varepsilon_1,\delta}$ for some $\varepsilon_1 > 0$ (see (4)).

2. To get expansions of the probabilities (1), we need estimates for $\Pi(u,B)$ and $H(u,A)$ in $\tilde{U}_\delta = \{u: |\mathrm{Im}\, u| \geq \delta,\ \mathrm{Re}\, u = 0\}$. In this case we use the uniform boundedness of the functions $\psi_{1\pm}(\lambda,u)$, $\psi_{1\pm}^{-1}(\lambda,u)$ for $u \in \tilde{U}_\delta$, $|\mathrm{Re}\,\lambda| \leq \delta$ (see [7]) as well as some results from [7]. Let

$$\Pi_b(u, x) = \Pi(u, [b, x)),\ x > b,\ \Pi_a(u, y) = \Pi(u, (y, -a]),\ y < -a.$$

L e m m a 7. Let conditions 1-3 hold. If $\sigma^2 = 0$, $\gamma > 0$ we assume in addition that

$$|\psi_2(\lambda)| \leqslant C_5 |\lambda|^{-r},\ 0 < r < 1,\ \lambda_- \leqslant \mathrm{Re}\,\lambda \leqslant \lambda_+,\ |\mathrm{Im}\,\lambda| > C_6.$$

Then, uniformly for $u \in \tilde{U}_\delta$ one has

$$|\Pi_b(u, x)| = O\left(\frac{e^{-\delta_1 x/2}}{|u|^{\gamma_2}}\right),\quad \gamma_2 > 0,\quad x > b.$$

Proof. From Lemma 1 for $c > 0$ we obtain

$$\begin{aligned}\Pi_b(u, x) &= -\frac{1}{2\pi i}\int\limits_{\mathrm{Re}\,\lambda=-c} \frac{[r_{u+}(\lambda)(1 - V_{u-}(\lambda))]^{[b,\infty)}}{\lambda r_{u+}(\lambda)} e^{-\lambda x} d\lambda \\ &= -[r_{u+}(\lambda)(1 - V_{u-}(\lambda))]_{\lambda=0}^{[b,\infty)} \\ &\quad - \frac{1}{2\pi i}\int\limits_{\mathrm{Re}\,\lambda=\delta_1/2} \frac{[r_{u+}(\lambda)(1 - V_{u-}(\lambda))]^{[b,\infty)}}{\lambda r_{u+}(\lambda)} e^{-\lambda x} d\lambda.\end{aligned} \tag{18}$$

Let

$$r_{u+}(\lambda) = \int_0^\infty e^{\lambda x} df(x, u).$$

Then, if the conditions of the Lemma are satisfied uniformly in $\tilde{U}_\delta$,

$$\left|\int_y^\infty df(x, u)\right| = O\left(\frac{e^{-\delta_1 y}}{|u|^{\gamma_2}}\right),\quad \gamma_2 > 0,\quad y > 0. \tag{19}$$

However, if for $\sigma^2 = 0$ and $\gamma > 0$ the condition of the Lemma is not satisfied, then, nevertheless,

$$\left|\int_y^\infty df(x, u)\right| = O\left(e^{-\delta_1 y}\right),\quad y > 0. \tag{20}$$

These estimates follow from the results of [7] by virtue of the continuity of f(x,u) in x, x > 0, for the processes considered (see [13]). Using (19), it is not hard to prove that the first summand in (18) can be estimated by a quantity $O(e^{-\delta_1 b}/|u|^{\gamma_2})$. Let us estimate the integral in (18):

$$\int_{\mathrm{Re}\,\lambda=\delta_1/2} \frac{[r_{u+}(\lambda)(1-V_{u-}(\lambda))]^{[b,\infty)}}{\lambda r_{u+}(\lambda)} e^{-\lambda x} d\lambda = \int_{\mathrm{Re}\,\lambda=\delta_1/2} \frac{1-V_{u-}(\lambda)}{\lambda} e^{-\lambda x} d\lambda$$

$$- \int_{\mathrm{Re}\,\lambda=\delta_1/2} \frac{[r_{u+}(\lambda)(1-V_{u-}(\lambda))]^{(-\infty,b)}}{\lambda r_{u+}(\lambda)} e^{-\lambda x} d\lambda = J_1 - J_2.$$

By Jordan's lemma, $J_1 = 0$. To estimate J_2, we shall examine three cases.

I. Let $\sigma^2 > 0$. Then (see (5))

$$\frac{1}{r_{u+}(\lambda)} = \frac{\psi_{1+}(\lambda, u)(\mu_+(u)-\lambda)}{\psi_{1+}(0, u)\mu_+(u)} = \frac{(u-\psi(\lambda))(\mu_+(u)-\lambda)}{(u-\psi_3(\lambda))\psi_{1+}(0, u)\mu_+(u)\psi_{1-}(\lambda, u)}$$

$$= \frac{\mu_+(u)-\lambda}{\psi_{1+}(0, u)\mu_+(u)\psi_{1-}(\lambda, u)} - \frac{2(\psi_4(\lambda)+u_1)}{\sigma^2(\mu_-(u)-\lambda)\psi_{1+}(0, u)\mu_+(u)\psi_{1-}(\lambda, u)},$$

$$J_2 = \int_{\mathrm{Re}\,\lambda=\delta_1/2} \frac{[r_{u+}(\lambda)(1-V_{u-}(\lambda))]^{(-\infty,b)}(\mu_+(u)-\lambda)}{\lambda\mu_+(u)\psi_{1+}(0, u)\psi_{1-}(\lambda, u)} e^{-\lambda x} d\lambda$$

$$- \frac{2}{\sigma^2}\int_{\mathrm{Re}\,\lambda=\delta_1/2} \frac{[r_{u+}(\lambda)(1-V_{u-}(\lambda))]^{(-\infty,b)}(\psi_4(\lambda)+u_1)}{\lambda(\mu_-(u)-\lambda)\psi_{1+}(0, u)\mu_+(u)\psi_{1-}(\lambda, u)} e^{-\lambda x} d\lambda = J_3 - J_4.$$

Since $[e^{-\lambda b} r_{u+}(\lambda)(1 - V_{u-}(\lambda))]^{(-\infty,0)} \to 0$ as $|\lambda| \to \infty$, $\mathrm{Re}\,\lambda > 0$ and $x - b > 0$, then

$$J_3 = \int_{\mathrm{Re}\,\lambda=\delta_1/2} \frac{\left[e^{-\lambda b} r_{u+}(\lambda)(1-V_{u-}(\lambda))\right]^{(-\infty,0)}(\mu_+(u)-\lambda)}{\lambda\mu_+(u)\psi_{1+}(0, u)\psi_{1-}(\lambda, u)} e^{-\lambda(x-b)} d\lambda = 0,$$

$$J_4 = \frac{2}{\sigma^2}\int_{\mathrm{Re}\,\lambda=\delta_1/2} \frac{(\psi_4(\lambda)+u_1)(1-V_{u-}(\lambda))}{\lambda\psi_{1+}(\lambda, u)\psi_{1-}(\lambda, u)(\mu_+(u)-\lambda)(\mu_-(u)-\lambda)} e^{-\lambda x} d\lambda$$

$$- \frac{2}{\sigma^2}\int_{\mathrm{Re}\,\lambda=\delta_1/2} \frac{[r_{u+}(\lambda)(1-V_{u-}(\lambda))]^{[b,\infty)}(\psi_4(\lambda)+u_1)}{\lambda(\mu_-(u)-\lambda)\psi_{1+}(0, u)\mu_+(u)\psi_{1-}(\lambda, u)} e^{-\lambda x} d\lambda = \frac{2}{\sigma^2}(J_5 - J_6),$$

$$|J_5| \leqslant C_7 e^{-\delta_1 x/2} \int\limits_{\operatorname{Re}\lambda=\delta_1/2} \frac{|\lambda|^{1-p}}{|\mu_+(u)-\lambda|\,|\mu_-(u)-\lambda|}|d\lambda|$$

$$= C_7 e^{-\delta_1 x/2} \int\limits_{-\infty}^{\infty} \frac{|i\mu+\delta_1/2|^{1-p}}{|i\mu+\delta_1/2-\mu_+(u)|\,|i\mu+\delta_1/2-\mu_-(u)|}d\mu$$

$$= C_7 e^{-\delta_1 x/2}|\mu_+(u)|^{1-p} \int\limits_{-\infty}^{\infty} \frac{\left|\mu-\dfrac{i\delta_1}{2|\mu_+(u)|}\right|^{1-p} d\mu}{\left|\mu - i\dfrac{\delta_1/2-\mu_+(u)}{|\mu_+(u)|}\right|\left|\mu - i\dfrac{\delta_1/2-\mu_-(u)}{|\mu_+(u)|}\right|}$$

$$\leqslant C_8 e^{-\delta_1 x/2}|u|^{-p/2}.$$

The last estimate follows from the fact that $|\arg \mu_\pm(u)| \to \pi/4$ as $|u| \to \infty$, $u \in \tilde{U}_\delta$ and $|\mu_\pm(u)| \geqq C_9|u|^{\frac{1}{2}}$ (see [7]). Now, we estimate the integral J_6. To this end, we use the following relation (an analog of which was considered in the proof of Lemma 5):

for $f(\lambda) \in \mathfrak{B}(0,\delta_1)$, $0 \leqq \varepsilon \leqq \delta_1$

$$\left[\frac{f(\lambda)}{\lambda-\varepsilon}\right]^{[b,\infty)} = \frac{[f(\lambda)]^{[b,\infty)}}{\lambda-\varepsilon} - \frac{e^{(\lambda-\varepsilon)b}}{\lambda-\varepsilon}[f(\lambda)]^{(-\infty,b)}_{\lambda=\varepsilon}.$$

This yields

$$|J_6| = \frac{2}{\sigma^2}\left| \int\limits_{\operatorname{Re}\lambda=\delta_1/2} \frac{\left[\dfrac{\psi_{1+}(0,u)\mu_+(u)(1-V_{u-}(\lambda))(\lambda-\delta_1)}{\psi_{1+}(\lambda,u)(\mu_+(u)-\lambda)}\right]^{[b,\infty)}(\psi_4(\lambda)+u_1)}{\lambda(\lambda-\delta_1)(\mu_-(u)-\lambda)\psi_{1+}(0,u)\mu_+(u)\psi_{1-}(\lambda,u)} e^{-\lambda x} d\lambda \right.$$

$$\left. + \int\limits_{\operatorname{Re}\lambda=\delta_1/2} \frac{\left[\dfrac{\psi_{1+}(0,u)\mu_+(u)(1-V_{u-}(\lambda))(\lambda-\delta_1)}{\psi_{1+}(\lambda,u)(\mu_+(u)-\lambda)}\right]^{(-\infty,b)}_{\lambda=\delta_1}(\psi_4(\lambda)+u_1)e^{-\lambda x}}{\lambda(\lambda-\delta_1)(\mu_-(u)-\lambda)\psi_{1+}(0,u)\mu_+(u)\psi_{1-}(\lambda,u)e^{-(\lambda-\delta_1)b}} d\lambda \right|$$

$$\leqslant C_{10} e^{-\delta_1 x/2} \int\limits_{\operatorname{Re}\lambda=\delta_1/2} \frac{|\lambda|^{1-p}}{|\lambda-\delta_1|\,|\mu_-(u)-\lambda|}|d\lambda| \leqslant C_{11} e^{-\delta_1 x/2}|u|^{-p/2}.$$

The proof of the Lemma for $\sigma^2 > 0$ is completed.

II. Let $\sigma^2 = 0$, $\gamma < 0$. Then

$$u - \psi_3(\lambda) = \gamma(\lambda - \mu_-(u)),$$

$$\frac{1}{r_{u+}(\lambda)}=\frac{\psi_{1+}(\lambda,u)}{\psi_{1+}(0,u)}=\frac{u-\psi(\lambda)}{u-\psi_3(\lambda)}\,\frac{1}{\psi_{1+}(0,u)\,\psi_{1-}(\lambda,u)}$$
$$=\frac{1}{\psi_{1+}(0,u)\,\psi_{1-}(\lambda,u)}+\frac{\psi_2(\lambda)}{\gamma(\lambda-\mu_-(u))\,\psi_{1+}(0,u)\,\psi_{\ -}(\lambda,u)},$$

$$J_2=\int\limits_{\mathrm{Re}\,\lambda=\delta_1/2}\frac{[r_{u+}(\lambda)(1-V_{u-}(\lambda))]^{(-\infty,b)}}{\lambda\psi_{1+}(0,u)\,\psi_{1-}(\lambda,u)}e^{-\lambda x}d\lambda$$
$$+\int\limits_{\mathrm{Re}\,\lambda=\delta_1/2}\frac{[r_{u+}(\lambda)(1-V_{u-}(\lambda))]^{(-\infty,b)}\psi_2(\lambda)}{\gamma\lambda(\lambda-\mu_-(u))\,\psi_{1+}(0,u)\,\psi_{1-}(\lambda,u)}e^{-\lambda x}d\lambda=J_7+J_8.$$

As in the previous case, $J_7 = 0$ for $x > b$. By condition 3, $|\psi_2(\lambda)| \leqq C_{12}$ and therefore

$$|J_8|\leqslant C_{13}e^{-\delta_1 x/2}\int\limits_{\mathrm{Re}\,\lambda=\delta_1/2}\frac{|d\lambda|}{|\lambda|\,|\lambda-\mu_+(u)|}\leqslant C_{14}e^{-\delta_1 x/2}\frac{|\ln|u||}{|u|}$$

(see [7]).

III. Let $\sigma^2 = 0$, $\gamma > 0$. In this case

$$u-\psi_3(\lambda)=\gamma(\lambda-\mu_+(u)),$$

$$\frac{1}{r_{u+}(\lambda)}=\frac{\psi_{1+}(\lambda,u)\,(\mu_+(u)-\lambda)}{\psi_{1+}(0,u)\,\mu_+(u)}=\frac{(u-\psi(\lambda))\,(\mu_+(u)-\lambda)}{(u-\psi_3(\lambda))\,\psi_{1+}(0,u)\,\mu_+(u)\,\psi_{1-}(\lambda,u)}$$
$$=\frac{\mu_+(u)-\lambda}{\psi_{1+}(0,u)\,\mu_+(u)\,\psi_{1-}(\lambda,u)}+\frac{\psi_2(\lambda)}{\gamma\psi_{1+}(0,u)\,\mu_+(u)\,\psi_{1-}(\lambda,u)}.$$

As in the preceding case, $J_2 = J_9 + J_{10}$, where

$$J_9=\int\limits_{\mathrm{Re}\,\lambda=\delta_1/2}\frac{[r_{u+}(\lambda)(1-V_{u-}(\lambda))]^{(-\infty,b)}(\mu_+(u)-\lambda)}{\lambda\psi_{1+}(0,u)\,\mu_+(u)\,\psi_{1-}(\lambda,u)}e^{-\lambda x}d\lambda=0.$$

By the conditions of the Lemma

$$|J_{10}|=\left|\int\limits_{\mathrm{Re}\,\lambda=\delta_1/2}\frac{[r_{u+}(\lambda)(1-V_{u-}(\lambda))]^{(-\infty,b)}\psi_2(\lambda)}{\gamma\lambda\psi_{1+}(0,u)\,\mu_+(u)\,\psi_{1-}(\lambda,u)}e^{-\lambda x}d\lambda\right|$$
$$\leqslant C_{15}e^{-\delta_1 x/2}|\mu_+(u)|^{-1}\int\limits_{\mathrm{Re}\,\lambda=\delta_1/2}\frac{|d\lambda|}{|\lambda|^{1+r}}\leqslant C_{16}e^{-\delta_1 x/2}|u|^{-1/2}.$$

The Lemma is completely proved.

In a similar way we prove

Lemma 8. Let conditions 1-3 be satisfied. If for $\sigma^2 = 0$, $\gamma < 0$ one has

$$|\psi_2(\lambda)| \leqslant C_{17}|\lambda|^{-r}, \quad 0<r<1, \quad \lambda_- \leqslant \operatorname{Re}\lambda \leqslant \lambda_+, \quad |\operatorname{Im}\lambda| > C_{18},$$

then uniformly in $u \in \tilde{U}_\delta$,

$$|\Pi_a(u, x)| = O\left(\frac{e^{-\delta_1 x/2}}{|u|^{\gamma_3}}\right), \quad \gamma_3 > 0, \quad x < -a.$$

Lemma 9. Let conditions 1-3 be satisfied. Then for any set $A = (-a_1, b_1) \subset (-a, b)$,

$$\left| uH(u, A) - u\int_0^\infty e^{-ut}\mathbf{P}(\xi(t) \in A)dt \right| = O\left(\frac{e^{-\delta_1 \min(a,b)}}{|u|^{\gamma_4}}\right)$$

uniformly in $u \in \tilde{U}_\delta$, $\gamma_4 > 0$.

Proof. Follows from relation (3) and Lemmas 1, 2, as well as a formula for the convolution of functions in V, if we note that in the representation

$$r_{u-}(\lambda) = \int_{-\infty}^0 e^{\lambda x} dg(x, u)$$

we have the estimate

$$\left| \int_{-\infty}^x dg(y, u) \right| = O\left(\frac{e^{\delta_1 x}}{|u|^{\gamma_5}}\right), \quad \gamma_5 > 0$$

uniformly in $u \in \tilde{U}_\delta$ if the conditions of Lemma 8 are satisfied, and the estimate $O(e^{\delta_1 x})$ if for $\sigma^2 = 0$, $\gamma < 0$ the conditions of the lemma are not satisfied. These estimates are analogous to estimates (19), (20) and are obtainable by the method used in [7].

2. Asymptotic expansions of the probabilities

By the inversion formula for the Laplace integral, for $c > 0$

$$\mathbf{P}(Y \in B, T < t) = \frac{1}{2\pi i}\int_{c-i\infty}^{c+i\infty} \frac{\Pi(u, B)\, e^{tu}}{u} du, \quad B \subset (-\infty, -a] \cup [b, \infty);$$

$$\mathbf{P}(\xi(t) \in A, T > t) = \mathbf{P}(\xi(t) \in A) - \frac{1}{2\pi i}\int_{c-i\infty}^{c+i\infty} \frac{\Phi(u, A)\, e^{tu}}{u} du,$$

$$A \subset (-a, b);$$

$$\Phi(u, A) = u \int_0^\infty e^{-ut} \mathbf{P}(\xi(t) \in A)\, dt - uH(u, A).$$

In these integrals, the integrands are analytic in the half-plane $\operatorname{Re} u > 0$, continuous and bounded, including the boundary $\operatorname{Re} u = 0$, with the exception of some small neighborhood of the point $u = 0$. Hence we take the imaginary axis as the contour of integration. We go around the point $u = 0$ lying on this axis, from the right along a circle of radius ε_3, $\varepsilon_3 > 0$ being an arbitrary sufficiently small number. Let K_1 denote this contour. If the conditions of Lemmas 7-9 are satisfied, it is easy to establish that

$$\left| \int_{\tilde{U}_\delta} \frac{\Pi(u, B)\, e^{tu}}{u}\, du \right| = O\left(e^{-\delta_1 \min(a,b)}\right), \quad B \subset (-\infty, -a] \cup [b, \infty);$$

$$\left| \int_{\tilde{U}_\delta} \frac{\Phi(u, A)\, e^{tu}}{u}\, du \right| = O\left(e^{-\delta_1 \min(a,b)}\right), \quad A \subset (-a, b).$$

Next, if $\varepsilon(u)$ is a function analytic in U_δ and continuous on the boundary of U_δ, then, by the arbitrariness of $\varepsilon_3 > 0$

$$\left| \int_{K_1 \setminus \tilde{U}_\delta} \frac{\varepsilon(u)\, e^{tu}}{u}\, du \right| \leqslant C \sup_{u \in U_\delta} |\varepsilon(u)| \ln t.$$

This estimate is obtained if we set $\varepsilon_3 = 1/t$.

Let $K_2 = K_1 \setminus \tilde{U}_\delta$. Consider the integral

$$\overline{J}(t) = \frac{1}{2\pi i} \int_{K_2} \frac{F(u)\, e^{(k_1 a + k_2 b)\lambda_-(u)} \Pi_1(u)\, e^{tu}}{u e^{(k_3 a + k_4 b)\lambda_+(u)}}\, du, \tag{21}$$

where $F(u) = \sum_{k=0}^{\infty} f_k u^{k/2}$ in a neighborhood of zero, $k_1, k_2, k_3, k_4 \geqq 0$. Note that in $U_{\varepsilon_1, \delta}$ we have the expansions (see [7])

$$\lambda_\pm(u) = \pm \alpha_1 \sqrt{u} + \alpha_2 u \pm \alpha_3 u^{3/2} + \ldots, \quad \alpha_1 = \sqrt{2/\psi''(0)}\,, \tag{22}$$

a root of a positive number being positive. Let K denote the contour obtained from the contour

$$\left\{ |\arg u| = \frac{3\pi}{4}, \quad u \in K^1_{\varepsilon_1, \delta} \right\}$$

by bending $K^1_{\varepsilon_1,\delta}$ inwards near the point $u = 0$,

$$K_3 = \{u : u \in K^1_{\varepsilon_1,\delta}, |\operatorname{Im} u| = \delta\}.$$

Since the integrand is analytic in $K^1_{\varepsilon_1,\delta}$, $K \cup K_3$ can be substituted for K_2 in (21). It follows from the expansions (22) that for sufficiently small $\varepsilon_1 > 0$ and $\delta > 0$ there exists a $\gamma_5 = \gamma_5(\varepsilon) > 0$ such that for $u \in K_3$

$$\operatorname{Re} \lambda_-(u) < -\gamma_5, \quad \operatorname{Re} \lambda_+(u) > \gamma_5.$$

This, in turn, implies that the integrand in (21) is uniformly bounded in K_3 by

$$O(\exp(-\gamma_5((k_1 + k_3)a + (k_2 + k_4)b)))$$

due to the uniform boundedness of $\Pi_1(u)$ (see Section 1). Thus,

$$\bar{J}(t) = J(t) + O(\exp(-\gamma_5((k_1 + k_3)a + (k_2 + k_4)b))),$$

where $J(t)$ is obtained from $\bar{J}(t)$ by substituting k for k_2.

For $M \geqq 1$ we have the relation (see [4])

$$\Pi_1(u) = \sum_{s=0}^{M-1} (1 - \mu^a(u) H_1(u)) (\mu^{b+a}(u) H_2(u))^s + \Pi_1(u) (\mu^{b+a}(u) \cdot H_2(u))^M.$$

Let

$$\tau_1 = a/t, \quad \tau_2 = b/t, \quad \tau_3 = 1/t, \quad \tilde{a}_1(u) = F(u)/u, \quad \tilde{a}_2(u) = \tilde{a}_1(u) H_1(u),$$

$$\begin{aligned} \hbar_{1s}(u) &= (\tau_1 s + \tau_2 s)(\lambda_-(u) - \lambda_+(u)) + (k_1\tau_1 + k_2\tau_2)\lambda_-(u) \\ &\quad - (k_3\tau_1 + k_4\tau_2) \times \lambda_+(u) + s\tau_3 \ln H_2(u) + u, \qquad 0 \leqslant s \leqslant M, \end{aligned}$$

$$\hbar_{2s}(u) = \hbar_{1s}(u) + \tau_1(\lambda_-(u) - \lambda_+(u)), \qquad 0 \leqslant s \leqslant M - 1.$$

Then

$$J(t) = \frac{1}{2\pi i}\left[\sum_{s=0}^{M-1} (J_{1s}(t) - J_{2s}(t)) + J_M(t)\right],$$

where

$$J_{js}(t) = \int_K \tilde{a}_j(u) e^{t\tilde{\hbar}_{js}(u)} du, \; s = 0, 1, 2, \ldots, \quad M-1, \; j = 1, 2$$

(in what follows, j takes on the values 1, 2), and

$$J_M(t) = \int_K \tilde{a}_1(u) \Pi_1(u) e^{t\tilde{\hbar}_{1M}(u)} du.$$

First we consider the case where $f_0 = 0$ in the expansion of the function $F(u)$ (see (21)). Note that this condition is not satisfied for $F_3(u,A)$ and $F_4(u,A)$ (Theorem 2). We make a substitution $z = \sqrt{u}$. Let Γ denote the image of K in the plane z, and let $\eta(z) = H_2(z)$, $\psi_-(z) = \lambda_-(u)$, $\psi_+(z) = \lambda_+(u)$, $h_{js} = \tilde{h}_{js}(u)$, $a_j(z) = 2z\tilde{a}_j(u)$, $s = 0,1,\ldots,M-1$, $h_{1M}(z) = \tilde{h}_{1M}(u)$. Then

$$J_{js}(t) = \int_{\Gamma} a_j(z)\, e^{th_{js}(z)}\, dz.$$

In this case $a_j(z)$, $h_{js}(z)$ are analytic in a neighborhood of zero, and the subsequent application of the modified saddle-point method from [2] is the same as in [4]. We omit the proofs.

Let

$$\begin{aligned} F_1(z, y_1, y_2, \ldots, y_5) &\equiv (y_1 + y_2)\left(\psi'_-(z) - \psi'_+(z)\right) \\ &+ y_3\eta'(z)/\eta(z) + y_4\left(k_1\psi'_-(z) - k_3\psi'_+(z)\right) \\ &+ y_5\left(k_2\psi'_-(z) - k_4\psi'_+(z)\right) + 2z = 0, \\ F_2(z, y_1, y_2, \ldots, y_5) &\equiv F_1(z, y_1, \ldots, y_5) + y_4\left(\psi'_-(z) - \psi'_+(z)\right) = 0. \end{aligned}$$

By the implicit function theorem, there exist solutions of these equations $z_1(y_1,\ldots,y_5)$, $z_2(y_1,\ldots,y_5)$, respectively, representable in a neighborhood $\Delta = \{|y_i| < \tau,\ i = 1,2,\ldots,5\}$ as convergent power series in $y_1,y_2,\ldots,y_5$. Note that

$$h'_{js}(z) = F_j(z, \tau_1 s, \tau_2 s, \tau_3 s, \tau_1, \tau_2).$$

Let $M - 1 = [\tau t/(a+b)]$. Then for $0 \leqq s \leqq M - 1$ the point $(\tau_1 s, \tau_2 s, \tau_3 s, \tau_1, \tau_2) \in \Delta$. For solutions of the equations $h'_{js}(z) = 0$, $s = 0,1,\ldots,M-1$, we have the expansion

$$\begin{aligned} z_{js} \equiv z_{js}(\tau_1 s, \tau_2 s, \tau_3 s, \tau_1, \tau_2) &= \alpha_1(s + j - 1 + (k_1 + k_3)/2)\tau_1 \\ &+ \alpha_1(s + (k_2 + k_4)/2)\tau_2 - s\eta_1\tau_3/2 + O((\tau_1 + \tau_2)^2 s^2). \end{aligned}$$

The points z_{js} are saddle-points of the functions $h_{js}(z)$, $0 \leqq s \leqq M-1$. In a neighborhood of zero,

$$\begin{aligned} h_{js}(z) = &-\alpha_1 z[(2(s + j - 1) + k_1 + k_3)\tau_1 + (2s + k_2 + k_4)\tau_2] \\ &+ s\eta_1 z + z^2[1 + O(s(\tau_1 + \tau_2))] + O(z^3), \end{aligned}$$

therefore,

$$\begin{aligned} h_{js}(z_{js}) = &-[\alpha_1(s + j - 1 + (k_1 + k_3)/2)\tau_1 + \alpha_2(s + (k_2 + k_4)/2)\tau_2 \\ &- s\eta_1\tau_3/2]^2 + O((\tau_1 + \tau_2)^3 s^3), \quad h''_{js}(z_{js}) = 2 + O(s(\tau_1 + \tau_2)). \end{aligned}$$

For this choice of M, $J_M(t) = O(e^{-\gamma_3 t})$, $\gamma_3 > 0$. A similar assertion is proved in [4], and hence we skip it.

Let

$$Q_{jsi} \equiv Q_{jsi}(k_1, k_2, k_3, k_4, F) = \frac{\sqrt{-1}}{2\pi} \sum_{r=0}^{i} q_{2i,r}^{j,s} \left(-h_{js}^{(0)}\right)^{-r-i-1/2} \Gamma(r+i+1/2),$$

where $q_{i,r}^{j,s}$ is the coefficient of z^i in the product

$$\frac{1}{r!}\left(a_{js}^{(0)} + a_{js}^{(1)} z + \dots\right)\left(h_{js}^{(1)} z + h_{js}^{(2)} z^2 + \dots\right)^r,$$

$$a_{is}^{(k)} = a_{js}^{(k)}(z_{js})/k!, \quad h_{js}^{(k)} = h_{js}^{(k+2)}(z_{js})/(k+2)!, \quad k = 0, 1, 2, \dots$$

Lemma 10. Let $a = o(t)$, $b = o(t)$, $a \to \infty$, $b \to \infty$, $(a+b)/\sqrt{t} \to \infty$ as $t \to \infty$. Then for an arbitrary integer $q \geqq 1$,

$$J(t) = \left[\sum_{i=1}^{q-1} t^{-i-1/2}\left(Q_{10i} e^{th_{10}(z_{10})} - Q_{20i} e^{th_{20}(z_{2s})}\right)\right.$$
$$\left. + t^{-q-1/2} O\left(e^{-\gamma[a(k_1+k_3)+b(k_2+k_4)]^2/t}\right)\right]\left(1 + O\left(e^{-\gamma(a+b)^2/t}\right)\right), \quad \gamma > 0.$$

Lemma 11. Let $a = x_1\sqrt{t}$, $b = x_2\sqrt{t}$, $0 < x_j < \infty$. Then for any integer $q \geqq 1$,

$$J(t) = \sum_{i=0}^{q-1} t^{-i-1/2} \sum_{r=0} t^{-r/2} p_{ri} + O\left(t^{-q-1/2}\right),$$

$$p_{ri} \equiv p_{ri}(k_1, k_2, k_3, k_4, F)$$
$$= \sum_{s=0}^{\infty} \left(q_r^{1,i}(s) \exp\left(-\alpha_1^2 [x_1(s + (k_1+k_3)/2) + x_2(s + (k_2+k_4)/2)]^2\right)\right.$$
$$\left. - q_r^{2i}(s) \exp\left(-\alpha_1^2 [x_1(s + 1 + (k_1+k_3)/2) + x_2(s + (k_2+k_4)/2)]^2\right)\right),$$

and the $q_r^{j,i}(s)$ are determined from the expansion

$$Q_{jsi} \exp\{th_{js}(z_{js}) + \alpha_1^2 [x_1(s + j - 1 + (k_1+k_3)/2)$$
$$+ x_2(s + (k_2+k_4)/2)]^2\} = \sum_{r=0}^{\infty} t^{-r/2} q_r^{j,i}(s).$$

Lemma 12. Let $a = x\sqrt{t}$, $b = o(a) \to \infty$, $t \to \infty$, $0 < x < \infty$. Then for any integer $q \geqq 1$,

$$J(t) = \sum_{i=0}^{q-1} t^{-i-1/2} \sum_{i_1, i_2=0}^{\infty} \left(\tau_1 \sqrt{t}\right)^{i_1} \left(1/\sqrt{t}\right)^{i_2} p_{i_1, i_2, i} + O\left(t^{-q-1/2}\right),$$

$$p_{i_1,i_2,i} \equiv p_{i_1,i_2,i}(k_1, k_2, k_3, k_4, F)$$
$$= \sum_{s=0}^{\infty} \Big(q^{1,i}_{i_1,i_2}(s) \exp\{-\alpha_1^2 x^2 (s + (k_1+k_3)/2)^2\}$$
$$- q^{2,i}_{i_1,i_2}(s) \exp\{-\alpha_1^2 x^2 (s + 1 + (k_2+k_4)/2)^2\}\Big),$$

and the $q^{j,i}_{i_1,i_2}$ are determined from the expansions

$$Q_{jsi} \exp\{th_{js}(z_{js}) + \alpha_1^2 x^2 (s + j - 1 + (k_1+k_3)/2)^2\}$$
$$= \sum_{i_1,i_2=0}^{\infty} (\tau_1 \sqrt{t})^{i_1} (1/\sqrt{t})^{i_2} q^{j,i}_{i_1,i_2}(s).$$

The case $b = x\sqrt{t}$, $a = o(b)$ can be examined similarly.

Theorem 2, Lemmas 10-12 and the estimates derived above yield complete asymptotic expansions of the probabilities $P(\xi(t) \in A, T > t)$, $A \subset (-a,b)$ in the range of deviations of the boundaries. As an example we give the case of normal deviations of the boundaries.

T h e o r e m 3. Let the set $A = (-a_1, b_1) \subset (-a,b)$ not depend on t, $a = x_1\sqrt{t}$, $b = x_2\sqrt{t}$, $0 < x_j < \infty$. Then for any integer $q \geq 1$,

$$\mathbf{P}(\xi(t) \in A, T > t) = \mathbf{P}(\xi(t) \in A)$$
$$- \sum_{i=0}^{q-1} t^{-i-1/2} \sum_{r=0}^{\infty} t^{-r/2} (p_{ri}(0, 1, 0, 1, F_3(\cdot, A))$$
$$- q_{ri}(0, 1, 0, 1, F_4(\cdot, A))) + O(t^{-q-1/2}),$$

where the difference between q_{ri} and p_{ri} is that in their definition, starting with (21), $H_3(u)$ replaces $H_1(u)$ and a and b switch places.

Turning again to the integral (21), we assume that $f_0 \neq 0$ in the expansion of $F(u)$. This condition is satisfied for $F_1(u,B)$ and $F_2(u,B)$. Upon the substitution $\sqrt{u} = z$, the functions $b_j(z) = za_j(z) = 2z^2\tilde{a}_j(u)$ and $h_{js}(z)$ are analytic in a neighborhood of zero and

$$J_{js}(t) = \int_{\Gamma} \frac{b_j(z)}{z} e^{th_{js}(z)} dz,$$
$$J(t) = \frac{1}{2\pi i} \left(\sum_{s=0}^{M-1} (J_{1s}(t) - J_{2s}(t)) + J_M(t) \right).$$

Further calculation of the integrals $J_{js}(t)$ follows that in [2], [6], [7]. Let

$$g_{js}(z, z_{js}) = h_{js}(z + z_{js}) - h_{js}(z_{js}) = z^2 \sum_{k=0}^{\infty} h_{js}^{(k)} z^k,$$
$$h_{js}^{(k)} = h_{js}^{(k)}(z_{js})/(k+2)!.$$

We change the contour of integration Γ in $J_{js}(t)$ in such a way that the endpoints of Γ do not change and the new contour Γ_1 contains a segment $(z_{js} - i\rho, z_{js} + i\rho)$ for some $\rho > 0$. Then

$$J_{js}(t) = \int_{z_{js}-i\rho}^{z_{js}+i\rho} \left(b_j(z)\, e^{th_{js}(z)}/z\right) dz + e^{th_{js}(z_{js})} O\left(e^{-\gamma_4 t}\right)$$
$$= \int_{z_{js}-i\rho}^{z_{js}+i\rho} \frac{b_j(z) - b_j(0)}{z} e^{th_{js}(z)}\, dz + \int_{-i\rho}^{i\rho} \frac{b_j(0)}{z + z_{js}} e^{th_{js}(z+z_{js})} dz$$
$$+ e^{th_{js}(z_{js})} O\left(e^{-\gamma_4 t}\right) = J'_{js}(t) + J''_{js}(t) + e^{th_{js}(z)} O\left(e^{-\gamma_4 t}\right).$$

We limit ourselves to the case $a = x_1\sqrt{t}$, $b = x_2\sqrt{t}$. Other cases involve no serious difficulties (see [6]).

Let $p_{ri}^{(2)}(k_1,k_2,k_3,k_4,F)$ be defined just as are the $p_{ri}(k_1,k_2,k_3,k_4,F)$, but with the function $\tilde{Q}_{jsi}$ obtained from Q_{jsi} by replacing $a_j(z)$ by $(b_j(z) - b_j(0))/z$. The expansion $(1/2\pi i) \sum_{s=0}^{M-1} J'_{js}(t)$ is obtained from Lemma 11 due to the fact that $(b_j(z) - b_j(0))/z$ is analytic in a neighborhood of zero. For $J''_{js}(t)$ one has the following relation (see [2], [6]):

$$\sum_{s=0}^{M-1} \left(J''_{1s}(t) - J''_{2s}(t)\right) = b_1(0) \sum_{s=0}^{M-1} \left(\int_{-i\rho}^{i\rho} \frac{e^{th_{1s}(z_{1s}) + tg_{1s}(z, z_{1s})}}{z + z_{1s}}\, dz \right.$$
$$\left. - \int_{-i\rho}^{i\rho} \frac{e^{th_{2s}(z_{2s}) - tg_{2s}(z, z_{2s})}}{z + z_{2s}}\, dz \right)$$
$$= 2\pi i b_1(0) \left[\sum_{s=0}^{M-1} \left(\Phi\left(|z_{1s}| \sqrt{2th_{1s}^{(0)}}\right) - \Phi\left(|z_{2s}| \sqrt{2th_{2s}^{(0)}}\right)\right) \right.$$
$$+ \sum_{i=1}^{q} t^{-i/2} \sum_{s=0}^{M-1} \left(e^{th_{1s}(z_{1s})} \Pi_{3i-1}^{(1)}\left(z_{1s}\sqrt{t}\right) \right.$$

$$- e^{th_{2s}(z_{2s})}\Pi^{(2)}_{3i-1}(z_{2s}\sqrt{t})\Big)\Big] + t^{-\frac{q+1}{2}} \sum_{s=0}^{M-1} \Big(r_{1sq}e^{th_{1s}(z_{1s})+C_{1s}tz_{1s}^3}$$

$$- r_{2sq}e^{th_{2s}(z_{2s})+C_{20}tz_{2s}^3}\Big) + O\left(e^{-\gamma_5 t}\right),\ \gamma_5 > 0, \tag{23}$$

where $q \geqq 1$ is an arbitrary integer, the $\Pi_i^{(j)}(u)$ are polynomials in u of degree i with coefficients admitting expansions in powers of $\tau_1 s$, $\tau_2 s$, $\tau_3 s$, τ_1, τ_2, $0 \leqq s \leqq M-1$, $b_1(0) = b_2(0)$, and

$$\Phi(t) = \frac{1}{\sqrt{2\pi}} \int_t^{\infty} e^{-x^2/2}dx, \quad |r_{jsq}| = O\left(1, |z_{js}\sqrt{t}|^{3q+2}\right)$$

for $A = x_1\sqrt{t}$, $b = x_2\sqrt{t}$, $0 < x_j < \infty$,

$$\sum_{s=0}^{M-1} \left(\Phi\left(|z_{2s}|\sqrt{2th_{2s}^{(0)}}\right) - \Phi\left(|z_{1s}|\sqrt{2th_{1s}^{(0)}}\right)\right)$$

$$= \frac{1}{\sqrt{2\pi}} \sum_{s=0}^{M-1}\left(\exp\left\{-\frac{\left(|z_{1s}|\sqrt{2th_{1s}^{(0)}}\right)^2}{2}\right\} \sum_{k=0}^{\infty} \frac{\left(|z_{1s}|\sqrt{2th_{1s}^{(0)}}\right)^{2k+1}}{(2k+1)!!}\right.$$

$$\left. - \exp\left\{-\frac{\left(|z_{2s}|\sqrt{2th_{2s}^{(0)}}\right)^2}{2}\right\} \sum_{k=0}^{\infty} \frac{\left(|z_{2s}|\sqrt{2th_{2s}^{(0)}}\right)^{2k+1}}{(2k+1)!!}\right)$$

$$= \frac{1}{\sqrt{2\pi}} \sum_{s=0}^{M-1} (\exp\{-\alpha_1^2[x_1(s+(k_1+k_3)/2) + x_2(s+(k_2+k_4)/2)]^2\}$$

$$\times \sum_{r=0}^{\infty} t^{-r/2} g_r^{(1)}(s) - \exp\{-\alpha_1^2[x_1(s+1-(k_1+k_3)/2) + x_2(s$$

$$+ (k_2+k_4)/2)]^2\} \sum_{r=0}^{\infty} t^{-r/2} g_r^{(2)}(s) = \frac{1}{\sqrt{2\pi}} \sum_{r=0}^{\infty} t^{-r/2}$$

$$\times \sum_{s=0}^{\infty} \left(g_r^{(1)}(s) \exp\{-\alpha_1^2[x_1(s+(k_1+k_3)/2) + x_2(s+(k_2+k_4)/2)]^2\}\right.$$

$$\left. - g_r^{(2)}(s) \exp\{-\alpha_1^2[x_1(s+1+(k_1+k_3)/2) + x_2(s+(k_2+k_4)/2)]^2\}\right)$$

$$+ O\left(e^{-\gamma_6 t}\right) = -\sum_{r=0}^{\infty} t^{-r/2} p_r(k_1, k_2, k_3, k_4) + O\left(e^{-\gamma_6 t}\right),\ \gamma_6 > 0, \tag{24}$$

$$\sum_{s=0}^{M-1} \Pi^{(j)}_{3i-1}(z_{js}\sqrt{t}) \exp\{th_{js}(z_{js}) + \alpha_1^2[x_1(s+j-1+(k_1+k_3)/2)$$

$$+ x_2(s+(k_2+k_4)/2)]^2\} = \sum_{r=0}^{\infty} t^{-r/2} g_{1,r}^{j,i}(s),$$

where the $g^{j,i}_{1,r}(s)$ are polynomials of x_1, x_2 and s of degree in each variable at most $3r$.

Let

$$\sum_{s=0}^{\infty}\Big(g^{1,i}_{1,r}(s)\exp\{-\alpha_1^2[x_1(s+(k_1+k_3)/2)+x_2(s+(k_2+k_4)/2)]^2\}$$
$$-g^{2,i}_{1,r}(s)\exp\{-\alpha_1^2[x_1(s+1+(k_1+k_3)/2)+x_2(s+(k_2+k_4)/2)]^2\}\Big) \quad (25)$$
$$= p^{(1)}_{ri}(k_1,k_2,k_3,k_4,F).$$

In (23),

$$h_{js}(z_{js})-C_{18+j}z^3_{js}=-z^2_{js}(1+O(s(\tau(t))))+C_{18+j}z^3_{js}$$
$$=-z^2_{js}(1+O(s\tau(t))),\qquad \tau(t)=\max(\tau_1,\tau_2).$$

Choosing a sufficiently small $\tau > 0$ can lead to

$$1+O(s\tau(t))\geqslant\gamma_7,\quad \gamma_7>0 \qquad \text{(see [6])}.$$

Thus, we have proved the following

T h e o r e m 4. If $a = x_1\sqrt{t}$, $b = x_2\sqrt{t}$, $0 < x_j < \infty$ and the conditions of Lemma 7 hold, then for $x > b$ for any integer $q \geqq 1$,

$$\mathbf{P}(b\leqslant Y<x,\ T<t)=2F_1(0,[b,x))\Bigg\{\sum_{r=0}^{\infty}t^{-r/2}p_r(0,0,0,1)$$
$$-\sum_{i=1}^{q}t^{-i/2}\Big(\sum_{r=0}^{\infty}t^{-r/2}p^{(1)}_{ri}(0,0,0,1,F_1(\cdot,[b,x)))\Big)$$
$$+\sum_{i=0}^{[(q-1)/2]}t^{-i-1/2}\Big(\sum_{r=0}^{\infty}t^{-r/2}p^{(2)}_{ri}(0,0,0,1,F_1(\cdot,[b,x)))\Big)\Bigg\}+O\big(t^{(-q+1)/2}\big),$$

the coefficients $p^{(1)}_{ri}(k_1,k_2,k_3,k_4,F)$, $p_r(k_1,k_2,k_3,k_4)$ are defined by formulas (24) and (25), and

$$p_0(0,0,0,1)=\frac{\alpha_1}{2\sqrt{\pi}}\sum_{s=0}^{\infty}\int_{-x_1}^{x_1}\exp\Big\{-\frac{\alpha_1^2}{4}[x+(2s+1)(x_1+x_2)]^2\Big\}dx.$$

Starting from (21), substituting $\Pi_2(u)$ for $\Pi_1(u)$, we can write expansions for $P(x < Y \leqq -a,\ T < t)$, $x < -a$.

The author takes this opportunity to express his gratitude to A.A. Borovkov and B.A. Rogozin for their valuable comments, and to V.I. Lotov for his continued interest in this work.

REFERENCES

[1] Borovkov, A.A. "Limit Theorems on the Distributions of Maxima of Sums of Bounded Lattice Random Variables. I; II." *Theory Probab. Applications*, vol.5, no.2; no.4 (1960): 125-55; 341-55. (English transl.)

[2] ______. New Limit Theorems in Boundary Value Problems for Sums of Independent Summands" (in Russian). *Sibirsk. matem. zhurnal*, vol.3, no.5 (1962): 645-94.

[3] Borovkov, A.A., and Rogozin, B.A. "Boundary Value Problems for Some Two-dimensional Random Walks." *Theory Probab. Applications*, vol.9, no.3 (1964): 361-88. (English transl.)

[4] Lotov, V.I. "Asymptotic Analysis of Distributions in Problems with Two Boundaries. I; II." *Theory Probab. Applications*, vol.24, no.3; no.4 (1979): 480-90; 869-76. (English transl.)

[5] ______. "On the Asymptotics of Distributions Relating to an Exit of a Nondiscrete Random Walk from an Interval" (in Russian). In press.

[6] ______. *Asymptotic Expansions in Two-sided Boundary Problems for Random Walks*. Kandidatskaya dissertatsiya. Moscow: Izd-vo Instituta Matematiki Akademii Nauk SSSR, 1977.

[7] Rogozin, B.A. "Distribution of the Maximum of a Process with Independent Increments" (in Russian). *Sibirsk. matemat. zh.*, vol.10, no.6 (1969): 1334-63.

[8] ______. "On Distributions of Functionals Related to Boundary Problems for Processes with Independent Increments." *Theory Probab. Applications*, vol.11, no.4 (1966): 580-91. (English transl.)

[9] Borovskikh, Yu.V. "Complete Asymptotic Expansions for the Resolvent of a Semi-continuous Process with Independent Increments with Absorption and Distributions of the Ruin Probability" (in Russian). In *Analiticheskie metody v teorii veroyatnostej*, pp. 10-21. Kiev: Naukova Dumka, 1979.

[10] Bratijchuk, N.S. "Complete Asymptotic Expansions for Distributions of the Exit Times of a Process with Independent Increments from a Strip" (in Russian). Kiev: Izd-vo Instituta Matematiki Akademii Nauk Ukr.SSR, 1981. Preprint 81.88.

[11] Borovkov, A.A. "On the First Passage Time for One Class of Processes with Independent Increments." *Theory Probab. Applications*, vol.10, no.2 (1965): 331-34. (English transl.)

[12] Zolotarev, V.M. "The First Passage Time of a Level and the Behavior at Infinity for a Class of Processes with Independent Increments." *Theory Probab. Applications*, vol.9, no.4 (1964): 653-62. (English transl.)

[13] Pecherskij (Pecherskii), E.A. "Some Identities Related to the Exit of a Random Walk from a Segment and a Semi-interval." *Theory Probab. Applications*, vol.19, no.1 (1974): 106-21. (English transl.)

[14] Lukacs, Eugene. *Characteristic Functions*. 2nd ed. rev. and enl. London: Griffin, 1970.

[15] Emery, D.J. "Exit Problem for a Spectrally Positive Process." *Advances in Applied Probability*, vol.5, no.3 (1973): 498-521.

A.A. Mogul'skij

PROBABILITIES OF LARGE DEVIATIONS FOR TRAJECTORIES OF RANDOM WALKS

1. Introduction

Let $(\xi_n)_{n=1}^{\infty}$ be a sequence of independent identically distributed random variables, $M\xi_1 = 0$, and let $M\exp\{\lambda\xi_1\} < \infty$ for $|\lambda| \leqq \delta$, $\delta > 0$. Let $s_n = s_n(t)$, $0 \leqq t \leqq 1$, denote a continuous polygonal line connecting the points $(i/n, S_i)$, $i = 0,1,\dots,n$, where $S_0 = 0$, $S_i = \xi_1 + \cdots + \xi_i$. For a function $H(t)$, $0 \leqq t \leqq 1$, we define the following subsets of the space $C(0,1)$ of continuous real functions $f(t)$, $0 \leqq t \leqq 1$:

$$A_H = \{f \in C(0,1): \inf_{0<t<1} \{f(t) - H(t)\} \leqslant 0\},$$
$$B_H = \{f \in C(0,1): \inf_{0<t<1} \{f(t) - H(t)\} > 0\}.$$

Following A.A. Borovkov [1], we call the problem of finding the asymptotics of the probability

$$\mathbf{P}(s \cdot r^{-1}(n) \in C),\ C \subseteq C(0,1), \tag{1.1}$$

when it tends to zero, the first and second boundary value problem, depending on whether the set C is of the form A_H, $\inf_{0\leqq t\leqq 1} H(t) > 0$, or of the form B_H, $H(0) < 0$, $\sup_{0\leqq t\leqq 1} H(t) > 0$. These problems originated with A.N. Kolmogorov [2], have been treated extensively (see [1-8] and the bibliographies therein). In the region of large deviations (when $r(n)/\sqrt{n} \to \infty$ as $n \to \infty$), the first and second boundary value problems are essentially different (we will touch on this later on). The first boundary problem has been solved almost completely [9-10]; the second one, however, only in the following special cases: 1) the boundary $H(t)$ is linear [8] and 2) the boundary $H(t)$ is different from $-\infty$ only at a finite number of points (see, for example, [10-12]) (then the first and secondary boundary problems reduce to finding probabilities of large derivations $r(n)$ bounded from above by a sequence $o(1)n/\ln^4 n$ is developed.

First, however, let us address the first and second boundary problems from the viewpoint of the work [1], wherein a "coarse" asymptotic of the probability (1.1) is obtained; namely for sets $C \subseteq C(0,1)$ from a sufficiently broad class, including sets of the form A_H and B_H, for deviations $r(n) = o(1)n$

$$\lim_{n\to\infty} \frac{n}{r^2(n)} \ln \mathbf{P}(s_n/r(n) \in C) = v(H_0), \tag{1.2}$$

where

$$v(H_0) = -\frac{1}{2}\int_0^1 \dot{H}_0^2(t)\,dt;$$

the function $H_0 \in \bar{C}$ is defined by the relation

$$v(H_0) = \sup\{v(f):\ f \in C,\ f(0) = 0\},\ H_0(0) = 0.$$

It is natural to call the function H_0 the most probable trajectory in the set C; thus, by (1.2), the principal term of the asymptotic of the probability (1.1) is given by the most probable trajectory H_0. It turns out that in the cases $C = A_H$ and $C = B_H$, the function H_0 determines "entirely" (to within an exponential factor) the probability (1.1). It is not hard to see that in the first boundary problem, the function H_0 consists of two linear pieces (here H_0 can be nonuniquely determined due to the fact that, as a rule, the set A_H is not convex); in the special cases 1 and 2 of the second boundary value problem, the function H_0 is linear or consists of a finite number of linear pieces, respectively. Thus, in the first boundary problem and in those particular cases of the second boundary problem, the fucntion H_0 consists of a finite number of linear pieces; the probability (1.1) differs only by an exponential factor from the sequence

$$\exp\left\{n\int_0^1 \Lambda\left(\frac{r}{n}\dot{H}_0(t)\right)dt\right\} = \exp\left\{\sum_{j=2}^{\infty} \Lambda_j \left(\frac{r}{n}\right)^j \int_0^1 (\dot{H}_0(t))^j\,dt\right\}, \tag{1.3}$$

where $\Lambda(\alpha) = \sum_{j=2}^{\infty} \Lambda_j \alpha^j$ is the so-called deviation function (the numbers Λ_j expressible in terms of the moments of the random variable ξ_1, form the so-called Cramér series). Our main restriction on the function H for the set B_H is the following: the function $H_0 = H_{0,H}$ consists of a finite number of pieces of two kinds: either linear pieces or pieces with a continuous second derivative bounded from zero. Under these assumptions, the probability (1.1) for the set $C = B_H$ differs only by an exponential factor from the sequence

$$\exp\left\{n\int_0^1 \Lambda\left(\frac{r}{n}\dot{H}_0(t)\right)dt + \left(\frac{r}{\sqrt{n}}\right)^{2/3} \times \sum_{j=0}^{\infty}\sigma_j\left(\frac{r}{n}\right)^j \int_0^1 |\ddot{H}_0(t)|^{2/3}(\dot{H}_0(t))^j\,dt\right\}, \tag{1.4}$$

where the numbers σ_j, $j = 0,1,\ldots,$ are expressed in terms of moments of the random variable ξ_1.

By the invariance principle, in the cases where the sequence $r = r(n)$ tends to infinity not very rapidly, the probability (1.1) is equivalent to the probability

$$\mathbf{P}(w/(r/\sqrt{n}) \in C), \tag{1.5}$$

where $w = w(t)$, $0 \leqq t \leqq 1$, is a Wiener process; from (1.3) and (1.4) one can see that in the first and second boundary problems this equivalence remains for deviations $r = o(n^{2/3})$ as well (see also [7]). Thus, for $C = B_H$ the probability (1.5) differs only by a power factor from the sequence

$$\exp\left\{\frac{r^2}{n}\Lambda_2\int_0^1 \dot{H}_0^2(t)\,dt + \left(\frac{r}{\sqrt{n}}\right)^{2/3}\sigma_0\int_0^1 |\ddot{H}_0(t)|^{2/3}\,dt\right\}.$$

To this we add that under broad assumptions the power terms in the second boundary problem for a Wiener process and polygonal lines coincide and have been determined (Theorem 2.2); the solution of the second boundary problem for a Wiener process has been obtained for the first time by this author in [13]-[15]. The results of this article are first reported in [16] and published in concise form in [17].

A few words about the methods used in [15] as well as herein are in order. The first procedure in both works is the application of an absolutely continuous measure transformation corresponding to the initial walk. At this stage, we separate the principle term of the required asymptotic -- as was done in most works on large deviations and goes back to Cramér [18]. The second procedure in [15] involves one more absolutely continuous transformation, to turn the Wiener process into an inhomogeneous recurrent Markov process. However, this procedure proved to be too complicated, for no reason, especially for random polygons. That is why in this article we use, at the second stage of the proof, appropriate estimates of the convergence rate of the invariance principle. We see that the limiting boundary is piecewise parabolic with the necessary degree of accuracy. In turn, this enables us to apply to the parabola the exact formulas obtained in Section 4.

2. The main theorem

Let $\phi(\lambda) = Me^{\lambda\xi}$; in problems on large deviations, the function

$$\Lambda(\alpha) = \Lambda_\xi(\alpha) = \inf_\lambda \{-\alpha\lambda + \ln \varphi(\lambda)\}, \quad -\infty < \lambda < \infty, \tag{2.1}$$

is of great significance, which, following A.A. Borovkov [1], we call the deviation function of the random variable ξ. Properties of this function were studied in [1]; let us briefly review them. For $M\xi = 0$, $M\xi^2 = \sigma^2 > 0$, the function $\Lambda(\alpha)$ is equal to 0 at a point $\alpha = 0$ and strictly less than 0 elsewhere; it is absolutely continuous for all α in the interval (α_-, α_+), where $\alpha_+ = \text{vrai sup } \xi$, $\alpha_- = \text{vrai inf } \xi$, and is equal to $-\infty$ outside this interval. The derivative $\dot{\Lambda}(\alpha)$ is nonincreasing, hence the deviation function is upward convex. Set

$$\lambda(\alpha) = -\dot{\Lambda}(\alpha)\,. \tag{2.2}$$

We see that the function $\lambda(\alpha)$ is the unique solution of the equation

$$\dot{\varphi}(\lambda)/\varphi(\lambda) = \alpha,$$

hence we have

$$\Lambda(\alpha) = -\alpha\lambda(\alpha) + \ln \varphi(\lambda(\alpha)). \tag{2.3}$$

In some neighborhood of $\alpha = 0$, the functions $\Lambda(\alpha)$ and $\lambda(\alpha)$ are analytic, with

$$\Lambda(\alpha) = -\frac{1}{2\sigma^2}\alpha^2 + \sum_{j=3}^{\infty} \Lambda_j \alpha^j, \quad \lambda(\alpha) = \frac{1}{\sigma^2}\alpha - \sum_{j=2}^{\infty} (j+1)\Lambda_{j+1}\alpha^j. \tag{2.4}$$

Next we consider the function

$$\sigma^2(\alpha) \equiv \frac{\partial^2}{\partial\lambda^2}\left(\frac{\varphi(\lambda + \lambda(\alpha))}{\varphi(\lambda(\alpha))} e^{-\lambda\alpha}\right)\Bigg|_{\lambda=0}. \tag{2.5}$$

This is also an analytic function, with

$$\sigma^2(\alpha) = -[\ddot{\Lambda}(\alpha)]^{-1} = \sigma^2 + \sum_{j=1}^{\infty} D_j\alpha^j. \qquad (2.6)$$

Finally, the probabilistic meaning of the deviation function becomes clear from the next assertion: let Δ_α be the interval $(\alpha-\Delta, \alpha+\Delta)$. Then

$$\lim_{\Delta\to 0}\lim_{n\to\infty} \frac{1}{n} \ln \mathbf{P}\left(\frac{\xi_1 + \dots \xi_n}{n} \in \Delta_\alpha\right) = \Lambda(\alpha),$$

where the random variables $\xi_1,\dots,\xi_n$ are independent and have the same distribution as the variable ξ, for which the deviation function is constructed. All of these properties of the function $\Lambda(\alpha)$ can be found in [1], [19]. As an example we consider the normal random variable with parameters $(0,\sigma^2)$, with the functions having the form

$$\Lambda(\alpha) = -\frac{\alpha^2}{2\sigma^2}, \quad \lambda(\alpha) = \frac{\alpha}{\sigma_2}, \quad \sigma^2(\alpha) = \sigma^2. \qquad (2.7)$$

By the Airy function [20] one means the unique solution of the equation

$$\ddot{V}(x) = xV(x), \ -\infty < x < \infty,$$

satisfying the condition

$$V(0) = \sqrt{\pi}3^{-1/3}/\Gamma(2/3), \ \dot{V}(0) = -\sqrt{\pi}3^{-1/3}/\Gamma(1/3).$$

Set

$$z(x) = V\left(2^{1/3}x + \lambda_0\right)\left(\int_0^\infty V^2(2^{1/3}t + \lambda_0)\,dt\right)^{-1/2}, \qquad (2.8)$$

where $\lambda_0 < 0$ is the greatest zero of $V(x)$.

Recall that $\xi,\xi_1,\dots$ is a sequence of independent identically distributed variables, $M\xi = 0$, $M\xi^2 = 1$, $M\exp\{\lambda\xi\} < \infty$ for $|\lambda| \leqq \delta$, $\delta > 0$, $S_i = \xi_1 + \dots + \xi_i$.

T h e o r e m 2.1. Let the function $H(t)$, $0 \leqq t \leqq 1$, satisfy the conditions: 1) $H(0) = 0$, 2) the third derivative $H^{(3)}(t)$ exists and is continuous, 3) $\sup_{0\leqq t\leqq 1} \ddot{H}(t) < 0$. Let the density $p(t) = \dot{P}(\xi < t)$ exist, be continuous and bounded. Let the sequence $\tau = r(n)$ of positive numbers be such that

$$\lim_{n\to\infty} r/\sqrt{n} = \infty, \quad \lim_{n\to\infty} r\ln^4 n/n = 0.$$

Then for any $x > 0$, $y > 0$, one has as $n \to \infty$:

$$\frac{\partial}{\partial y} \mathbf{P}\left(S_i + \left(\frac{n^2}{3}\right) > rH(i/n), \quad i = 1, \ldots, n, \quad S_n + x\left(\frac{n^2}{r}\right)^{1/3}\right.$$

$$\left. < rH(1) + y\left(\frac{n^2}{r}\right)^{1/3}\right) = |\ddot{H}(0)|^{1/6} |\ddot{H}(1)|^{1/6} z(x|\ddot{H}(0)|^{1/3}) z(y|\ddot{H}(1)|^{1/3})$$

$$\times \exp\left\{n\int_0^1 \Lambda\left(\frac{r}{n}\dot{H}(t)\right)dt - x\left(\frac{n^2}{r}\right)^{1/3}\lambda\left(\frac{r}{n}\dot{H}(0)\right) + y\left(\frac{n^2}{r}\right)^{1/3}\lambda\left(\frac{r}{n}\dot{H}(1)\right)\right\}$$

$$\times \exp\left\{\left(\frac{r}{\sqrt{n}}\right)^{2/3} \frac{\lambda_0}{2^{1/3}} \int_0^1 |\ddot{H}(t)|^{2/3} \sigma^2\left(\frac{r}{n}\dot{H}(t)\right)dt\right\}(1 + o(1)),$$

where the functions z, Λ, λ, σ^2 are defined by formulas (2.8), (2.1), (2.2), (2.5), respectively.

It is worthwhile to explain that we have chosen the normalization $(n^2/r)^{1/3}$ for the beginning and the "end" of the walk, because it is precisely in a strip about the function rH, of width $C(n^2/r)^{1/3}$, where C is sufficiently great, that the trajectories of the walk most crucial for our study of the probability considered are concentrated.

3. An absolutely continuous transformation

Let random variables $\eta_i = \eta_i(\omega)$, $i = 1,2$, $\omega \in \Omega$, be given on a probability space (Ω, F, P), and let the indicator A of the event $A \in F$ be denoted by $I_A = I_A(\omega)$. Also, by $M'(\eta_1; A, \eta_2=x)$ we denote

$$d/dx\mathbf{M}(\eta_1 I_A I\{\eta_2 < x\});$$

if $\eta_1 \equiv 1$, then we write $P'(A, \eta_2=x)$ instead of $M'(1; A, \eta_2=x)$. In particular, we consider the function

$$\begin{aligned} P_n(x, y) &= P_n(H; x, y) \\ &\equiv \mathbf{P}'(S_i + x > rH(i/n), \ i = 1, \ldots, n, \ S_n + x = rH(1) + y). \end{aligned} \tag{3.1}$$

Let $P = P_n$ be the measure on the Borel σ-algebra $\mathcal{B}$ of events of $C(0,1)$ corresponding to the random polygon $s_n = s_n(t)$ connecting the points $(i/n, S_i)$, $i = 0,1,\ldots,n$. We define an absolutely continuous transformation of this measure as follows: for any $A \in \mathcal{B}$ set

$$\tilde{\mathbf{P}}'(s_n + x \in A,\ s_n(1) + x = rH(1) + y)$$

$$\equiv \prod_{i=1}^{n} \varphi^{-1}(\lambda(\alpha_i))\, \mathbf{M}'\left(e^{\sum_{i=1}^{n} \lambda(\alpha_i)\xi_i};\ s_n + x \in A,\ s_n(1) + x = rH(1) + y\right),$$

where $\alpha_i = \alpha_{i,n} \equiv r(H(i/n) - H((i-1)/n))$, $i = 1,\dots,n$. Now, consider along with the variables $\xi_1,\dots,\xi_n$ the independent random variables $\tilde{\xi}_1,\dots,\tilde{\xi}_n$, where the distribution of $\tilde{\xi}_i$ is uniquely defined by the Laplace transformation:

$$\tilde{\varphi}_i(\lambda) = \mathbf{M}e^{\lambda\tilde{\xi}_i} \equiv \varphi(\lambda + \lambda(\alpha_i))/\varphi(\lambda(\alpha_i)).$$

Let $\tilde{s}_n = \tilde{s}_n(t)$ be the random polygon connecting the points $(i/n, \tilde{S}_i)$, $i = 0,1,\dots,n$, where $\tilde{S}_0 = 0$, $\tilde{S}_i = \tilde{\xi}_1 + \cdots + \tilde{\xi}_i$. Then it is obvious that to this polygon in $C(0,1)$ there corresponds the distribution coinciding with $\tilde{P}$. We agree to denote the mathematical expectation with respect to the measure $\tilde{P}$ by the letter with a tilde (i.e., $\tilde{M}$).

Using this transformation, we can write the function in question in the form

$$\mathbf{P}_n(x, y) = \exp\left\{\sum_{i=1}^{n} [\ln \varphi(\lambda(\alpha_i)) - \alpha_i\lambda(\alpha_i)]\right\} \tilde{\mathbf{M}}'\left(\exp\left\{-\sum_{i=1}^{n} \lambda(\xi_i)(\xi_i - \alpha_i)\right\};\right.$$

$$\left.\inf_{1<i<n-1} (S_i - rH(i/n + x) > 0,\ S_n - rH(1) + x = y\right).$$

Next we set

$$X_i = \xi_i - \alpha_i,\ i = 1, 2, \dots, n;\ U_i = X_1 + \dots + X_i,$$

so that $U_i + x = S_i - rH(i/n) + x$, which together with formula (2.3) yield

$$\mathbf{P}_n(x, y) = \exp\left\{\sum_{i=1}^{n} \Lambda(\alpha_i)\right\} \tilde{\mathbf{M}}'\left(\exp\left\{-\sum_{i=1}^{n} \lambda(\alpha_i) X_i\right\};\right.$$

$$\left.\inf_{1<i<n-1} (U_i + x) > 0,\quad U_n + x = y\right).$$

For arbitrary scalars $x_0, x_1, \dots, x_n$, $y_1, \dots, y_n$, one has

$$\sum_{i=1}^{n} x_i y_i = y_n \sum_{i=1}^{n} x_i - y_1 x_0 - \sum_{j=1}^{n-1} (y_{j+1} - y_j) \sum_{i=0}^{j} x_i.$$

Setting $x_0 = x$, $y_i = \lambda(\alpha_i)$, $x_i = X_i$, $i = 1,\dots,n$, we get the following relation on the event $\{U_n + x = y\}$ for $y > 0$:

$$-\sum_{i=1}^{n}\lambda(\alpha_i)X_i=\lambda(\alpha_1)x-\lambda(\alpha_n)y+\sum_{j=1}^{n-1}\beta_j(U_j+x),$$

where $\beta_j \equiv \lambda(\alpha_{j+1}) - \lambda(\alpha_j)$, $j = 1,\dots,n-1$, hence

$$\mathbf{P}_n(x,y)=\exp\left\{\sum_{i=1}^{n}\Lambda(\alpha_i)+\lambda(\alpha_1)x-\lambda(\alpha_n)y\right\}$$
$$\times\widetilde{\mathbf{M}}'\left(\exp\left\{\sum_{j=1}^{n-1}\beta_j(U_j+x)\right\};\quad \inf_{1\leqslant j\leqslant n-1}(U_j+x)>0,\quad U_n+x=y\right).$$

Set

$$k=k(n)\equiv\left[\left(\frac{n^2}{r}\right)^{2/3}\right] \tag{3.2}$$

and in what follows we use the letter k solely for (3.2). Then, obviously,

$$P_n(x\sqrt{k},y\sqrt{k})=\frac{1}{\sqrt{k}}M_n(x,y)$$
$$\times\exp\left\{\sum_{j=1}^{n}\Lambda(\alpha_j)+\lambda(\alpha_1)x\sqrt{k}-\lambda(\alpha_n)y\sqrt{k}\right\}, \tag{3.3}$$

where

$$M_n(x,y)\equiv\widetilde{\mathbf{M}}'\left(e^{\sum_{i=1}^{n-1}\beta_j(U_j+x\sqrt{k})};\quad \inf_{1\leqslant j\leqslant n-1}(U_j+x\sqrt{k})>0,\quad U_n/\sqrt{k}+x=y\right).$$

For brevity, we omit the tilde in the notation for the mathematical expectation with respect to the measure $\tilde{P}$. This will cause no misunderstanding if we bear in mind that the measure P is used for the random variables ξ_i, S_j and $\tilde{P}$ for X_i, U_j.

4. Exact formulas for a Wiener process

The following formula is the starting point of this section.

$$\frac{\partial}{\partial\beta}\mathbf{P}(w(t)+\alpha>sH(t);\quad 0\leqslant t\leqslant T,\quad w(T)+\alpha<sH(1)+\beta)$$
$$=\exp\left\{-\frac{s^2}{2}\int_0^T\dot{H}^2(t)dt+s\dot{H}(0)\alpha-s\dot{H}(T)\beta\right\} \tag{4.1}$$
$$\times\frac{\partial}{\partial\beta}\mathbf{M}\left(e^{s\int_0^T\ddot{H}(t)(w(t)+\alpha)dt};\quad \inf_{0<t<T}(w(t)+\alpha)>0,\quad w(T)+\alpha<\beta\right),$$

where $\alpha > 0$, $\beta > 0$, $s > 0$, $T > 0$, the function $H(t)$, $0 \leqq t \leqq T$, has a continuous second derivative $\ddot{H}(t)$, $H(0) = 0$, $\sup_{0\leqq t\leqq T} \ddot{H}(t) < 0$. Formulas (4.1) can be derived, for example, from formula (3.3); to do this, one needs to choose the initial normal random variables, with parameters $(0,1)$, set $x = \alpha\sqrt{n}/\sqrt{k}$, $y = \beta\sqrt{n}/\sqrt{k}$, $r = s\sqrt{n}$ in (3.3), use formulas (2.7) and pass to the limit in n. For $\rho > 0$, $t > 0$, we set

$$T_\rho(t; x, y) = \mathbf{M}'\left(e^{-\rho\int_0^t (w(u)+x)\,du} ; \inf_{0<u<t} (w(u)+x) > 0, \quad w(t)+x = y \right);$$

for the functions $\phi \in L_2(0,\infty)$ we define the operator

$$T_{\rho,t}\varphi(x) = \int_0^\infty T_\rho(t; x, y)\,\varphi(y)\,dy.$$

It is seen that the family $(T_{\rho,t})_{t\geqq 0}$ is a homogeneous semigroup of self-conjugate completely continuous operators in $L_2(0,\infty)$. Hence by the Hilbert-Schmidt theorem (see, for example, [21]) the kernel of $T_{\rho,t}$ can be written in the form

$$T_\rho(t; x, y) = \sum_{j=1}^{\infty} e^{\lambda_j(\rho)t} \varphi_j(\rho; x)\,\varphi_j(\rho; y), \tag{4.2}$$

where $e^{\lambda_j(\rho)}$ and $\phi_j(\rho, x)$ are the jth eigenvalue and the jth eigenvector of $T_{\rho,1}$, respectively. It follows from

$$e^{\lambda_j(\rho)t}\varphi_j(\rho; x) = \mathbf{M}\left(e^{-\rho\int_0^t (w(u)+x)\,du} \varphi_j(\rho; w(t)+x); \inf_{0<u<t} (w(u)+x) > 0 \right)$$

that the function $\phi_j(\rho; x)$ and the number $\lambda_j(\rho)$ satisfy the equation

$$\frac{1}{2}\ddot{\varphi}_j(\rho; x) = (\lambda_j(\rho) + \rho x)\,\varphi_j(\rho; x), \quad x > 0, \tag{4.3}$$

and the boundary conditions

$$\varphi_j(\rho;\ 0) = \varphi_j(\rho;\ \infty) = 0. \tag{4.4}$$

Next, since the Airy function $V(x)$ satisfies the equation

$$\ddot{V}(x) = xV(x), \ -\infty < x < \infty,$$

and the boundary conditions

$$V(0) = \sqrt{\pi}/(3^{1/3}\Gamma(2/3)), \quad V(\infty) = 0,$$

then one can see that the function $\phi_j(\rho, x)$ and the number $\lambda_j(\rho)$ are expressible in terms of the function $V(x)$ and its zeros, respectively.

To this end, let $\mu_1 < \mu_2 < \cdots < \mu_n < \cdots$ denote all the zeros of $V(x)$; it is well known [20] that they are all simple, negative and countable in number. Note in passing that as $x \to \infty$,

$$V(x) = \frac{1}{2} x^{-1/4} e^{-s}(1 + o(1)),\ \dot{V}(x) = -\frac{1}{2} x^{1/4} e^{-s}(1 + o(1)),$$
$$\mu_n = -\left(\frac{3}{8}(4n-1)\pi\right)^{2/3}(1 + o(1)),\quad \sup_{-\infty < x < \infty} \{|V(x)| + |\dot{V}(x)|\} < \infty, \tag{4.5}$$

where $s = (2/3x)^{3/2}$. For $j = 1,2,\ldots,\ x > 0$, set

$$\lambda_j = \mu_j/2^{1/3},\ z_j(x) = c_j V(2^{1/3}x + \mu_j), \tag{4.6}$$

where the numbers $c_j > 0$ are such that $\int_0^\infty z_k^2(x)\,dx = 1$. Then the number $\lambda_j(\rho)$ and the function $\phi_j(\rho;x)$ have the form

$$\lambda_j(\rho) = \lambda_j \rho^{2/3},\ \varphi_j(\rho;\ x) = \rho^{1/6} z_j(\rho^{1/3}x),\ j = 1,\ 2,\ \ldots, \tag{4.7}$$

where the numbers λ_j and the functions $z_j(x)$ are defined in (4.6); this is easily seen by checking that the numbers and functions thus determined satisfy equation (4.3) and the conditions (4.4). Now, using (4.1), (4.2) and (4.7), one can write an exact formula for the case $\ddot{H}(t) = \text{const}$ (i.e., $H(t)$ is a parabola).

T h e o r e m 4.1. Let $H(0) = 0$, $\ddot{H}(t) = c < 0$ for $0 \leqq t \leqq T$. Then for any $x > 0$, $y > 0$, $T > 0$, $s > 0$,

$$\frac{\partial}{\partial y} \mathbf{P}(w(t) + x > sH(t);\ 0 \leqslant t \leqslant T,\ w(T) + x < sH(T) + y)$$
$$= \exp\left\{-\frac{s^2}{2}\int_0^T \dot{H}^2(t)\,dt + s\dot{H}(0)\,x - s\dot{H}(T)\,y\right\}$$
$$\times \sum_{j=1}^{\infty} e^{\lambda_j (sc)^{2/3} T} (sc)^{1/3} z_j\left((sc)^{1/3}x\right) z_j\left((sc)^{1/3}y\right),$$

where the numbers λ_j and the functions $z_j(x)$ are defined in (4.6).

5. The lemma on approximation

In this section, along with the function $H(t)$ of Theorem 2.1 we construct a piecewise parabolic function $G(t)$ such that certain functionals of both functions are close with a given accuracy. For the given $0 < \Delta \leqq \frac{1}{4}$, let $m = m(\Delta) \equiv \max\{i = 3j:\ (i+1)\Delta \leqq 1\}$; let $\delta_1, \delta_2, \ldots, \delta_m, \delta_{m+1}$ denote the intervals $(0,\Delta)$, $(\Delta,2\Delta)$, ..., $(m\Delta-\Delta,\ m\Delta)$, $(m\Delta,1)$, respectively.

L e m m a 5.1. Let $H(0) = 0$ and let for $0 < a \leqq A < \infty$,

$$-A \leqslant \ddot{H}(t) \leqslant -a, \ |H^{(3)}(t)| \leqslant A, \ |\dot{H}(t)| \leqslant A \quad \text{for} \quad 0 \leqslant t \leqslant 1.$$

Then one can construct a function $G(t)$, $0 \leqq t \leqq 1$, $G(0) = 0$, $\dot{G}(0) = \dot{H}(0)$, such that for $t \in \delta_i$ one has $\ddot{G}(t) = a_i$, $i = 1,2,\ldots,m+1$, and for some $C < \infty$ depending only on a and A, the following relations hold:

1. $\sup\limits_{0<t<1} |\ddot{H}(t) - \ddot{G}(t)| \leqslant C\Delta$,

2. $\sup\limits_{0<t<1} |\dot{H}(t) - \dot{G}(t)| \leqslant C\Delta^2$,

3. $\sup\limits_{0<t<1} |H(t) - G(t)| \leqslant C\Delta^3$,

4. $\left| \int\limits_0^1 (\dot{H}^2(t) - \dot{G}^2(t))\, dt \right| \leqslant C\Delta^4$,

5. $\left| \int\limits_0^1 (\dot{H}^i(t) - \dot{G}^i(t))\, dt \right| \leqslant i(i-1)\, C^{i-2}\Delta^3, \quad i = 3, 4, \ldots,$

6. $\left| \int\limits_0^1 (|\ddot{H}(t)|^{2/3} - |\ddot{G}(t)|^{2/3})\, dt \right| \leqslant C\Delta^2$.

Proof. We denote by $f(t) = O(\Delta^i)$ functions of t (including constants) which admit the estimate $|f(t)| \leqq C\Delta^i$, where the number $C < \infty$ depends only on a and A; by C we denote various constants depending only on a and A.

To construct the function $G(t)$, it suffices to find numbers $a_1,\ldots,a_{m+1}$, $a_i = \ddot{G}(t)$ for $t \in \delta_i$. We find the numbers $a_1,\ldots,a_m$ from $m/3$ systems of equations (and initial conditions $G(0) = 0$, $\dot{G}(0) = \dot{H}(0)$):

$$\mathrm{I}_i:\ G(3i\Delta) = H(3i\Delta),$$

$$\mathrm{II}_i:\ \dot{G}(3i\Delta) = \dot{H}(3i\Delta),$$

$$\mathrm{III}_i:\ \int\limits_{3(i-1)\Delta}^{3i\Delta} (\dot{G}^2(t) - \dot{H}^2(t))\, dt = 0,$$

$i = 1,2,\ldots, m/3$. We find the number a_{m+1} from the condition $\mathrm{I}_{m/3+1}$: $G(1) = H(1)$.

First we prove the existence of a solution of the I_i, II_i, III_i, $i = 1,\ldots, m/3$; obviously, it suffices to consider the equations $\mathrm{I}_1 - \mathrm{III}_1$. We rewrite these equations in a more convenient form:

$$\mathrm{I}_1'\colon \Phi_1 \equiv 5a_1 + 3a_2 + a_3 = 2\frac{1}{\Delta^2}\int\limits_0^{3\Delta}\int\limits_0^{t} \ddot{H}(u)\,du\,dt = 9\ddot{H}(0) + O(\Delta),$$

$$\mathrm{II}_1'\colon \Phi_2 \equiv a_1 + a_2 + a_3 = \frac{1}{\Delta}\int\limits_0^{3\Delta} \ddot{H}(u)\,du = 3\ddot{H}(0) + O(\Delta),$$

$$\mathrm{III}_1'\colon \Phi_3 \equiv 7a_1^2 + 4a_2^2 + a_3^2 + 9a_1a_2 + 3a_1a_3 + 3a_2a_3 = \frac{3}{\Delta^3}\int\limits_0^{3\Delta} \dot{H}^2(t)\,dt$$

$$= 27\ddot{H}^2(0) + O(\Delta).$$

Solving this system, we see that

$$a_j = \ddot{H}(0) + O(\Delta),\ j = 1,\ 2,\ 3. \tag{5.1}$$

Since the equation $\mathrm{I}_{m/3+1}$ also has the solution

$$a_{m/3+1} = \ddot{H}(m\Delta) + O(\Delta), \tag{5.2}$$

all the numbers $a_1,\ldots,a_{m+1}$ are found and the function $G(t)$ is constructed. It remains to see that properties 1-6 hold.

Set

$$\Delta_1 = \delta_1 + \delta_2 + \delta_3,\ \ldots,\ \Delta_{m/3} = \delta_{m-2} + \delta_{m-1} + \delta_m,\ \Delta_{m/3+1} = \delta_{m+1}.$$

Obviously, it suffices to prove that the following relations hold:

1′. $\sup\limits_{t\in\Delta_i} |\ddot{H}(t) - \ddot{G}(t)| \leqslant C\Delta,\quad i = 1,\ldots, m/3+1,$

2′. $\sup\limits_{t\in\Delta_i} |\dot{H}(t) - \dot{G}(t)| \leqslant C\Delta^2,\quad i = 1,\ldots, m/3+1,$

3′. $\sup\limits_{t\in\Delta_i} |H(t) - G(t)| \leqslant C\Delta^3,\quad i = 1,\ldots, m/3+1,$

4′. $\left|\int\limits_{\Delta_{m/3+1}} (\dot{H}^2(t) - \dot{G}^2(t))\,dt\right| \leqslant C\Delta^4,$

5′. $\left|\int\limits_{\Delta_i} (\dot{H}^j(t) - \dot{G}^j(t))\,dt\right| \leqslant C^{j-1}j(j-1)\Delta^4,\quad i = 1,\ldots, m/3+1,\ j \geqslant 3,$

6′. $\left|\int\limits_{\Delta_i} (|\ddot{H}(t)|^{2/3} - |\ddot{G}(t)|^{2/3})\,dt\right| \leqslant C\Delta^3,\quad i = 1,\ldots, m/3,$

7′. $\left|\int\limits_{\Delta_{m/3+1}} (|\ddot{H}(t)|^{2/3} - |\ddot{G}(t)|^{2/3})\,dt\right| \leqslant C\Delta^2.$

Relation 1' follows from formulas (5.1) and (5.2); relation 2' from relation 1'; 3' from 2'; 7' from 1'. Hence it remains to prove relations 4'-6'. Let us start with 4': let

$$L = \dot{H}(m\Delta) = \dot{G}(m\Delta),\ \dot{G}_0(t) = \dot{G}(t) - L,\ \dot{H}_0(t) = \dot{H}(t) - L\ .$$

Then

$$\int_{m\Delta}^{1} (\dot{G}^2(t) - \dot{H}^2(t))\,dt = \int_{m\Delta}^{1} (\dot{G}_0^2(t) - \dot{H}_0^2(t))\,dt + 2L\int_{m\Delta}^{1} (\dot{H}_0(t) - \dot{G}_0(t))\,dt\}.$$

Since

$$\int_{m\Delta}^{1} (H_0(t) - G_0(t))\,dt = 0$$

is satisfied by $I_{m/3+1}$, we obtain

$$\int_{m\Delta}^{1} (\dot{H}^2(t) - \dot{G}^2(t))\,dt = \int_{m\Delta}^{1} \left(\dot{H}_0^2(t) - \dot{G}_0^2(t)\right)dt.$$

It is obvious, next, that for $m\Delta \leqq t \leqq 1$,

$$\dot{H}_0^2(t) = (\ddot{H}(m\Delta)\,t)^2 + O(\Delta^3),\ \dot{G}_0^2(t) = (\ddot{H}(m\Delta)\,t)^2 + O(\Delta^3),$$

hence relation 4' is proved:

$$\int_{m\Delta}^{1} (\dot{H}^2(t) - \dot{G}^2(t))\,dt = \int_{m\Delta}^{1} O(\Delta^3)\,dt = O(\Delta^4)\ .$$

To prove 5' we need relations (5.1) and (5.2) only, thus it suffices to prove 5' for $i = 1$. Let $L = \dot{H}(0) = \dot{G}(0)$, $\dot{H}_0(t) = \dot{H}(t) - L$, $\dot{G}_0(t) = \dot{G}(t) - L$. Then

$$\dot{H}^s(t) = (\dot{H}_0(t) + L)^s = \sum_{i=0}^{s} C_s^i L^{s-i}\dot{H}_0^s(t),$$

$$\dot{G}^s(t) = (\dot{G}_0(t) + L)^s = \sum_{i=0}^{s} C_s^i L^{s-i}\dot{G}_0^s(t)$$

Hence

$$A_s \equiv \left|\int_0^{3\Delta} (\dot{H}^s(t) - \dot{G}^s(t))\,dt\right| \leqslant \sum_{i=0}^{s} C_s^i A^{s-i} B_i,$$

where

$$B_i \equiv \left|\int_0^{3\Delta} (\dot{H}_0^i(t) - \dot{G}_0^i(t))\,dt\right|.$$

Since

$$B_0 = 0,\ B_1 = H_0(3\Delta) - G_0(3\Delta) = 0,$$

then

$$A_s \leqslant \sum_{i=2}^{s} C_s^i A^{s-i} B_i. \tag{5.3}$$

Now, to estimate B_i, we note that for $0 \leqq t \leqq 3\Delta$,

$$\dot{G}_0(t) = t\ddot{H}(0) + O(\Delta^2), \quad \dot{H}_0(t) = t\ddot{H}(0) + O(\Delta^2);$$

$$B_i = \int_0^{3\Delta} \sum_{j=0}^{i} C_i^j \left[(t\ddot{H}(0))^j O^{i-j}(\Delta^2) - (t\ddot{H}(0))^j O^{i-j}(\Delta^2) \right] dt$$

$$\leqslant \sum_{j=0}^{i-1} C_i^j \frac{1}{j+1} |\ddot{H}(0)|^j \Delta^{j+1} O^{i-j}(\Delta^2),$$

where $|\ddot{H}(0)| \leqq A$. Continuing the calculation, we obtain

$$B_i \leqslant \Delta \cdot \Delta^2 \sum_{j=0}^{i-1} C_i^j \frac{1}{j+1} A^j \Delta^j O^{i-1-j}(\Delta^2) \leqslant 2\Delta^3 \left| \sum_{j=0}^{i-1} C_{i-1}^j A^j O^{i-1-j}(\Delta^2) \right|.$$

We used in the last inequality, the obvious relation

$$\frac{1}{j+1} C_i^j \leqslant 2C_{i-1}^j, \quad 0 \leqslant j \leqslant i-1,$$

and thus obtained

$$B_i \leqslant 2\Delta^3 |A\Delta + O(\Delta^2)|^{i-1}$$

yielding

$$B_i \leqslant \Delta^3 (C\Delta)^{i-1}, \ i = 2, 3, \ldots \tag{5.4}$$

Substituting (5.4) into (5.3) yields

$$A_s \leqslant \sum_{i=2}^{s} C_s^i \Delta^3 (C\Delta)^{i-1} A^{s-i} = \Delta^4 C \sum_{i-2=0}^{s-2} C_s^i (C\Delta)^{i-2} A^{s-2-(i-2)}.$$

Next, using the obvious inequality

$$C_s^i \leqslant s(s-1) C_{s-2}^{i-2}, \quad 0 \leqslant i-2 \leqslant s-2,$$

we obtain

$$A_s \leqslant \Delta^4 C s(s-1) \sum_{i-2=0}^{s-2} C_{s-2}^{i-2} (C\Delta)^{i-2} A^{(s-2)-(i-2)} = \Delta^4 C s(s-1)(A + C\Delta)^{s-2}.$$

Relation 5' is proved.

Finally, let us prove 6'. It also suffices to check it for $i = 1$. Set

$$\ddot{H}(t) = \ddot{H}(0) + \varepsilon(t),$$

where $|\varepsilon(t)| = O(\Delta)$. Hence

$$|\ddot{H}(t)|^{2/3} = |\ddot{H}(0)|^{2/3} + \frac{2}{3}\varepsilon(t)|\ddot{H}(0)|^{-1/3} + O(\Delta^2),$$

$$\int_0^{3\Delta} |\ddot{H}(t)|^{2/3} dt = 3\Delta|\ddot{H}(0)|^{2/3} + \frac{2}{3}|\ddot{H}(0)|^{-1/3}\int_0^{3\Delta}\varepsilon(t)\,dt + O(\Delta^3). \tag{5.5}$$

Next we set $a_i = \ddot{H}(0) + \varepsilon_i$, $i = 1,2,3$, so that $\varepsilon_i = O(\Delta)$. Then

$$\int_0^{3\Delta} |\ddot{G}(t)|^{2/3} dt = \sum_{i=1}^{3}\Delta|a_i|^{2/3} = |\ddot{H}(0)|^{2/3}3\Delta + \frac{2}{3}|\ddot{H}(0)|^{-1/3} \times \Delta(\varepsilon_1+\varepsilon_2+\varepsilon_3) + O(\Delta^3). \tag{5.6}$$

Now we take advantage of relation II_1:

$$\dot{H}(3\Delta) = \dot{G}(3\Delta),$$

where

$$\dot{H}(3\Delta) = \dot{H}(0) + 3\Delta\ddot{H}(0) + \int_0^{3\Delta}\varepsilon(t)\,dt,$$

$$\dot{G}(3\Delta) = \dot{H}(0) + 3\Delta\ddot{H}(0) + \Delta(\varepsilon_1+\varepsilon_2+\varepsilon_3),$$

to obtain

$$\int_0^{3\Delta}\varepsilon(t)\,dt = \Delta(\varepsilon_1+\varepsilon_2+\varepsilon_3),$$

Hence comparing (5.5) and (5.6) yields the required estimate in 6':

$$\left|\int_0^{3\Delta}|\ddot{H}(t)|^{2/3}dt - \int_0^{3\Delta}|\ddot{G}(t)|^{2/3}dt\right| = O(\Delta^3).$$

Lemma 5.1 is proved.

6. Proof of the main Theorem 2.1

Let

$$s = s(n) \equiv \min\{[\ln^2 n],\ [(r/\sqrt{n})^{1/9}]\}.$$

We set the number Δ from Lemma 5.1 equal to ks/n, where $k = k(n)$ is given by (3.2). Thus

$$\Delta = \left(\frac{\sqrt{n}}{r}\right)^{2/3}\min\left\{[\ln^2 n], \left[\left(\frac{r}{\sqrt{n}}\right)^{1/9}\right]\right\}(1+o(1)). \tag{6.1}$$

Let $m = m(\Delta) \equiv \max\{i = 3j\colon (i+1)\Delta \leq 1\}$ and $\delta_1,\ldots,\delta_{m+1}$ be the intervals $(0,\Delta),(\Delta,2\Delta),\ldots,(m\Delta,1)$, respectively. By Lemma 5.1, one can construct a function $G(t)$, $0 \leq t \leq 1$, which on each of the intervals δ_i is a piece of a parabola and, in particular, satisfies the relations

$$\sup_{0<t<1} |G(t) - H(t)| \leqslant C\Delta^3, \quad G(0) = H(0), \quad G(1) = H(1).$$

In this section, we denote by C constants that depend only on the function H and, possibly, on the distribution of the random variable ξ_1. Let $\varepsilon = C\Delta^3$. It is not hard to see that for all $x > 0$, $y > 0$, the following inequalities hold:

$$\mathbf{P}_n(G;\ \sqrt{k}x^+,\ \sqrt{k}y^+) \geqslant \mathbf{P}_n(H;\ \sqrt{k}x,\ \sqrt{k}y) \geqslant \mathbf{P}_n(G;\ \sqrt{k}x^-,\ \sqrt{k}y^-), \tag{6.2}$$

where $x^{\pm} = x \pm r\varepsilon/\sqrt{k}$, $y^{\pm} = y \pm r\varepsilon/\sqrt{k}$, the function $P_n(H;\ x,y)$ is defined by formula (3.1). Hence, to find the asymptotics of $P_n(H;\ \sqrt{k}\,x,\ \sqrt{k}\,y)$ it suffices to determine the behavior of the outer terms in (6.2) and see that they are asymptotically equivalent. Part of the calculations for H have already been made; these transformations have given formula (3.3). We start from this formula and assume that all the transformations preceding (3.3) have been made with respect to the function G rather than H. Therefore,

$$\mathbf{P}_n\left(G;\ \sqrt{k}\,x^{\pm},\ \sqrt{k}\,y^{\pm}\right) = k^{-1/2} \exp\left\{\sum_{j=1}^{n} \Lambda(\alpha_j)\right\} \exp\{\lambda(\alpha_1)\sqrt{k}\,x^{\pm}$$
$$- \lambda(\alpha_n)\sqrt{k}\,\lambda(\alpha_n)\}\, \mathbf{M}_n\left(x^{\pm},\ y^{\pm}\right),$$

where the numbers α_i and β_i are constructed from G. Using the properties of the functions $\Lambda(\alpha)$ and $\lambda(\alpha)$ listed in Section 2, as well as the smoothness properties of G, we easily see that

$$\sum_{i=1}^{n} \Lambda(\alpha_i) + \lambda(\alpha_1)\sqrt{k}\,x^{\pm} - \lambda(\alpha_n)\sqrt{k}\,y^{\pm}$$
$$= n\int_0^1 \Lambda\left(\frac{r}{n}\dot{G}(t)\right)dt + x\sqrt{k}\,\lambda\left(\frac{r}{n}\dot{G}(0)\right) - y\sqrt{k}\,\lambda\left(\frac{r}{n}\dot{G}(1)\right) + o(1).$$

Now, it follows from Lemma 5.1 that

$$x\sqrt{k}\,\lambda\left(\frac{r}{n}\dot{G}(0)\right) - y\sqrt{k}\,\lambda\left(\frac{r}{n}\dot{G}(1)\right) = x\sqrt{k}\,\lambda\left(\frac{r}{n}\dot{H}(0)\right)$$
$$- y\sqrt{k}\,\lambda\left(\frac{r}{n}\dot{H}(1)\right) + o(1). \tag{6.3}$$

Since by virtue of (2.4)

$$n\int_0^1 \Lambda\left(\frac{r}{n}\dot{G}(t)\right)dt = -\frac{r^2}{2n}\int_0^1 \dot{G}^2(t)\,dt + n\sum_{i=3}^{\infty}\Lambda_i\left(\frac{r}{n}\right)^i\int_0^1 \dot{G}^i(t)\,dt,$$

then using assertions 4 and 3 of Lemma 5.1 and the choice of Δ (see (6.1)), we easily obtain that

$$n\int_0^1 \Lambda\left(\frac{r}{n}\dot{G}(t)\right)dt = n\int_0^1 \Lambda\left(\frac{r}{n}\dot{H}(t)\right)dt + o(1). \tag{6.4}$$

By (6.3) and (6.4), to prove Theorem 2.1 it suffices to show that

$$\mathbf{M}_n(x^{\pm}, y^{\pm}) = |\ddot{H}(0)|^{1/6}|\ddot{H}(1)|^{1/6}z_1(x|\ddot{H}(0)|^{1/3})$$
$$\times z_1(y|\ddot{H}(1)|^{1/3})\exp\left\{\left(\frac{r}{\sqrt{n}}\right)^{2/3}\lambda_1\int_0^1 |\ddot{H}(t)|^{2/3}\sigma^2\left(\frac{r}{n}\dot{H}(t)\right)dt\right\}\times(1+o(1)), \tag{6.5}$$

where λ_1 and $z_1(x)$ are defined in (4.6). For this we need to introduce an elaborate notation. Set

$$U_{i,j} \equiv X_i + \cdots + X_j,$$

so that $U_j = U_{1,j}$, where the sums $U_j = X_1 + \cdots + X_j$ were defined in Section 3. For $j = 1,2,\ldots,n-1$, we write

$$\gamma_j = \gamma_{j,n} \equiv k\sqrt{k}\beta_{j,n};\quad \gamma_n = \gamma_{n,n} = 0,$$

where the numbers β_j are defined for the function G as in Section 2:

$$\beta_j \equiv \lambda(\alpha_{j+1}) - \lambda(\alpha_j),\quad \alpha_j \equiv r(G(j/n) - G((j-1)/n)).$$

Here

$$\gamma_j = \ddot{G}(j/n) + o(1) = \ddot{H}(j/n) + o(1),\quad j = 1,\ldots,n-1.$$

Let us next define the family of functions of $x > 0$, $y > 0$:

$$R^{(n)}(i,j;x,y) = \mathbf{M}'\Big(e^{-\frac{1}{k}\sum_{m=i}^{j}\gamma_m(U_{i,m}/\sqrt{k}+x)};$$
$$\inf_{i<m<j}(U_{i,m} + x\sqrt{k}) > 0,\ \frac{U_{i,j}}{\sqrt{k}} + x = y\Big),$$

and the family of operators from $L_2(0,\infty)$ to $L_2(0,\infty)$:

$$R^{(n)}(i,j)\varphi(x) = \int_0^{\infty} R^{(n)}(i,j;x,y)\varphi(y)\,dy.$$

Obviously, the function $M_n(x,y)$ considered can be written in the form

$$\mathbf{M}_n(x, y) = R^{(n)}(1, n; x, y);$$

and for any $1 \leqq i \leqq j < s \leqq n$ we have

$$R^{(n)}(i, s)\varphi(x) = R^{(n)}(i, j) \cdot R^{(n)}(j+1, s)\varphi(x).$$

Thus, the function $M_n(x,y)$, being the kernel of a certain operator, happens to be the convolution of kernels of n operators $R^{(n)}(1,1), \ldots, R^{(n)}(n,n)$. Each of the operators $R^{(n)}(i,i)$ converges as $n \to \infty$ to the identity operator, and the operator $R^{(n)}(1,n)$ converges to the null operator. By the invariance principle, the superposition of any k consecutive operators $R^{(n)}(i+1, i+1), \ldots, R^{(n)}(i+k, i+k)$ (the number k is given by (3.2)) converges as $n \to \infty$ to some nonsingular operator, which is defined with respect to a Wiener process. In this case, the estimate of the rate of convergence, of order $(\ln^4 k/\sqrt{k})$, makes it possible to pass from sums to the Wiener process.

Set

$$\sigma^2(i,j) \equiv \mathbf{M}(X_i + \ldots + X_j)^2 = \sigma^2(\alpha_i) + \ldots + \sigma^2(\alpha_j),$$

where the function $\sigma^2(\alpha)$ is defined by (2.5). The family of functions

$$T_\rho(t; x, y) = \mathbf{M}'\left(e^{-\rho\int_0^t (w(u)+x)du}; \quad \inf_{0 \leqslant u \leqslant t} (w(u) + x) > 0, \quad w(t) + x = y\right)$$

was introduced and studied in Section 4. We define the operator corresponding to this family to be

$$T_t\varphi(x) = T_{\rho,t}\varphi(x) \equiv \int_0^\infty T_\rho(t; x, y)\,\varphi(y)\,dy.$$

For $n = 1, 2, \ldots$ set

$$N = N(n) \equiv \sup\{i \geqslant 1 \colon (i+1)k \leqslant n\} \qquad (6.6)$$

and separate the operators introduced earlier into groups of k: $R_1^{(n)} \equiv R^{(n)}(1,k)$, $R_2^{(n)} \equiv R^{(n)}(k+1, 2k)$, $\ldots$, $R_{N+1}^{(n)} \equiv R^{(n)}(Nk+1, n)$; the operator $R_{N+1}^{(n)}$ consists of $k' = n - Nk$ elements, where k' varies from k to $2k-1$. Let $R_i^{(n)}(x,y)$ denote the kernel of $R_i^{(n)}$. Observe that by the construction of the function G (the choice of Δ by formula (6.1)), the numbers $\gamma_{i,n}$ for $k(i-1)+1 \leqq j \leqq ki$ coincide.

Set

$$\gamma^i = \gamma_n^i \equiv \gamma_{(i-1)k+1,n} = \ldots = \gamma_{ik,n}, \quad i = 1, \ldots, N;$$

$$\gamma^{N+1} = \gamma_n^{N+1} \equiv \gamma_{Nk+1,n} = \ldots = \gamma_{n-1,n}.$$

Also, introduce the operators to approximate $R_1^{(n)}, \ldots, R_{N+1}^{(n)}$:

$$T_i^{(n)} \equiv T_{\gamma_n^i, B_i^2}, \quad i = 1, \ldots, N+1,$$

$$B_i^2 \equiv \sigma^2(k(i-1)+1, ki)/k, \quad i = 1, \ldots, N;$$

$$B_{N+1}^2 \equiv \sigma^2(kN+1, n)/k.$$

Let $T_i^{(n)}(x,y)$ denote the kernel of $T_i^{(n)}$.

Consider next the functions

$$a_x^\pm = a_x^\pm(\cdot) \equiv R_1^{(n)}(x^\pm, \cdot) \in L_2(0, \infty),$$

$$b_y^\pm = b_y^\pm(\cdot) \equiv R_{N+1}^{(n)}(\cdot, y^\pm) \in L_2(0, \infty),$$

where $x^\pm$ and $y^\pm$ are defined by (6.2):

$$A^{(n)} \equiv R_2^{(n)} \ldots R_N^{(n)}.$$

Setting

$$a_x = a_x(\cdot) \equiv T_1^{(n)}(x, \cdot) \in L_2(0, \infty),$$

$$b_y = b_y(\cdot) \equiv T_{N+1}^{(n)}(\cdot, y) \in L_2(0, \infty),$$

$$A^{(0)} \equiv T_2^{(n)} \ldots T_N^{(n)},$$

we see that the functions considered can be represented as

$$\mathbf{M}_n(x^\pm, y^\pm) = (a_x^\pm, A^{(n)} b_y^\pm).$$

Lemmas 6.1 - 6.3 will be proved in Section 7. By $\|\cdot\|$ we denote the norm in $L_2(0,\infty)$.

L e m m a 6.1. There is a $C < \infty$ such that for $1 \leqq i \leqq N+1$,

$$\| T_i^{(n)} - R_i^{(n)} \| \leqslant C \ln^4 k / \sqrt{k}.$$

L e m m a 6.2. For any $x > 0$, $y > 0$

$$\lim_{n\to\infty} \| a_x^\pm - a_x \| = 0, \quad \lim_{n\to\infty} \| b_y^\pm - b_y \| = 0.$$

L e m m a 6.3. For any vector $\phi \in L_2(0,\infty)$

$$\lim_{n\to\infty} \| A^{(0)}\varphi / (\| T_2^{(n)} \| \ldots \| T_N^{(n)} \|) - \psi_0 (\psi_1, \varphi) \| = 0.$$

with

$$\psi_0 = \psi_0(x) \equiv \| T_1^{(n)} \| \, | \gamma_n^1 |^{1/6} z_1 (| \gamma_n^1 |^{1/3} x),$$

$$\psi_1 = \psi_1(x) \equiv \| T_{N+1}^{(n)} \| \, | \gamma_n^{N+1} |^{1/6} z_1 (| \gamma_n^{N+1} |^{1/3} x),$$

where the functions $z_1(x)$ are defined by formula (4.6).

We use these lemmas to complete proving Theorem 2.1. For any $\phi \in L_2(0,\infty)$, obvious calculations yield

$$A^{(n)}\varphi = A^{(0)}\varphi + \psi,$$

where

$$\psi = \sum_{i=2}^{N} R_2^{(n)} \dots R_{i-1}^{(n)} \left(R_i^{(n)} - T_i^{(n)}\right) T_{i+1}^{(n)} \dots T_N^{(n)} \varphi.$$

To estimate the norm of ψ, note that by Lemma 6.1, for $1 \leqq i \leqq N+1$ one has

$$\left| \| R_i^{(n)} \| - \| T_i^{(n)} \| \right| \leqslant C \ln^4 k / \sqrt{k};$$

therefore,

$$\| \psi \| \leqslant \| \varphi \| \sum_{i=2}^{N} \| R_2^{(n)} \| \dots \| R_{i-1}^{(n)} \| \| R_i^{(n)} - T_i^{(n)} \| \| T_{i+1}^{(n)} \| \dots \| T_N^{(n)} \|$$

$$\leqslant \| \varphi \| NC \ln^4 k / \sqrt{k} \cdot \prod_{i=2}^{N} \left(\| T_i^{(n)} \| + C \ln^4 k / \sqrt{k} \right).$$

which, obviously, implies that

$$\| \psi \| \Big/ \prod_{i=2}^{N} \| T_i^{(n)} \| \leqslant CN \ln^4 k / \sqrt{k} \cdot \exp \{ CN \ln^4 k / \sqrt{k} \}.$$

Since for the sequences $r = r(n) = o(n / \ln^4 n)$ one has

$$N \ln^4 k / \sqrt{k} = n \ln^4 k / k \sqrt{k} \cdot (1 + o(1)) = o(1),$$

we obtain

$$\| \psi \| = o(1) \| T_2^{(n)} \| \dots \| T_N^{(n)} \|. \tag{6.7}$$

It follows from (6.7) by Lemma 6.3, that for any $\phi \in L_2(0,\infty)$,

$$\lim_{n \to \infty} \left\| A^{(n)} \varphi \Big/ \prod_{i=2}^{N} \| T_i^{(n)} \| - \psi_0 (\psi_1, \varphi) \right\| = 0.$$

which together with Lemma 6.2 enables us to assert that for $x,y > 0$ one has

$$\lim_{n \to \infty} \left| \mathbf{M}_n (x^{\pm}, y^{\pm}) \Big/ \prod_{i=2}^{N} \| T_i^{(n)} \| - (a_x, \psi_0)(\psi_1, b_y) \right| = 0.$$

We have thus obtained

$$\mathbf{M}_n(x^{\pm}, y^{\pm}) = | \ddot{H}(0) |^{1/6} | \ddot{H}(1) |^{1/6} z_1 (x \,|\, \ddot{H}(0) |^{1/3})$$

$$\times z_1 \left(y \,|\, \ddot{H}(1) |^{1/3} \right) \| T_1^{(n)} \| \dots \| T_{N+1}^{(n)} \| (1 + o(1)),$$

hence to prove Theorem 2.1 it suffices to convince ourselves that

$$\prod_{i=1}^{N+1} \| T_i^{(n)} \| = \exp\left\{\left(\frac{r}{\sqrt{n}}\right)^{2/3} \lambda_1 \int_0^1 | \ddot{H}(t) |^{2/3} \sigma^2 \left(\frac{r}{n} \dot{H}(t)\right) dt\right\} \times (1 + o(1)). \tag{6.8}$$

Since

$$\ln \| T_{N+1}^{(n)} \| = \lambda_1 (\gamma_n^{N+1})^{2/3} \frac{\sigma^2 (Nk+1, n)}{k},$$

$$\ln \| T_i^{(n)} \| = \lambda_1 (\gamma_n^i)^{2/3} \frac{\sigma^2 ((i-1)k+1, ik)}{k}, \qquad i = 1, \ldots, N,$$

one has

$$\ln \left(\prod_{i=1}^{N+1} \| T_i^{(n)} \|\right) = \lambda_1 \frac{1}{k} \left[\sum_{i=1}^{N} (\gamma_n^i)^{2/3} \sigma^2 ((i-1)k + 1, ik)\right.$$

$$\left. + (\gamma_n^{N+1})^{2/3} \sigma^2 (kN + 1, n)\right] = \lambda_1 \frac{n}{k} \frac{1}{n} \sum_{i=1}^{n-1} (\gamma_{i,n})^{2/3} \sigma^2 (i, i) + \lambda_1 \frac{1}{k} (\gamma_{n-1,n})^{2/3} \sigma^2 (n, n)$$

$$= \lambda_1 n \frac{\sqrt{k}}{n} \sum_{i=1}^{n-1} \left[\lambda\left(r\left(G\left(\frac{i+1}{n}\right) - G\left(\frac{i}{n}\right)\right)\right) - \lambda\left(r\left(G\left(\frac{i}{n}\right) - G\left(\frac{i-1}{n}\right)\right)\right)\right]^{2/3}$$

$$\times \sigma^2 (i, i) + o(1).$$

Using now the smoothness properties of G and the properties (2.4) of the function $\lambda(\alpha)$ and (2.6) of the function $\sigma^2(\alpha)$, we easily obtain that

$$\ln \left(\prod_{i=1}^{N+1} \| T_i^{(n)} \|\right) = \lambda_1 \left(\frac{r}{\sqrt{n}}\right)^{2/3} \int_0^1 | \ddot{G}(t) |^{2/3} \sigma^2 \left(\frac{r}{n} \dot{G}(t)\right) dt + o(1).$$

Next, by the estimates of Lemma 5.1 and the choice of Δ we are assured that

$$\left(\frac{r}{\sqrt{n}}\right)^{2/3} \int_0^1 | \ddot{G}(t) |^{2/3} \sigma^2 \left(\frac{r}{n} \dot{G}(t)\right) dt$$

$$= \left(\frac{r}{\sqrt{n}}\right)^{2/3} \int_0^1 | \ddot{H}(t) |^{2/3} \sigma^2 \left(\frac{r}{n} \dot{H}(t)\right) dt + o(1),$$

i.e., (6.8) holds true. Theorem 2.1 is proved.

7. Proof of Lemmas 6.1 - 6.3

We shall use the following assertion [11].

T h e o r e m 7.1. Let $Y_1,\dots,Y_k$ be independent identically distributed random variables, $MY_1 = 0$, $MY_1^2 = 1$, $\beta_3 \equiv M|Y_1|^3 < \infty$, such that the density $\frac{d}{dt}P(Y_1 < t)$ is uniformly bounded by the number A. Then

$$\sup_x \left| \frac{d}{dx} \mathbf{P}\left(Y_1 + \dots + Y_k < x\sqrt{k}\right) - (2\pi)^{-1/2} e^{-\frac{x^2}{2}} \right| \leqslant C \frac{\beta_3}{\sqrt{k}} \max(1, A^3),$$

where $C < \infty$ is an absolute constant.

Using the notation adopted in the preceding sections, we formulate the following asertion.

L e m m a 7.1. We can find $0 < c \leqq C < \infty$ such that for all $1 \leqq i \leqq N+1$, $n = 1,2,\dots$, $x > 0$, $y > 0$ we have

$$|R_i^{(n)}(x, y)| \leqslant Ce^{-c(x+y)}, \quad |T_i^{(n)}(x, y)| \leqslant Ce^{-c(x+y)}.$$

Proof. By $0 < c \leqq C < \infty$ we denote constants not depending on n and i, $1 \leqq i \leqq N+1$. We prove the first inequality for $i = 1$; for the remaining i, the proof is, obviously, similar. By $k/2$ we denote the $[k/2]$-integer part of $k/2$. Since

$$R_1^{(n)}(x, y) \equiv R^{(n)}(1, k; x, y)$$
$$= \int_0^\infty R^{(n)}(1, k/2; x, z)\, R^{(n)}(k/2+1, k; z, y)\, dz,$$

we have by the Cauchy inequality that

$$|R_1^{(n)}(x, y)| \leqslant I_1^{1/2} I_2^{1/2}, \tag{7.1}$$

where

$$I_1 \equiv \int_0^\infty (R^n(1, k/2; x, z))^2 dz, \quad I_2 \equiv \int_0^\infty (R^{(n)}(k/2+1, k; z, y))^2 dz.$$

To estimate the integral I_1, we check that the following estimate holds:

$$R^{(n)}(1,\ k/2; x, y) \leqslant Ce^{c\frac{r^3}{n^3}(x+y)} \tag{7.2}$$

To do this we turn to the initial random variable $\xi_1,\dots,\xi_n$:

$$R^{(n)}(1, k/2; x, y)=\exp\left\{-\sum_{i=1}^{k/2}\Lambda(\alpha_i)-x\sqrt{k}\lambda(\alpha_1)+y\sqrt{k}\lambda(\alpha_{k/2})\right\}$$
$$\times\mathbf{P}'\Big(S_i+\sqrt{k}x>rG(i/n);\quad i=1,\dots,k/2;\quad S_{k/2}/\sqrt{k}+x$$
$$=\frac{r}{\sqrt{k}}G((k/2)/n)+y\Big)\leqslant\exp\left\{-\sum_{i=1}^{k/2}\Lambda(\alpha_i)-x\sqrt{k}\lambda(\alpha_1)+y\sqrt{k}\lambda(\alpha_{k/2})\right\}$$
$$\times\mathbf{P}'\left(S_{k/2}/\sqrt{k}+x=\frac{r}{\sqrt{k}}G((k/2)/n)+y\right).$$

Next we go over to the new random variables $\bar{X}_1, \dots, \bar{X}_{k/2}$ and their sums $\bar{U}_j = \bar{X}_1 + \dots + \bar{X}_j$, setting

$$\mathbf{M}e^{\lambda\bar{X}_i}\equiv\varphi(\lambda+\lambda(\bar{\alpha}_i))\,e^{-\lambda\bar{\alpha}_i}/\varphi(\lambda(\bar{\alpha}_i)),$$

where

$$\bar{\alpha}_1=\dots=\bar{\alpha}_{k/2}\equiv rG((k/2)/n)/(k/2).$$

Thus, the random variables $\bar{X}_1, \dots, \bar{X}_{k/2}$ are identically distributed. We obtain

$$\mathbf{P}'\left(S_{k/2}/\sqrt{k}+x=\frac{r}{\sqrt{k}}G((k/2)/n)+y\right)=\mathbf{P}'(\bar{U}_{k/2}/\sqrt{k}+x=y)$$
$$\times\exp\left\{\sum_{i=1}^{k/2}\Lambda(\bar{\alpha}_i)+x\sqrt{k}\lambda(\bar{\alpha}_1)-y\sqrt{k}\lambda(\bar{\alpha}_{k/2})\right\}. \tag{7.4}$$

Using the smoothness properties of $G(t)$, it is not hard to show that

$$\left|\sum_{i=1}^{k/2}\Lambda(\alpha_i)-\sum_{i=1}^{k/2}\Lambda(\bar{\alpha}_i)\right|+x\sqrt{k}|\lambda(\alpha_1)-\lambda(\bar{\alpha}_1)|+y\sqrt{k}|\lambda(\alpha_{k/2})-\lambda(\bar{\alpha}_{k/2})|$$
$$\leqslant C+c\frac{r^3}{n^3}(x+y);$$

hence by (7.3) and (7.4) we have

$$|R^{(n)}(1, k/2; x, y)|\leqslant Ce^{c\frac{r^3}{n^3}(x+y)}\mathbf{P}'\left(\frac{\bar{U}_{k/2}}{\sqrt{k}}+x=y\right).$$

To obtain estimate (7.2), it suffices to prove that

$$\mathbf{P}'(\bar{U}_{k/2}/\sqrt{k}+x=y)\leqslant C. \tag{7.5}$$

When the random variables $\xi_1,\dots,\xi_n$ are bounded, the density of the transformed random variables $\bar{X}_1, \dots, \bar{X}_{k/2}$ is uniformly bounded, and the estimate (7.5) follows from Theorem 7.1. In the general case, one can only assert that for the density $\bar{p}(t)$ of the $\bar{X}_1$ one has

$$\bar{p}(t) \leqslant C \exp\left\{c \frac{r}{n} |t|\right\}. \tag{7.6}$$

Choose two numbers $a > 0$, $b > 0$ such that

$$\alpha \equiv \max(a, b) = \frac{n}{rc} \frac{1}{6} \ln k, \quad \mathbf{M}(\bar{X}_1; -b < \bar{X}_1 < a) = 0,$$

where the number $c > 0$ is given in formula (7.6). Introduce the following events:

$$A_0 = \left\{ \min_{1 \leqslant i \leqslant k/2} \bar{X}_i > -b, \max_{1 \leqslant i \leqslant k/2} \bar{X}_i < a \right\},$$

$$A_1 = \left\{ \min_{1 \leqslant i \leqslant k/2} \bar{X}_i < -b, \min_{1 \leqslant i \leqslant k/2} |\bar{X}_i| \leqslant \ln^2 k \right\} \cup \left\{ \max_{1 \leqslant i \leqslant k/2} \bar{X}_i > a, \min_{1 \leqslant i \leqslant k/2} |\bar{X}_i| \leqslant \ln^2 k \right\},$$

$$A_j = \{(j-1) \ln^2 k \leqslant \min_{1 \leqslant i \leqslant k/2} |\bar{X}_i| < j \ln^2 k\}, \quad j = 2, 3, \ldots$$

It is clear that

$$\mathbf{P}'(\bar{U}_{k/2}/\sqrt{k} + x = y) \leqslant \sum_{j=0}^{\infty} \mathbf{P}'(A_j; \bar{U}_{k/2}/\sqrt{k} + x = y).$$

We estimate each summand on the right of the last inequality. By virtue of Theorem 7.1, since the density of $\bar{X}_i$ admits the estimate

$$\bar{p}(t) \leqslant C e^{c \frac{r}{n} \alpha} \leqslant C k^{\frac{1}{6}}; \quad -b < t < a,$$

on the event A_0 we have

$$\mathbf{P}'(A_0; \bar{U}_{k/2}/\sqrt{k} + x = y) \leqslant C.$$

The density of the smallest element in modulus admits the estimate

$$\bar{p}(t) \leqslant C e^{-c \frac{r}{n} j \ln^2 k}$$

on the event A_j, $j = 1, 2, \ldots$. Hence for $j = 2, 3, \ldots$ one has

$$\mathbf{P}'(A_j; \bar{U}_{k/2}/\sqrt{k} + x = y)$$

$$\leqslant \frac{k}{2} \int_{-\infty}^{\infty} \mathbf{P}'\left(\bar{U}_{k/2-1}/\sqrt{k} + x = z, \min_{1 \leqslant i \leqslant k/2-1} |\bar{X}_i| \geqslant (j-1) \ln^2 k\right) \times \mathbf{P}'\left(\bar{X}_{k/2}/\sqrt{k} + z = y; |\bar{X}_{k/2}| < j \ln^2 k\right) dz$$

$$\leqslant Cke^{c\frac{r}{n}j\ln^2 k}\mathbf{P}(\min_{1\leq i\leq k/2-1}|\overline{X}_i|\geqslant (j-1)\ln^2 k)$$

$$\leqslant Cke^{c\frac{r}{n}j\ln^2 k-c(j-1)(k/2-1)\ln^2 k}\leqslant Cke^{-cjk}.$$

For $j = 1$ one has the estimate

$$\mathbf{P}'\left(A_1;\ \overline{U}_{k/2}/\sqrt{k}+x=z\right)\leqslant Ck^2e^{-c\ln^5 k},$$

which is established similarly. Hence

$$\sum_{j=1}^{\infty}\mathbf{P}'\left(A_j;\ \overline{U}_{k/2}/\sqrt{k}+x=y\right)\leqslant Ck\left(ke^{-c\ln^5 k}+\sum_{j=2}^{\infty}e^{-cjk}\right)\leqslant C$$

and relation (7.5) has been obtained; estimate (7.2) has also been obtained. Substituting (7.2) into the integral I_1, we have

$$|I_1|\leqslant C\int_0^{\infty}R^{(n)}(1,\ k/2;\ x,\ z)\,e^{c\frac{r^3}{n^3}(x+z)}dz$$

$$\leqslant Ce^{c\frac{r^3}{n^3}x}\mathrm{M}e^{-\frac{\rho}{k}\sum_{j=1}^{k/2}(U_j/\sqrt{k}+x)+c\frac{r^3}{n^3}U_{k/2}/\sqrt{k}}$$

$$=Ce^{\left(c\frac{r^3}{n^3}-\rho(k/2)/k\right)x}\prod_{j=1}^{k/2}\mathrm{M}e^{\left(\frac{k-i}{k\sqrt{k}}\rho+c\frac{r^3}{n^3}\right)X_i}\leqslant Ce^{-\frac{\rho}{4}x}.$$

Thus, for the first integral I_1 in (7.1) we have obtained the estimate

$$|I_1|\leqslant Ce^{-cx}.$$

Turning now to the walk

$$U_k=y,\quad U_{k-1},\quad \ldots,\quad U_{k/2+1},$$

we see that the integral I_2 is of the same nature as the integral I_1, and a similar estimate is valid for it:

$$|I_2|\leqslant Ce^{-cy}.$$

Hence, by (7.1) assertion one of Lemma 7.1 is proved. Since the proof of assertion 2 is similar (and much simpler), Lemma 7.1 is proved.

The following assertion will be proved in Section 9.

L e m m a 7.2. We can find a number $C < \infty$ such that for all $1 \leqq i \leqq N+1$, $n = 1,2,..,$ $x > 0$, $y > 0$, one has

$$\left|R_i^{(n)}(x,\ y)-T_i^{(n)}(x,\ y)\right|\leqslant C\ln^3 k/\sqrt{k}.$$

We prove Lemmas 6.1 - 6.3, using Lemma 7.1 and 7.2.

Proof of Lemma 6.1. Since

$$\| R_i^{(n)} - T_i^{(n)} \|^2 \leqslant I, \qquad I \equiv \int_0^\infty \int_0^\infty (R_i^{(n)}(x, y) - T_i^{(n)}(x, y))^2 \, dxdy,$$

it suffices to estimate the integral I. For $a > 0$ let

$$I_1 \equiv \int_0^{a \ln k} \int_0^{a \ln k} (R_i^{(n)}(x, y) - T_i^{(n)}(x, y))^2 \, dxdy, \; I_2 \equiv I - I_1.$$

By Lemma 7.2,

$$|I_1| \leqslant Ca^2 \ln^2 k \ln^6 k/k,$$

and by Lemma 7.1,

$$|I_2| \leqslant Ce^{-ac \ln k};$$

taking a sufficiently large, we get the necessary estimate:

$$|I| = |I_1| + |I_2| \leqslant C \ln^8 k/k.$$

Lemma 6.1 is proved. Lemma 6.2 is an obvious corollary of Lemmas 7.1 and 7.2.

Proof of Lemma 6.3. Along with the operator (see (4.2) and (4.3))

$$T_{\rho,t}\varphi(x) = \sum_{j=1}^{\infty} e^{\lambda_j \rho^{2/3} t} \varphi_j(\rho; x) \; (\varphi_j(\rho; \cdot), \varphi(\cdot))$$

we consider the operator

$$L_{\rho,t}\varphi(x) \equiv e^{\lambda_1 \rho^{2/3} t} \varphi_1(\rho; x) (\varphi_j(\rho; \cdot), \varphi(\cdot));$$

and along with the operator $T_1^{(n)}, \ldots, T_{N+1}^{(n)}$, we consider the operators $L_1^{(n)}, \ldots, L_{N+1}^{(n)}$, which we define as

$$L_i^{(n)} \equiv L_{\rho_i, t_i},$$

where $\rho_i = \rho_i^{(n)}$, $t_i = t_i^{(n)}$ are the parameters defining the operator $T_i^{(n)} = T_{\rho_i, t_i}$. By the construction of G with respect to Δ (see (6.1)), every consecutive s operators $T_1^{(n)}, \ldots, T_s^{(n)}$; $T_{s+1}^{(n)}, \ldots, T_{2s}^{(n)}$; $\ldots$; $T_{(m-1)s+1}^{(n)}, \ldots, T_{ms}^{(n)}$ and, finally, the remaining $T_{ms+1}^{(n)}, \ldots, T_N^{(n)}$ (the number $m = m(\Delta)$ is defined at the beginning of Section 4) coincide. The last (m+1)*th* group of operators consists of $s' = N - sm$ elements so that $s/2 \leqq s' \leqq 2s$. The same is true for the operators $L_1^{(n)}, \ldots, L_N^{(n)}$. Set

$$A_j \equiv ((L^{(n)}_{js+1})^s - (T^{(n)}_{js+1})^s) \| T^{(n)}_{js+1} \|^{-s}, \quad j = 1, \ldots, m;$$
$$A_{m+1} \equiv ((L^{(n)}_{ms+1})^{s'} - (T^{(n)}_{ms+1})^{s'}) \| T^{(n)}_{ms+1} \|^{-s'}.$$

Note that the norms of the operators $L_i^{(n)}$ and $T_i^{(n)}$ coincide and are equal to $\|T_i^{(n)}\| = \exp\{\lambda_1 \rho_i^{2/3} t_i\}$; the norms of the A_j are equal to

$$\|A_j\| = e^{(\lambda_2 - \lambda_1)\rho^{2/3}_{js+1} t_{js+1} s}, \quad 1 \leqslant j \leqslant m, \quad \|A_{m+1}\| = e^{(\lambda_2 - \lambda_1)\rho^{2/3}_{ms+1} t_{ms+1} s'}.$$

Thus, by the estimate $\sup_{1 \leq j \leq m+1} \|A_j\| \leq Ce^{-cs}$, we have that for any function $\phi \in L_2(0,\infty)$,

$$\lim_{n\to\infty} \left\| \frac{T_2^{(n)} \ldots T_N^{(n)} \varphi}{\|T_2^{(n)}\| \ldots \|T_N^{(n)}\|} - \frac{L_2^{(n)} \ldots L_N^{(n)} \varphi}{\|T_2^{(n)}\| \ldots \|T_N^{(n)}\|} \right\| = 0 \tag{7.7}$$

(here the same arguments are used as in deriving (6.7)).

It is obvious that

$$\frac{L_2^{(n)} \ldots L_N^{(n)} \varphi(x)}{\prod_{j=2}^{N} \|T_j^{(n)}\|} = \varphi_1(\rho_1; x) \prod_{j=1}^{m} (\varphi_1(\rho_{js}; \cdot),$$
$$\varphi_1(\rho_{js+1}; \cdot))(\varphi_1(\rho_{ms+1}; \cdot), \varphi(\cdot)) . \tag{7.8}$$

By the choice of G (and the smoothness of H) one has

$$\max_{1 < j \leqslant m} | \rho_{js} - \rho_{js+1} | \leqslant C\Delta.$$

Next, since

$$(\varphi_1(\rho; \cdot), \varphi_1(\rho; \cdot)) \equiv 1,$$

then

$$\frac{\partial}{\partial \tilde{\rho}} (\varphi_1(\tilde{\rho}; \cdot), \varphi_1(\rho; \cdot)) \Big|_{\tilde{\rho}=\rho} = 0,$$

and hence

$$\max_{1 < j < m} | (\varphi_1(\rho_{js}; \cdot), \varphi_1(\rho_{js+1}; \cdot)) - 1 | \leqslant C\Delta^2.$$

By the last inequality it follows from (7.8) that

$$\lim_{n\to\infty} \left\| \frac{L_2^{(n)} \ldots L_N^{(n)} \varphi(x)}{\prod_{j=2}^{N} \|T_j^{(n)}\|} - \varphi_1(\rho_1; x)(\varphi_1(\rho_{sm+1}; \cdot), \varphi(\cdot)) \right\| = 0,$$

which together with (7.7) proves Lemma 6.3.

8. Solution of the second boundary value problem

For the given function $H(t)$, $0 \leqq t \leqq 1$, by $B_H \subseteq C(0,1)$ we denoted (see Section 1) the set of functions $f \in C(0,1)$ which lie above $H(t)$ for all $0 \leqq t \leqq 1$. The most probable trajectory $H_0(t)$ in B_H is uniquely determined from the relation $H_0(0) = 0$,

$$\int_0^1 \dot{H}_0^2(t)\,dt = \inf\left\{\int_0^1 \dot{f}^2(t)\,dt\colon f(0) = 0,\ f \in B_H\right\}. \tag{8.1}$$

Let there exist q intervals $\Delta_i = (t_{2i-2}, t_{2i-1})$, $i = 1,\dots,q$, such that $H_0(t) > H(t)$ for $t \in \bigcup_{i=1}^{q} \Delta_i$; for $t \notin \bigcup_{i=1}^{q} \Delta_i$, one has $H_0(t) = H(t)$. We denote the length of Δ_i by $|\Delta_i| = t_{2i-1} - t_{2i-2}$ and the distance between neighboring intervals Δ_i and Δ_{i+1} by $\delta_i = t_{2i} - t_{2i-1}$ (it is assumed that $t_1 < t_2 < \dots < t_{2q-1}$). Theorem 2.1, together with the lemmas used to prove it, enables one to solve sufficiently completely the second boundary problem in the region of large deviations $r = r(n) = o(n/\ln^4 n)$. The asymptotic of the probability sought "piles up" from the "asymptotics" on individual intervals, the form of the preceding and subsequent intervals being significant in determining the constants. Naturally, there are many subcases. We formulate the result only in one particular case -- the remaining special cases are similar (within the assumptions of Theorem 2.1. Recall that $(\xi_i)_{i=1}^{\infty}$ are independent identically distributed random variables, $M\xi_1 = 0$, $M\xi_1^2 = 1$, $M\exp\{\lambda\xi_1\} < \infty$ for $|\lambda| \leqq \delta$, $\delta > 0$, $S_i = \xi_1 + \dots + \xi_i$,

$$\Lambda(\alpha) = \inf_{\lambda}\{-\alpha\lambda + \ln Me^{\lambda\xi_1}\},\quad \sigma^2(\alpha) = -\left[\frac{d^2}{d\alpha^2}\Lambda(\alpha)\right]^{-1},$$

the scalar sequence $r = r(n)$ satisfies the conditions:

$$\lim_{n\to\infty} r/\sqrt{n} = \infty,\quad \lim_{n\to\infty} r\ln^4 n/n = 0. \tag{8.2}$$

T h e o r e m 8.1. Let the density $p(t) = \dot{P}(\xi_1 < t)$ be continuous and bounded. Furthermore, let the function $H(t)$ satisfy the following conditions: 1) $H(0) < 0$; 2) $t_{2q-1} = 1$; 3) $|\Delta_i| > 0$, $i = 1,\dots,q$; 4) $\delta_i > 0$, $i = 1,\dots,q-1$; 5) the function $H^{(3)}(t)$ is continuous for $t \notin \bigcup_{i=1}^{q} \Delta_i$; 6) the function $H(t)$ is continuous in some

neighborhood of the set $[0,1] \setminus \bigcup_{i=1}^{q} \Delta_i$; 7) $\inf \{\ddot{H}(t) : t \notin \bigcup_{i=1}^{q} \Delta_i\} < 0$.
Then, as $n \to \infty$,

$$\mathbf{P}(S_i > rH(i/n);\ i = 1, \ldots, n)$$
$$= \Lambda^{q-1} \prod_{i=1}^{q-1} |\Delta_i|^{-1/2} \prod_{i=1}^{2q-2} |\ddot{H}(t_i)|^{-1/6} \left(\frac{r}{\sqrt{n}}\right)^{-(q-1)/3}$$
$$\times \exp\left\{n \int_0^1 \Lambda\left(\frac{r}{n} H_0(t)\right) dt + \left(\frac{r}{\sqrt{n}}\right)^{2/3} \frac{\lambda_0}{2^{1/3}} \int_0^1 |\ddot{H}_0(t)|^{2/3}\right.$$
$$\left.\times \sigma^2\left(\frac{r}{n} \dot{H}_0(t)\right) dt\right\} \times (1 + o(1)),$$

where

$$\Lambda = 2^{-17/20} \sqrt{\pi} \left[\int_0^\infty (V^2(2^{1/3}x + \lambda_0))\, dx\right]^{-1},$$

and λ_0 is the largest zero of the Airy function $V(x)$ (see [20]).

REMARK 8.1. Analogous assertions hold as well in the lattice case where the common distribution of the summands ξ_i has support at points of the form (jh), $-\infty < j < \infty$. The proof in the absolutely continuous case differs from that in the lattice case only in those points where the local limit theorem is used. In particular, if in Theorem 8.1 the condition for the existence of a smooth density is replaced by the lattice condition, then the assertion remains unchanged.

REMARK 8.2. For a homogeneous process with independent increments $\xi(t)$, $0 \leqq t < \infty$, assertions concerning the behavior of the probability $P(\xi(nt) > rH(t);\ 0 \leqq t \leqq 1)$ are just like the corresponding assertions for sums both in the absolutely continuous case and the lattice case. The functions $\Lambda(\alpha)$, $\lambda(\alpha)$ and $\sigma(\alpha)$ are constructed with respect to the random variable $\xi(1)$. Proofs for the sums follow those for the processes; in particular, it is convenient to obtain an absolutely continuous transformation for a process by passing to the limit, just as was done for the Wiener process in Section 4.

Let $F_n(t)$, $0 \leqq t \leqq 1$, be an empirical distribution function constructed on the basis of a simple sample of size n from uniform distribution on $[0,1]$. It is well known (see, for example, [1]) that the process $F_n(t) - t$ has the same distribution as the process $n^{-1}\eta(nt)$,

$0 \leqq t \leqq 1$, under the condition that $n^{-1}\eta(n) = 0$, where $\eta(t)$ is the Poisson process wtih drift: $M\exp\{\lambda\eta(t)\} = \exp\{t(e^{\lambda} - 1 - \lambda)\}$. It is not hard to see that for $-1 < \alpha < \infty$, one has

$$\Lambda_{\eta(1)}(\alpha) = \alpha - (\alpha + 1)\ln(\alpha + 1),$$
$$\lambda_{\eta(1)}(\alpha) = \ln(\alpha + 1), \ \sigma^2_{\eta(1)}(\alpha) = \alpha + 1.$$

Hence by remarks 8.1 and 8.2 the following theorem holds true.

T h e o r e m 8.2. Let the function $H(t)$ satisfy the conditions $H(0) < 0$, $H(1) = 0$ and let the function $H_0(t)$ be defined from the relations $H_0(0) = H_0(1) = 0$,

$$\int_0^1 \dot{H}_0^2(t)\,dt = \inf\left\{\int_0^1 \dot{f}^2(t)\,dt : f(0) = f(1) = 0,\right.$$

$$\left. f(t) \geqslant H(t);\ 0 \leqslant t \leqslant 1\right\}.$$

Furthermore, let conditions 1) - 7) of Theorem 8.1 be satisfied and let the scalar sequence $r = r(n)$ satisfy (8.2). Then as $n \to \infty$

$$\mathbf{P}(n(F_n(t) - t) > rH(t);\ 0 \leqslant t \leqslant 1) = \Lambda^{q-1}\prod_{i=1}^{q}|\Delta_i|^{-1/2}$$

$$\times \prod_{i=1}^{2q-2}|\ddot{H}(t_i)|^{-1/6}\left(\frac{r}{\sqrt{n}}\right)^{-(q-1)/3}$$

$$\exp\left\{-n\int_0^1\left(\frac{r}{n}\dot{H}_0(t) + 1\right)\ln\left(\frac{r}{n}\dot{H}_0(t) + 1\right)dt\right\}$$

$$\times \exp\left\{\left(\frac{r}{\sqrt{n}}\right)^{2/3}\frac{\lambda_0}{2^{1/3}}\int_0^1|\ddot{H}_0(t)|^{2/3}\left(\frac{r}{n}\dot{H}_0(t) + 1\right)dt\right\} \times (1 + o(1)),$$

where λ_0, Λ are given in Theorem 8.1.

REMARK 8.3. Cramér's condition $Me^{\lambda\xi_1} < \infty$ for $|\lambda| \leqq \delta$, $\delta > 0$, can be weakened by replacing it, for instance, by the condition $Me^{\lambda|\xi_1|^{\beta}} < \infty$ for $|\lambda| \leqq \delta$, $\delta > 0$, $\beta > 0$. Using the truncation technique, one can next derive analogs of Theorems 2.1 and 8.1, in which parts of (2.4) and of (2.6) represent the functions $\Lambda(\alpha)$, $\lambda(\alpha)$ and $\sigma^2(\alpha)$ and the sequence $r = r(n)$ is bounded above by $o(1)n^{1/(2-\beta)}$, $0 < \beta < 1$.

REMARK 8.4. The proofs of Theorems 2.1 and 8.1 "almost" omit the fact that the summands $\xi_1, \xi_2, \ldots, \xi_n$ are identically distributed; moreover, this property is gone at the first stage of the proof. Hence it is possible to obtain by the method suggested the appropriate assertions for differently distributed summands as well as for the series $\xi_{1,n}, \ldots, \xi_{n,m}$, $n = 1,2,\ldots$.

Proof of Theorem 8.1. Is basically prepared by the proof of Theorem 2.1 and requires no additional propositions. Hence we give it in less detail. For brevity, we assume that $q = 2$; proving the case $q > 2$ is exactly the same. Thus, for $q = 2$, the interval $[0,1]$ partitions into three segments: $\Delta_1 = [0,t_1)$, $\delta = [t_1,t_2]$ and $\Delta_2 = (t_2,1]$, where $0 < t_1 < t_2 < 1$, and $H_0(t) > H(t)$ for $t \in \Delta_1 \cup \Delta_2$; $H_0(t) = H(t)$ for $t \in \delta$. The principal transformation has to be performed with respect to the $H_0(t)$ defined by formula (8.1). Let $i_1 = \sup\{i \geq 1: i < nt_1\}$, $i_2 = \inf\{i \leq n: t > nt_2\}$; upon transformation, the random variables $\xi_1,\ldots,\xi_n$ become $X_1,\ldots,X_n$, now differently distributed; since $\dot{H}_0(t) = \dot{H}_0(0)$ for $0 \leq t < t_1$ and $\dot{H}_0(t) = \dot{H}_0(1)$ for $t_2 > t \geq 1$, the $X_1,\ldots,X_{i_1}$ have common distribution, and the $X_{i_2},\ldots,X_n$ have the same property. Just as earlier, we denote the sums by $U_{ij} = X_i + \cdots + X_j$. Let $s = s(n) \equiv r/\sqrt{n}$. Also, let

$$A_n(x_1) \equiv \mathbf{P}'\left(\frac{U_{1i}}{\sqrt{n}} > s\left(H(i/n) - \frac{i}{i_1} H(i_1/n)\right);\ i = 1, \ldots, i_1,\ \frac{U_{1,i_1}}{\sqrt{n}} = x_1\right),$$

$$B_n(x_2) \equiv \mathbf{P}\left(\frac{U_{i_2,i}}{\sqrt{n}} + \frac{\sqrt{k}}{\sqrt{n}} x_2 > s(H(j/n) - H(i_2/n));\ i = i_2, \ldots, n\right),$$

$$C_n(x_1, x_2) \equiv \mathbf{M}'\left(e^{\sqrt{k}\sum_{j=i_1+1}^{i_2-1} \beta_j\left(\frac{U_{i_1+1,j}}{\sqrt{k}} + x_1\right)};\right.$$

$$\left.\inf_{i_1+1 < j \leq i_2+1}\left\{\frac{U_{i_1+1,j}}{\sqrt{k}} + x_1\right\} > 0,\ \frac{U_{i_1+1,i_2-1}}{\sqrt{k}} + x_1 = x_2\right).$$

Upon transformation, the probability sought can be represented

$$\mathbf{P}(S_i > rH(i/n);\ i = 1, \ldots, n) = \exp\left\{n \sum_{i=1}^{n} \Lambda(\alpha_i)\right\}$$

$$\times \int_0^\infty \int_0^\infty A_n(x_1)\, C_n(x_1, x_2)\, B_n(x_2)\, dx_1 dx_2. \tag{8.3}$$

By Theorem 2.1, one has

$$C_n(x_1, x_2) = \varphi_1(|\ddot{H}(t_1)|; x_1)\cdot\varphi_1(|\ddot{H}(t_2)|; x_2)$$

$$\times \exp\left\{s^{2/3}\lambda_1 \int_{t_1}^{t_2} |\ddot{H}(t)|^{2/3}\sigma^2(s\dot{H}(t))\,dt\right\} \times (1 + o(1)), \tag{8.4}$$

where $\phi_1(\rho;x) = \rho^{1/6}z(\rho^{1/3}x)$. Next, using Theorem 7.1, one can show that

$$A_n(x_1) = \mathbf{P}'\left(w(t) > s\left(H(t) - \frac{t}{t_1}H(t_1)\right);\quad 0 \leqslant t \leqslant t_1,\quad w(t_1)\cdot s^{1/3} = x_1\right)$$

$$\times(1 + o(1)) = A(|\ddot{H}(t_1)|; x_1)\cdot\frac{s^{-1/3}}{\sqrt{2\pi t_1}}\times(1 + o(1)), \tag{8.5}$$

where

$$A(\rho; x) \equiv \mathbf{P}\left(w(t) + x > -\frac{1}{2}\rho t^2;\ 0 \leqslant t < \infty\right).$$

One can show in the same way that

$$B_n(x_2) = \mathbf{P}\left(w(t) - w(t_2) + \frac{\sqrt{k}}{\sqrt{n}}x_2 > s(H(t) - H(t_2));\quad t_2 \leqslant t \leqslant 1\right)$$

$$\times(1 + o(1)) = A(|\ddot{H}(t_2)|; x_2)\times(1 + o(1)). \tag{8.6}$$

Now, using (8.6), (8.5) and (8.4) and passing to the limit in (8.3) (which is appropriate by virtue of Lemma 7.1), we obtain the relation

$$\sqrt{2\pi t_1}s^{1/3}\exp\left\{-n\int_0^1 \Lambda(s\dot{H}_0(t))\,dt - s^{2/3}\lambda_1\int_0^1 |\ddot{H}_0(t)|^{2/3}\sigma^2(s\dot{H}_0(t))\,dt\right\}$$

$$\times\mathbf{P}(S_i > rH(i/n);\ 1 \leqslant i \leqslant n) = \int_0^\infty \varphi_1(|\ddot{H}(t_1)|; x)A(|\ddot{H}(t_1)|; x)\,dx$$

$$\times\int_0^\infty \varphi_1(|\ddot{H}(t_2)|; x)\,A(|\ddot{H}(t_2)|; x)dx\times(1 + o(1)). \tag{8.7}$$

It is obvious that

$$A(\rho;\ x) = A(1;\ \rho^{1/3}x).$$

Hence the right side in (8.7) has the form

$$|\ddot{H}(t_1)|^{-1/6}|\ddot{H}(t_2)|^{-1/6}\left(\int_0^\infty z(x)\,A(1; x)\,dx\right)^2, \tag{8.8}$$

where the function $z(x)$ is defined by (2.8),

$A(1;x) = P(\omega(t)+x > -t^2/2;\ 0 \leq t < \infty)$. It remains to determine the integral in (8.8), which we denote by $\Lambda_0^{1/2}$; it is seen that

$$\Lambda_0^{1/2} = \lim_{T\to\infty} \int_0^\infty z(x)\,\mathbf{P}\Big(w(t) > -x - \frac{1}{2}t^2;\ 0 \leqslant t \leqslant T\Big)\,dx.$$

Hence, using the exact formula from Theorem 4.1, we obtain

$$\Lambda_0^{1/2} = \lim_{T\to\infty} e^{\lambda_1 T - T^3/6} \int_0^\infty z(y)\,e^{Ty}\,dy.$$

Using formulas (4.5), by the Laplace method we easily find that

$$\Lambda_0^{1/2} = \sqrt{2\pi}\,2^{-5/3}\left(\int_0^\infty V^2\left(2^{1/3}x + \lambda_0\right)dx\right)^{-1/2},$$

with the $V(x)$ from (2.8). Theorem 8.1 is proved.

9. Technical lemmas

The objective of this section is to prove Lemma 7.2. Here we shall use the earlier notation. In particular, by Lemma 5.1, along with the function H, the function G was constructed -- it was done on the basis of the $\Delta = \Delta(n)$ defined by (6.1). Using the function G and the sequence $r = r(n)$ we constructed the random variables $X_{1,n}, \ldots, X_{n,n}$; $MX_{i,n} = 0$, $MX_{i,n}^2 = \sigma_{i,n}^2$, $i = 1,\ldots,n$. In this section, we have to deal with a set of $k = k(n) \equiv [(n^2/r)^{2/3}]$ variables taken one by one:

$$X_{\varkappa+1,\,n}, \ldots, X_{\varkappa+k,\,n},\ \varkappa = (j-1)k,\ j = 1, \ldots, N+1, \tag{9.1}$$

where the number $N = N(n)$ is defined by (6.6) and behaves like n/k as $n \to \infty$. Since, as a rule, n is not a multiple of k, the last group contains k' elements rather than k, where $k \leq k' \leq 2k-1$. It is clear that all of the assertions obtained in this section for a group of k elements are also valid for a group of k' elements.

For brevity, we denote the arbitrary set (9.1) by $Y_1,\ldots,Y_k$ and the sums by $V_{i,j} = Y_i + \cdots + Y_j$. Also, let

$$B^2(i,j) \equiv k^{-1}\left(\mathbf{M}Y_i^2 + \ldots + \mathbf{M}Y_j^2\right),$$

$$\mathbf{M}(i,j;x,y) = \mathbf{M}_\rho(i,j;x,y)$$

$$\equiv \mathbf{M}'\left(e^{-\rho^{\frac{1}{k}}\sum_{m=i}^{j}(V_{i,m}/\sqrt{k}+x)};\ \inf_{i\leqslant m\leqslant j}(V_{i,m}+x\sqrt{k})>0,\ \frac{V_{i,j}}{\sqrt{k}}+x=y\right).$$

As before by C we denote various constants not depending on n and κ. In the new notation, Lemma 7.2 has the following form:

L e m m a 9.1. For some $C < \infty$ not depending on n and κ, for all $x > 0$, $y > 0$, one has

$$|M(1,\ k;\ x,\ y)-M(B^2(1,\ k);\ x,\ y)|\leqslant C\ln^3 k/\sqrt{k},$$

where the function $M(t; x,y) = M_\rho(t; x,y)$ is defined in Section 4.

Proof. The families $M(i,j; x,y)$ and $M(B^2(i,j); x,y)$ have the semigroup property. Hence

$$I\equiv M(1,\ k;\ x,\ y)-M(B^2(1,\ k);\ x,\ y)$$

$$=\int_0^\infty (M(1,\ k/2;\ x,\ z)\cdot M(k/2+1,\ k;\ z,\ y)$$

$$-M(B^2(1,\ k/2);\ x,\ z)-M(B^2(k/2+1,\ k);\ z,\ y))\,dz,$$

where we use $k/2$ to denote $[k/2]$. Let

$$I_1\equiv\int_0^\infty (M(1,\ k/2;\ x,\ z)$$

$$-M(B^2(1,\ k/2);\ x,\ z))M(B^2(k/2+1,\ k);\ z,\ y)\,dz,$$

$$I_2\equiv\int_0^\infty M(1,\ k/2;\ x,\ z)(M(k/2+1,\ k;\ z,\ y)$$

$$-M(B^2(k/2+1,\ k),\ z,\ y))\,dz\ .$$

Then, obviously,

$$I = I_1 + I_2 . \tag{9.2}$$

We use integration by parts to estimate I_1:

$$I_1=-\int_0^\infty L(k;\ x,\ z)\ \frac{\partial}{\partial z}\ M(B^2(1,\ k/2);\ z,\ y)\,dz,$$

where

$$L(k;\ x,\ z)\equiv\int_0^z (M(1,\ k/2;\ x,\tilde{z})-M(B^2(1,\ k/2);\ x,\tilde{z}))\,d\tilde{z}.$$

Hence

$$|I_1| \leqslant \sup_{x,y>0} |L(k;x,y)| \int_0^\infty \left|\frac{\partial}{\partial z} M(B^2(1,k/2);z,y)\right| dz.$$

By formula (4.2) and relations (4.5) we have

$$\int_0^\infty \left|\frac{\partial}{\partial z} M(B^2(1,k/2);z,y)\right| dz \leqslant C,$$

hence

$$|I_1| \leqslant C \sup_{x,y>0} |L(k;x,y)|. \tag{9.3}$$

Here we interrupt the proof of Lemma 9.1 to prove a certain intermediate assertion. Set

$$F^+(i,j;x,y) = \int_0^y M(i,j;x,\tilde{y})\,d\tilde{y}, \quad F^+(t;x,y) = \int_0^y M(t;x,\tilde{y})\,d\tilde{y},$$

$$F^-(i,j,x,y) = \int_0^x M(i,j;\tilde{x},y)\,d\tilde{x}, \quad F^-(t;x,y) = \int_0^x M(t;\tilde{x},y)\,d\tilde{x}.$$

L e m m a 9.2. For some $C < \infty$ not depending on n and κ, for all $x > 0$, $y > 0$ we have

$$|F^{\pm}(1,k;x,y) - F^{\pm}(B^2(1,k);x,y)| \leqslant C \ln k/\sqrt{k}.$$

The proof is based on a theorem of A.I. Sakhanenko in [22], where a result obtained in [23] is extended to the case of nonidentically distributed summands.

T h e o r e m 9.1. [22]. Let $Z_1,\ldots,Z_k$ be independent random variables with zero mean, and for some numbers $s > 0$ and $L < \infty$ let the following inequality hold:

$$\mathbf{M}|Z_i|^3 \exp\{s|Z_i|\} \leqslant L\mathbf{M}Z_i^2, \quad 1 \leqslant i \leqslant k. \tag{9.4}$$

Then the random variables $Z_1,\ldots,Z_k$ can be defined on a common probability space wtih Gaussian independent random variables $T_1,\ldots,T_k$ for which $MT_i = MZ_i$, $MT_i^2 = MZ_i^2$, $i = 1,\ldots,k$, so that for all $x > 0$,

$$\mathbf{P}\left(\max_{1\leqslant i\leqslant k}\left|\sum_{j<i} Z_j - \sum_{j<i} T_j\right| > C(N/t + x)\right) \leqslant \exp\{-tx\},$$

where

$$t = \min(s,\ 1/L),\ N = 1 + \ln\left(t^2 \sum_{i=1}^{k} \mathbf{M}Z_i^2 + 1\right),$$

and C is an absolute constant.

It is clear that condition (9.4) is satisfied for the random variables $Y_1,\ldots,Y_k$ with constants s and L not depending on n and ξ. Hence, by Theorem 9.1, we assume that the random variables $Y_1,\ldots,Y_k$ are defined on a common probability space with the Gaussian random variables $W_1,\ldots,W_k$; we assume the latter to be the increments of a Wiener process:

$$W_i = w(kB^2(1,\ i)) - w(kB^2(1,\ i-1)),\ i = 1,\ \ldots,\ k,$$

where $B^2(1,0) = 0$. For $\varepsilon > 0$, set

$$\mathbf{P}(\varepsilon) = \mathbf{P}_k(\varepsilon) \equiv \mathbf{P}\left(\sup_{0 \leqslant t \leqslant kB^2(1,k)} |v_k(t) - w(t)| > \varepsilon\sqrt{k}\right),$$

where $v_k(t)$, $0 \leqq t \leqq B^2(1,k)$, is the polygon connecting the points $(kB^2(1,i),\ V_{1,i})$, $i = 0,1,\ldots,k$, where $B^2(1,0) = 0$, $V_{1,0} = 0$. Now, let $W_k(t)$, $0 \leqq t \leqq kB^2(1,k)$, be the polygon connecting the points $(kB^2(1,i),\ w(kB^2(1,k)))$, $i = 0,1,\ldots,k$. Then, obviously,

$$\begin{aligned}\sup_{0 \leqslant t \leqslant kB^2(1,k)} |v_k(t)/\sqrt{k} - w_k(t)/\sqrt{k}| &\leqslant \max_{1 \leqslant i \leqslant k} |V_{1,i}/\sqrt{k} \\ &\quad - w(k,\ B^2(1,\ i))/\sqrt{k}| \\ &\quad + \max_{1 \leqslant i \leqslant k} \sup_{kB^2(1,i-1) \leqslant t \leqslant kB^2(1,i)} |w(t) - w_k(t)|/\sqrt{k}.\end{aligned}$$

Since

$$|kB^2(i,\ i) - 1| = |kB^2(1,\ i) - kB^2(1,\ i-1) - 1| \leqslant Cr/n,$$

then

$$\mathbf{P}_k(\varepsilon) \leqslant k\mathbf{P}\left(\sup_{0 \leqslant t \leqslant 2}\left|w(t) - \frac{t}{2}w(2)\right| > \sqrt{k}\varepsilon/2\right) + \mathbf{P}_k^1(\varepsilon/2), \tag{9.5}$$

where

$$\mathbf{P}_k^1(\varepsilon) \equiv \mathbf{P}\left(\max_{1 \leqslant i \leqslant k} |V_{1,k} - w(k,\ B^2(1,\ i))|/\sqrt{k} > \varepsilon\right).$$

It is easy to see now that

$$\begin{aligned}e^{-2\rho\varepsilon}F^+(B^2(1,\ k);\ x-\varepsilon,\ y-2\varepsilon) - \mathbf{P}_k(\varepsilon) &\leqslant F^+(1,\ k;\ x,\ y) \\ &\leqslant e^{2\rho\varepsilon}F^+(B^2(1,\ k);\ x+\varepsilon,\ y+2\varepsilon) + \mathbf{P}_k(\varepsilon).\end{aligned} \tag{9.6}$$

In (9.6) we take the number $\varepsilon = C(N/t + \ln k/t)/\sqrt{k}$, where the C, t and N are from the formulation of Theorem 9.1. By (9.5) and Theorem 9.1, we get $P_k(\varepsilon) \leqq C\exp\{-\ln k\} = C/k$. Next, since

$$|1 - \exp\{\pm 2\rho\varepsilon\}| \leqslant C \ln k/\sqrt{k}$$

and

$$\sup_{x>0,y>0} |(\partial^2/\partial x\partial y)F^+(B^2(1, k); x, y)| \leqslant C$$

(this follows from (4.2) and (4.5)), one has

$$|F^+(B^2(1, k); x \pm \varepsilon, y \pm 2\varepsilon) - F^+(B^2(1, k); x, y)| \leqslant C \ln k/\sqrt{k},$$

and the part of Lemma 9.2 related to the functions F^+ is proved. Since the proof for the F^- is exactly the same, Lemma 9.2 is proved.

We return to proving Lemma 9.1 that we have interrupted at formula (9.3). By Lemma 9.2, we have

$$|L(k; x, y)| \leqslant C \ln k/\sqrt{k},$$

and therefore we have derived the estimate

$$|I_1| \leqslant C \ln k/\sqrt{k}. \tag{9.7}$$

We proceed to estimate the integral I_2 in formula (9.2). For brevity, set

$$A(k; z, y) = M(k/2 + 1, k; z, y) - M(B^2(k/2 + 1, k); z, y),$$

so that

$$I_2 = \int_0^\infty M(1, k/2; x, z)\, A(k; z, y)\, dz.$$

On trajectories of the walk

$$x,\ V_{1,1}/\sqrt{k} + x,\ V_{1,2}/\sqrt{k} + x,\ \ldots,\ V_{1,k}/\sqrt{k} + x$$

we introduce the fucntional

$$\eta = \eta(x) \equiv \inf\{i \geqslant 0 : V_{1,i}/\sqrt{k} + x \geqslant a/\sqrt{\ln k}\}$$

which is the first instant of reaching the level $a/\sqrt{\ln k}$.

Then

$$I_2 = I_{2,1} + I_{2,2}, \tag{9.8}$$

where

$$I_{2,1} \equiv \int_{z=0}^{\infty} \int_{t=a/\sqrt{\ln k}}^{\infty} \sum_{i=0}^{k/4} \mathrm{M}'\left(e^{-\frac{\rho}{k}\sum_{j=1}^{i}(V_{1,j}/\sqrt{k}+x)} \right. ;$$

$$\inf_{1\leq j\leq i}(V_{1,j}/\sqrt{k}+x)>0, \qquad \eta=i,$$

$$\left.V_{1,i}/\sqrt{k}+x=t\right) M(i+1,\ k/2;\ t,\ z)\, A(k;\ z,\ y)\, dz,\ dt;$$

$$I_{2,2}\equiv\int_0^\infty \mathbf{M}'\left(e^{-\frac{\rho}{k}\sum_{j=1}^{k/2}(V_{1,j}/\sqrt{k}+x)};\right.$$

$$\left.\inf_{1\leq j\leq k/2}(V_{1,j}/\sqrt{k}+x)>0,\quad V_{1,k/2}/\sqrt{k}+x=z,\ \eta>k/4\right) A(k;\ z,\ y)\, dz.$$

Now, we estimate the integral $I_{2,2}$; we have

$$|I_{2,2}|\leq \mathbf{P}\left(\inf_{1\leq i\leq k/4}(V_{1,i}/\sqrt{k}+x)>0,\ \eta>k/4\right)$$

$$\times\sup_{x,y>0} M(k/4+1,\ k/2;\ x,\ y)\int_0^\infty |A(k;\ z,\ y)|\, dz.$$

By Lemma 7.1,

$$\sup_{x,y>0}|M(k/2+1,\ k/2;\ x,\ y)|\leq C;$$

furthermore, it is obvious that

$$\int_0^\infty |A(k;\ z,\ y)|\, dz\leq\int_0^\infty M(k/2+1,\ k;\ z,\ y)dy$$

$$+\int_0^\infty M(B^2(k/2+1,\ k);\ z,\ y)\, dy\leq C,$$

and hence

$$|I_{2,2}|\leq C\mathbf{P}\left(\inf_{1\leq i\leq k/4}(V_{1,i}/\sqrt{k}+x)>0,\ \sup_{1\leq i\leq k/4}(V_{1,i}/\sqrt{k}+x)\leq a/\sqrt{\ln k}\right).$$

It is known (see, for example [24]) that for $0 < \varepsilon \leqq 1$, we have the inequality

$$\mathbf{P}\left(\inf_{0\leq t\leq 1}(w(t)+x)>0,\ \sup_{0\leq t\leq 1}(w(t)+x)<\varepsilon\right)\leq Ce^{-\frac{\pi^2}{4\varepsilon^2}};$$

using this inequality and the estimate of the convergence rate in the invariance principle for a strip [25], we are in a position to choose an $a > 0$ not depending on n and κ, such that one has

$$|I_{2,2}| \leqq C/\sqrt{k}\ . \tag{9.9}$$

Now let us turn to the summand $I_{2,2}$ in (9.8), in which the $a > 0$ is fixed in accordance with (9.9). For brevity, set

$$c_i(x, t) \equiv \mathbf{M}\left(e^{-\frac{\rho}{k}\sum_{j=1}^{i}(V_{1,j}/\sqrt{k}+x)};\right.$$

$$\left.\inf_{1\leqslant j\leqslant i}(V_{1,j}/\sqrt{k}+x)>0,\quad \eta=i,\ V_{1,i}/\sqrt{k}+x=t\right),$$

so that

$$I_{2,1}=\int_{z=0}^{\infty}\int_{t=a/\sqrt{\ln k}}^{\infty}\sum_{i=0}^{k/4}c_i(x, t)\,M(i+1, k/2;\, t, z)\,A(k;\, z, y)\,dzdt.$$

Let $m = m(k) \equiv [k/\ln^3 k]$; in addition, set

$$N^+(i, j;\, x, y)\equiv\mathbf{M}'\left(e^{-\frac{\rho}{k}\sum_{p=i}^{j}(V_{i,p}/\sqrt{k}+x)};\quad V_{i,j}/\sqrt{k}+x=y\right),$$

$$N^-(i, j;\, x, y)\equiv N^+(i, j;\, x, y)-M(i, j;\, x, y).$$

Then we have the relation

$$I_{2,1} = I^+_{2,1} + I^-_{2,1}, \tag{9.10}$$

where

$$I^{\pm}_{2,1}\equiv\int_{z=0}^{\infty}\int_{u=0}^{\infty}\int_{t=a/\sqrt{\ln k}}^{\infty}\sum_{i=0}^{k/4}c_i(x, t)\,N^{\pm}(i+1, i+m;\, t, u)$$

$$\times\, M(i+m+1, k/2;\, u, z)\,A(k;\, z, y)\,dzdudt.$$

L e m m a 9.3. For $t \geqq a/\sqrt{\ln k}$, one has

$$\sup_{0\leqslant i\leqslant k/4}\int_0^{\infty}N^-(i+1, i+m;\, t, u)\,du\leqslant C\ln^{5/2}k/\sqrt{k}.$$

Proof. Let

$$A^{(m)}(t)\equiv\int_0^{\infty}N^-(i+1, i+m;\, t, u)\,du\,.$$

Since

$$A^{(m)}(t)=\mathbf{M}\left(e^{-\rho\left(\frac{m}{k}\right)^{3/2}\frac{1}{m}\sum_{j=i+1}^{i+m}\left(\frac{V_{i+1,j}}{\sqrt{m}}+t\frac{\sqrt{k}}{\sqrt{m}}\right)};\ \inf_{i+1\leqslant j\leqslant i+m}\left(\frac{V_{i+1,j}}{\sqrt{m}}+t\frac{\sqrt{k}}{\sqrt{m}}\right)<0\right),$$

then arguing just as in the proof of Lemma 9.2, we can show that

$$|A^{(m)}(t) - A(t)| \leqslant C\frac{\ln m}{\sqrt{m}} = C\frac{\ln^{5/2}k}{\sqrt{k}}, \tag{9.11}$$

where

$$A(t) \equiv \mathbf{M}\left(e^{-\rho\left(\frac{m}{k}\right)^{3/2}\int_0^{B^2}\left(w(u)+t\frac{\sqrt{k}}{\sqrt{m}}\right)du};\ \inf_{0\leqslant u\leqslant B^2}\left(w(u)+t\frac{\sqrt{k}}{\sqrt{m}}\right)<0\right),$$

$$B^2 \equiv B^2(i+1,\ i+m)\frac{k}{m} \leqslant C.$$

Next, by the Cauchy inequality,

$$A(t) \leqslant \mathbf{M}^{1/2}\left(e^{-2\rho\left(\frac{m}{k}\right)^{3/2}\int_0^{B^2}\left(w(u)+t\frac{\sqrt{k}}{\sqrt{m}}\right)du}\right)$$

$$\times\, \mathbf{P}^{1/2}\left(\inf_{0\leqslant u\leqslant B^2}\left(w(u)+t\sqrt{\frac{k}{m}}\right)<0\right) \leqslant C\mathbf{P}^{1/2}\left(\inf_{0\leqslant u\leqslant B^2}\left(w(u)+t\sqrt{\frac{k}{m}}\right)<0\right).$$

Setting t equal to the smallest value of $a/(\ln k)^{\frac{1}{2}}$, for $t \geqq a/(\ln k)^{\frac{1}{2}}$ we obtain

$$A(t) \leqslant C\mathbf{P}^{1/2}\left(\inf_{0\leqslant u\leqslant c} w(u) < -a\ln k\right).$$

Since (see, for example [1])

$$\mathbf{P}\left(\inf_{0\leqslant u\leqslant c} w(u) < -a\ln k\right) \leqslant Ce^{-\frac{a^2}{4}\ln^2 k},$$

for $t \geqq a/(\ln k)^{\frac{1}{2}}$ we have

$$A(t) \leqq C/k,$$

which together with (9.11) prove Lemma 9.3.

Now we return to formula (9.10). By Lemma 9.3,

$$|I_{2,1}^{-}| \leqslant C\ln^{5/2}k/\sqrt{k}, \tag{9.12}$$

hence it remains to estimate $I_{2,1}^{+}$. Let $\phi(\sigma^2,t) \equiv (\sqrt{2\pi}\,\sigma)^{-1}\exp\{-t^2/(2\sigma^2)\}$ be the density of normal distribution with parameters $(0,\sigma^2)$.

L e m m a 9.4. For certain numbers $a = a(x,k)$ and $b = b(x,k)$ satisfying the conditions

$$0 < (1+C)^{-1} \leqslant a,\ b \leqslant (1+C) < \infty,$$

for all $0 \leqq i \leqq k/4$, $x > 0$, $y > 0$, we have

$$\left| N^{+}(i+1,\, i+m;\, x,\, y) - a\varphi\left(b;\, \frac{\sqrt{k}}{\sqrt{m}}(x-y)\right)\right| \leqslant C\frac{\ln^3 k}{\sqrt{k}}\, e^{c\frac{r^3}{n^3}(x+y)},$$

where, as earlier, the constants c, $C < \infty$ do not depend on n and κ.

The proof of Lemma 9.4 follows the arguments used in deriving estimate (7.2) (see the proof of Lemma 7.1), based on the passage to identically distributed random variables and a subsequent application of Theorem 7.1; hence we omit it.

Let us go back to formula (9.10) and estimate the $I^{+}_{2,1}$. We have

$$|I^{+}_{2,1}| \leq J^{+} + J^{-} , \tag{9.13}$$

where

$$J \equiv \int\limits_{z=0}^{\infty}\int\limits_{u=0}^{\infty}\int\limits_{t=a/\sqrt{\ln k}}^{\infty}\sum_{i=0}^{k/4} c_i(x,\, t)\, L^{\pm}(t,\, u)$$

$$\times M(i+m+1,\, k/2;\, u,\, z)\, A(k;\, z,\, y)\, dzdudt,$$

$$L^{+}(t,\, u) \equiv a\phi(b;\, (t-u)\sqrt{k}/\sqrt{m}) ,$$
$$L^{-}(t,\, u) \equiv |L^{+}(i,\, u) - N^{+}(i+1,\, i+m;\, t,\, u)| .$$

By Lemmas 9.4 and 7.1,

$$|J^{-}| \leq C \ln^3 k/\sqrt{k} , \tag{9.14}$$

hence it remains to estimate J^{+}. Integrating by parts over u yields

$$J^{+} = -\int\limits_{z=0}^{\infty}\int\limits_{u=0}^{\infty}\int\limits_{t=a/\sqrt{\ln k}}^{\infty}\sum_{i=0}^{k/4} c_i(x,\, t)\frac{\partial}{\partial u} L^{+}(t,\, u)$$

$$\times F^{-}(i+m+1,\, k/2;\, u,\, z)\, A(k;\, z,\, y)\, dzdtdu,$$

where the function $F^{-}(i,j;\, x,y)$ is defined before Lemma 9.2. Next, we integrate by parts over z:

$$J^{+} = \int\limits_{z=0}^{\infty}\int\limits_{u=0}^{\infty}\int\limits_{t=a/\sqrt{\ln k}}^{\infty}\sum_{i=0}^{k/4} c_i(x,\, t)\frac{\partial}{\partial u} L^{+}(t,\, u)$$

$$\times \frac{\partial}{\partial z} F^{-}(i+m+1,\, k/2;\, u,\, z)\, B(k;\, z,\, y)\, dzdudt, \tag{9.15}$$

where

$$B(k;\, z,\, y) \equiv \int\limits_{0}^{z} A(k;\, \tilde{z},\, y)\, d\tilde{z}.$$

Lemma 9.5. For $0 \leqq i \leqq k/4$,

$$\sup_{u\geqslant 0}\int_0^\infty \left|\frac{\partial}{\partial z}F^-(i+m+1,\,k/2;\,u,\,z)\right|dz\leqslant C.$$

Proof. Set

$$d=i+m,\ k/2-d=p,$$

$$\eta_1=-\frac{\rho}{k}\sum_{j=d+1}^{k/2}V_{d+1,j}/\sqrt{k},\qquad \eta_2=\inf_{d+1<j<k/2}V_{d+1,j}/\sqrt{k},$$

$$\eta_3=V_{d+1,k/2}/\sqrt{k},$$

so that

$$M(d+1,\,k/2;\,u,\,z)=\mathbf{M}'\left(e^{\eta_1-\rho\frac{p}{k}u};\ \eta_2+u>0,\ \eta_3+u=z\right).$$

For any $\delta > 0$, we have

$$\left|\int_0^u \mathbf{M}'\left(e^{\eta_1-\rho\frac{p}{k}\tilde{u}};\ \eta_2+\tilde{u}>0,\ \eta_3+\tilde{u}=z\right)d\tilde{u}-\right.$$

$$\left.-\int_0^u \mathbf{M}'\left(e^{\eta_1-\rho\frac{p}{k}\tilde{u}};\ \eta_2+\tilde{u}>0,\ \eta_3+\tilde{u}=z+\delta\right)d\tilde{u}\right|$$

$$\leqslant\left|\int_{-\delta}^{0}\mathbf{M}'\left(e^{\eta_1-(\tilde{u}+\delta)\rho\frac{p}{k}};\ \eta_2+\tilde{u}+\delta>0,\ \eta_3+\tilde{u}=z\right)d\tilde{u}\right|$$

$$+\left|\int_{u-\delta}^{u}\mathbf{M}'\left(e^{\eta_1-(\tilde{u}+\delta)\rho\frac{p}{k}};\ \eta_2+\tilde{u}+\delta>0,\ \eta_3+\tilde{u}=z\right)d\tilde{u}\right|$$

$$+\left|\int_0^u \mathbf{M}'\left(e^{\eta_1-\rho\frac{p}{k}\tilde{u}};\ \eta_2+\tilde{u}>0,\ \eta_3+\tilde{u}=z\right)d\tilde{u}\right.$$

$$\left.-\int_0^u \mathbf{M}'\left(e^{\eta_1-\rho\frac{p}{k}\tilde{u}-\rho\delta\frac{p}{k}};\ \eta_2+\tilde{u}>0,\ \eta_3+\tilde{u}=z\right)d\tilde{u}\right|$$

$$+\left|\int_0^u \mathbf{M}'\left(e^{\eta_1-\rho\frac{p}{k}(\tilde{u}+\delta)};\ \eta_2+\tilde{u}>0,\ \eta_3+\tilde{u}=z\right)d\tilde{u}-\right.$$

$$\left.-\int_0^u \mathbf{M}'\left(e^{\eta_1-\rho\frac{p}{k}(\tilde{u}+\delta)};\ \eta_2+\tilde{u}+\delta>0,\ \eta_3+\tilde{u}=z\right)d\tilde{u}\right|.$$

Dividing the inequality by δ and letting δ tend to zero, yields

$$\left|\frac{\partial}{\partial z}\int_0^u M\left(d+1, k/2; \tilde{u}, z\right) d\tilde{u}\right| \leqslant \mathbf{P}'(\eta_3 = z) + \mathbf{P}'(\eta_3 + u = z)$$

$$+ \rho\frac{p}{k}\int_0^u \mathbf{M}'\left(e^{\eta_1 - \frac{p}{k}\tilde{u}};\ \eta_2 + \tilde{u} > 0,\ \eta_3 + \tilde{u} = z\right) d\tilde{u}$$

$$+ \mathbf{P}'(\eta_2 \geqslant -u;\ \eta_3 - \eta_2 = z).$$

Integrating the resulting inequality over z, we obtain

$$\int_0^\infty \left|\frac{\partial}{\partial z} F^-(d+1, k/2; u, z)\right| dz \leqslant 1 + 1$$

$$+ \rho\frac{p}{k}\int_0^u \mathbf{M}'\left(e^{\eta_1 - \rho\frac{p}{k}\tilde{u}};\ \eta_2 + \tilde{u} > 0\right) d\tilde{u} + \mathbf{P}(\eta_2 \geqslant -u)$$

$$\leqslant 1 + 1 + C + 1 = C;$$

the third summand in the middle part of the last inequality is estimated by the assertion of Lemma 7.1. Lemma 9.5 is proved.

Returning to formula (9.15), by Lemma 9.2 we have that

$$|B(k; z, y)| \leqslant C \ln k/\sqrt{k};$$

by Lemma 9.5 that

$$|J^+| \leqslant C \int_{u=0}^\infty \int_{t=a/\sqrt{\ln k}}^\infty \sum_{i=0}^{k/4} c_i(x, t)\frac{\partial}{\partial u} L^+(t, u)\, dt du \times \ln k/\sqrt{k};$$

Since

$$\int_0^\infty \left|\frac{\partial}{\partial u} L^+(t, u)\right| du \leqslant C\sqrt{k}/\sqrt{m} = C \ln^{3/2} k,$$

we obtain

$$|J^+| \leqq C \ln^{5/2} k/\sqrt{k} \ . \tag{9.16}$$

Combining formulas (9.2), (9.8), (9.10), (9.12) yields

$$|I| \leqslant |I_1| + |I_{2,2}| + |I_{2,1}^-| + |I^+| + |I^-|;$$

Hence by (9.7), (9.9), (9.12), (9.14), (9.16),

$$|I| \leqq C \ln^3 k/\sqrt{k} \ .$$

Lemma 9.1 is proved.

REFERENCES

[1] Borovkov, A.A. "Boundary-Value Problems for Random Walks and Large Deviations in Function Spaces." *Theory Probab. Applications*, vol.12, no.4 (1967): 575-95. (English transl.)

[2] Kolmogorov, A.N. "Eine Verallgemeinrung des Laplace-Liapounoffschen Satzes." *Izvestiya Akademii Nauk SSSR, Otd. matematich. i estestv. nauk* 15 (1931): 959-62.

[3] Borovkov, A.A., and Korolyuk, V.S. "On the Results of Asymptotic Analysis in Problems with Boundaries." *Theory Probab. Applications*, vol.10, no.2 (1965): 236-46. (English transl.)

[4] Groenboom, P., Oosterhoff, J., and Ruymgart, F.H. "Large Deviations Theorems for Empirical Probability Measures." *The Annals of Probability*, vol.7, no.4 (1979): 553-86.

[5] Bahadur, R.R., and Zabell, S.L. "Large Deviations of the Sample Mean in General Vector Spaces." *The Annals of Probability*, vol.7, no.4 (1979): 587-621.

[6] Borovkov, A.A., and Mogul'skii, A.A. "Probabilities of Large Deviations in Topological Spaces. I, II." *Siberian Math. Journal*, vol.19, no.5 (1978): 697-709; vol.21, no.5 (1980): 653-64. (English transl.)

[7] Borovkov, A.A. "Rate of Convergence and Large Deviations in Invariance Principle." *Proceedings of the International Congress of Mathematicians*, Helsinki, 1978, pp. 725-31 (vol.2). Helsinki: Academia Scientiarum Technica, 1980.

[8] ______. "New Limit Theorems in Boundary-Value Problems for Sums of Independent Summands" (in Russian). *Sibirskij matematich. zhurnal*, vol.3, no.5 (1962): 645-95.

[9] ______. "Some Results of Analysis of Large Deviations in Boundary Problems." *Soviet Math. Doklady*, vol.4, no.2 (1963): 959-62. (English transl.)

[10] ______. "Analysis of Large Deviations in Boundary-Value Problems with Arbitrary Boundaries. I, II" (in Russian). *Sibirskij matematich. zhurnal*, vol.5, no.2 (1964): 253-89; vol.5, no.4 (1964): 750-67.

[11] Petrov, V.V. *Sums of Independent Random Variables*. New York Berlin Heidelberg Tokyo: Springer-Verlag, 1975. (English transl.)

[12] Borovkov, A.A., and Rogozin, B.A. "On the Multidimensional Central Limit Theorem." *Theory Probab. Applications*, vol.10, no.1 (1965): 55-62. (English transl.)

[13] Mogul'skii, A.A. "On the Large Deviation Probabilities for a Wiener Process." *Theory Probab. Applications*, vol.25, no.3 (1980): 654-55. (English transl.)

[14] ______. "Large Deviations for a Wiener Process" (in Russian). In *Tezisy dokladov XII-ogo Evropejskogo soveshchaniya statistikov* (Abstracts of the Twelfth European Conference of Statisticians), p. 164. Varna, 1979.

[15] ______. "Large Deviations for a Wiener Process." In *Predel'nye teoremy i smezhnye voprosy* (Limit Theorems and Related Problems), pp. 25-49. Novosibirsk: Nauka, 1980.

[16] Mogul'skij, A.A. "Large Deviations for Trajectories of Random Walks." In *Tezisy dokladov Tret'ej Vil'nyuskoj konferentsii po teorii veroyatnostej* (Abstracts of the Third Vilnius Conference on Probability Theory), pp. 214-15. Vilnius, 1981.

[17] ______. *Veroyatnosti bol'shikh uklonenij dlya sluchajnykh bluzhdanij* (Probabilities of Large Deviations for Random Walks). Novosibirsk: Izd-vo Instituta Matematiki Sibirskogo Otdeleniya Akademii Nauk SSSR, 1980.

[18] Cramér, H. "On a New Limit Theorem of Probability Theory" (in Russian). *Uspekhi matematich. nauk* 10 (1944): 166-84.

[19] Borovkov, A.A. *Teoriya veroyatnostej* (Probability Theory). Moscow: Nauka, 1976.

[20] Yakovleva, G.D. *Tablitsy funktsij Ehjri i ikh proizvodnykh* (Tables of Airy Functions and Their Derivatives). Moscow: Nauka, 1969.

[21] Kolmogorov, A.N., and Fomin, S.V. *Elements of the Theory of Functions and Functional Analysis.* 2 vol. Rochester, New York: Graylock Press, 1957-61. (English transl.)

[22] Sakhanenko, A.I. "Unimprovable Estimates of the Rate of Convergence in the Invariance Principle" (in Russian). In *Tezisy dokladov kollokviuma po neparametrich. statistich. otsenivaniyu* (Abstracts of the Seminar on Nonparameter Statist. Estimation), p. 141. Budapest, 1980.

[23] Komlós, J., Major, P., and Tusnády, G. "An Approximation of Partial Sums of Independent RV'-s, and the Sample DF. II." *Z. Wahrscheinlichkeitstheorie verw. Gebiete*, B.34, H.1 (1976): 33-58.

[24] Mogul'skii, A.A. "Small Deviations in a Space of Trajectories." *Theory Probab. Applications*, vol.19, no.4 (1974): 726-36. (English transl.)

[25] Nagaev, S.V. "On the Speed of Convergence in a Boundary Problem. I." *Theory Probab. Applications*, vol.15, no.2 (1970): 163-86. (English transl.)

Part 2

Limit Theorems for Random Processes of Particular Types

I.S. Borisov

ON THE CONVERGENCE RATE IN THE CENTRAL LIMIT THEOREM

1. Introduction. Formulation of the main results

Let $\{X_k\}$ be a sequence of independent identically distributed random variables with values in an arbitrary measurable space $(\mathcal{X},\mathcal{A})$ and distribution $P(\cdot)$ on $\mathcal{A}$. Let $P_n(\cdot)$ denote the empirical distribution constructed from the sample $X_1,X_2,\ldots,X_n$:

$$P_n(A) = n^{-1}\sum_{i=1}^{n} I_{X_i}(A),$$

where $I_{X_i}(A) = 1$ if $X_i \in A$ and $I_{X_i}(A) = 0$ otherwise.

Consider the sequence of (so-called) empirical measures:

$$Q_n(\cdot) = n^{1/2}(P_n(\cdot) - P(\cdot)), \quad n = 1, 2, \ldots, \tag{1}$$

which are convenient to interpret as random processes given on the parametric set $\mathcal{A}$. By virtue of the multivariate central limit theorem, one can assert that the finite-dimensional distributions $Q_n(\cdot)$ converge weakly as $n \to \infty$ to the corresponding finite-dimensional distributions of a Gaussian random process $G(A)$, $A \in \mathcal{A}$, with zero mean and covariance

$$\mathbf{E}G(A)G(B) = P(A \cap B) - P(A)P(B). \tag{2}$$

Of considerable interest is a stronger form of convergence of Q_n to G, which implies a weak convergence of the distributions for a sufficiently broad class of functionals of the processes considered.

Let $\mathcal{B}$ be an arbitrary subclass of sets in $\mathcal{A}$. We say that for the sequence $\{Q_n\}$ the central limit theorem holds uniformly in $\mathcal{B}$, if on some probability space it is possible to construct random processes $\{\tilde{Q}_n\}$ as well as a Gaussian process G such that

$$\widetilde{Q}_n \overset{\mathcal{D}}{=} Q_n, \quad n = 1, 2, \ldots, \tag{3}$$

and, furthermore, as $n \to \infty$,

$$\sup_{A \in \mathcal{B}} | \widetilde{Q}_n (A) - G(A) | \to 0 \tag{4}$$

with probability 1, where "$\mathcal{D}$" in (3) and in what follows denotes equality of finite-dimensional distributions.

Observe that in this formulation, it is not required that the supremum in (4) be a measurable mapping, since convergence with probability 1 can be defined for sequences of arbitrary functions of an elementary outcome.

REMARK. A similar definition is given in [1]. However, the notation "$\mathcal{D}$" used therein means the coincidence of distributions on the σ-algebra generated by balls in the uniform topology of some function space. It is necessary, of course, to require that the processes Q_n be measurable with respect to the σ-algebra. Note also that under sufficiently broad assumptions concerning $\mathcal{B}$ and P, these two definitions are equivalent (also see the remark to Theorem 2).

Set

$$d(A, B) = P(A \backslash B) + P(B \backslash A).$$

The pair $(\mathcal{A}, d)$ can be viewed as a metric space with pseudometric $d(\cdot)$. It will be assumed throughout that $\mathcal{B}$ is a totally bounded set in $(\mathcal{A}, d)$.

Let $N_1(\varepsilon) = N_1(\varepsilon, \mathcal{B}, P)$ denote the smallest number of elements in the so-called ε-net with embedding (see [2]) for $\mathcal{B}$, i.e., $N_1(\varepsilon)$ is the smallest natural number for which we can find sets $A_1, A_2, \ldots, A_{N_1(\varepsilon)} \in \mathcal{A}$, such that for any $B \in \mathcal{B}$ one can choose a pair of sets A_{i_1} and A_{i_2}, $i_1, i_2 \leqq N_1(\varepsilon)$, with the following properties:

$$A_{i_1} \subseteq B \subseteq A_{i_2}, \quad d(A_{i_1}, A_{i_2}) \leqslant \varepsilon.$$

Following P. Dudley ([2]), we call $H_1(\varepsilon) = \log N_1(\varepsilon)$ the metric entropy.

It is shown in [2] that for the central limit theorem to hold uniformly in $\mathcal{B}$ it suffices that

$$\sum_{k=1}^{\infty} [2^{-k} H_1(2^{-k})]^{1/2} < \infty. \tag{5}$$

It is also established in [2] that condition (5) is generally not a necessary one. However, in [3] an example illustrates that relation (5) is nevertheless necessary for the central limit theorem to hold uniformly in $\mathcal{B}$. In other words, condition (5), in some sense, is unimprovable.

Our objective is to estimate the convergence rate in (4) when (5) is satisfied.

T h e o r e m 1. Let condition (5) be satisfied. Then on some probability space one can construct d-separable random processes $\tilde{Q}_n(\cdot)$, $n = 1,2,\ldots,$ and $G(\cdot)$, satisfying for any $x \geqq 0$, $r \geqq 0$, $M = 0,1,\ldots$ and $y \geqq 2r + 5$ the inequality

$$\mathbf{P}\Big\{\sup_{A\in\mathcal{B}} |\tilde{Q}_n(A) - G(A)| \geqslant Cn^{-1/2}\gamma^*(2^{-M}, \mathcal{B}, P)(\log n + x)$$
$$+ (4y+1)\Big(\sum_{k>M} [2^{-k} H_1(2^{-k})]^{1/2} + 5\cdot 2^{-M/2}\Big)\Big\} \leqslant Ke^{-\lambda x} + N_1(2^{-M})^{1-r}$$
$$\times [R_1(y,r) + I(n,M,H_1) R_2(y,r)] + N_1(2^{-M})^3 (1 - I(n,M,H_1))$$
$$\times \exp\Big\{-yn^{1/2} \sum_{k>M} [2^{-k} H_1(2^{-k})]^{1/2}\Big\}, \tag{6}$$

where

$$I(n,M,H_1) = \begin{cases} 1 & \text{if } \max(1, H_1(2^{-M-1})) < n2^{-M+1}, \\ 0 & \text{otherwise.} \end{cases}$$

$$R_1(y,r) = 3\left[\exp\left\{\frac{(y-r-2)^2 - (r+3)^2 + 1}{2(1+y)}\right\} - 1\right]^{-1};$$

$$R_2(y,r) = (2\pi)^{-1/2} \int_y^{\infty} [\exp\{(t^2-5)/2 - r\} - 1]^{-1} dt;$$

C, K, λ being absolute positive constants; $\gamma^*(\cdot)$ being a functional (for more detail, see Section 2) satisfying the inequality

$$\gamma^*(\varepsilon, \mathcal{B}, P) \leqslant N_1(\varepsilon). \tag{7}$$

REMARK 1. Separable random processes with respect to the pseudometric $d(\cdot)$ possess all the properties of usual separable (with respect to metrics) processes (see [4]). In particular, the supremum in (6) is a random variable.

REMARK 2. Relation (6) continues to hold if the function $N_1(\cdot)$ is replaced by any of its majorants for which (5) holds. This modification is often more convenient for obtaining estimates of the convergence rate in (4).

Theorem 1 can be generalized as follows.

T h e o r e m 2. Let condition (5) be satisfied. Then on some probability space one can construct the sample $X_{1n}, X_{2n}, \ldots, X_{nn}$, $n = 1,2,\ldots,$ coinciding in distribution with the sample $X_1, X_2, \ldots, X_n$ for each n and also a d-separable Gaussian process $G(\cdot)$, for which relation (6) holds if the supremum is replaced by some random variable Λ_n satisfying the inequality

$$\sup_{A\in\mathscr{B}} |\tilde{Q}_n(A) - G(A)| \leqslant \Lambda_n$$

with probability 1, where the random process $\tilde{Q}_n(\cdot)$ is defined in (1) for the sample $X_{1n}, \ldots, X_{nn}$.

REMARK. The processes $Q_n(\cdot)$ and $\tilde{Q}_n(\cdot)$ defined in (1) need no longer be d-separable. Hence the supremum in (4) need not be a random variable.

If $\mathcal{B}$ is "empirically measurable for P" (see [1]), then the notation "$\mathcal{D}$" as defined in the central limit theorem may be interpreted just as in [1], i.e., as the coincidence of probability distributions of the $Q_n(\cdot)$ and $\tilde{Q}_n(\cdot)$ on the σ-algebra generated by balls in the uniform topology of some function space. However, as Theorem 2 demonstrates, proving the central limit theorem when (5) is satisfied requires no measurability conditions. In that case, the class of functionals of the $Q_n(\cdot)$ converging in distribution to the corresponding functionals of the $G(\cdot)$ becomes broader.

In (6), set $r = 1$, $x = 2\lambda^{-1}\log n$, $y = 3\log n$. Then the right side of (6) will obviously have order of smallness $O(n^{-2})$. Furthermore, let

$$M \equiv M(n) = \max\left\{M \geqslant 0\colon \sum_{k>M} [2^{-k}H_1(2^{-k})]^{1/2} \geqslant N_1(2^{-M})\, n^{-1/2}\right\}.$$

It follows from the Borel-Cantelli lemma, that on the corresponding probability space, as $n \to \infty$,

$$\sup_{A\in\mathscr{B}} |\tilde{Q}_n(A) - G(A)| = O\left(\log n \sum_{k>M(n)} [2^{-k}H_1(2^{-k})]^{1/2}\right) \qquad (8)$$

with probability 1, the constant in the definition of the $O(\cdot)$ being absolute.

We cite another estimate, which in some cases is sharper than (8). In (6), we set

$$x = 2\lambda^{-1}\log n,\ y = 6 + 2r,\ r = 1 + 2\log n/\log N_1(2^{-M}),$$

and as $M = \tilde{M}(n)$ we take the smallest nonnegative integer that minimizes the expression

$$\Delta_n(M) = \max\left\{\left(1 + \frac{\log n}{H_1(2^{-M})}\right)\sum_{k>M}[2^{-k}H_1(2^{-k})]^{1/2},\ \frac{\gamma^*(2^{-M}, \mathcal{B}, P)\log n}{\sqrt{n}}\right\}$$

on the set

$$\{M : \max(1,\quad H_1(2^{-M-1})) < n2^{1-M}\}.$$

It is not hard to see that in this case (6) yields the inequality

$$\mathbf{P}\left\{\sup_{A\in\mathcal{B}}|\tilde{Q}_n(A) - G(A)| > C_0\Delta_n(\tilde{M}(n))\right\} \leqslant Cn^{-2}, \tag{9}$$

where C_0, C are absolute positive constants. As in (8), one can derive an estimate of the convergence rate in (4).

As an example we consider the case where $X = R^m$, A is the σ-algebra of Borel subsets in R^m, $\mathcal{B} = \{\{z \in R^m: z < t\};\ t \in R^m\}$ is the set of open "octants" in R^m (the relationship $z < t$ is understood coordinatewise). In other words, in this example we mean the approximation of empirical distribution functions in R^m). It is shown in [5] that for any distribution P in R^m

$$H_1(2^{-k}) \leqslant \hat{H}_1(2^{-k}) \equiv mk\log 2,\ \gamma^*(2^{-M}, \mathcal{B}, P) \leqslant 2^{(m-1)M}. \tag{10}$$

Noting Remark 2 to Theorem 1, one can substitute the function $\hat{H}_1(\cdot)$ for $H_1(\cdot)$ in the expression for $\Delta_n(M)$.

Set

$$M \equiv \hat{M}(n) = [(2m-1)^{-1}(\log_2 n - \log_2\log_2 n)] - 1,$$

where $[\cdot]$ is the integer part of a number. If $n \geqq 2^{3(2m-1)}$, then $\hat{M}(n) > 0$. In addition, it is easy to see that $m(M+1) < n2^{1-M}$. Hence

$$\inf_{\substack{M=0,1,\ldots\\ \max(1,\hat{H}_1(2^{-M-1}))<n2^{1-M}}} \Delta_n(M) \leqslant \Delta_n(\hat{M}(n)).$$

It follows now from (10) that for $n \geq 2^{3(2m-1)}$ one has

$$\hat{H}^{-1}(2^{-M}) \sum_{k>M} [2^{-k}\hat{H}_1(2^{-k})]^{1/2} \leqslant C_1 (mM2^M)^{-1/2} \leqslant C_2 n^{-\frac{1}{2(2m-1)}},$$

$$n^{-1/2}\gamma^*(2^{-M}, \mathscr{B}, P) \leqslant C_3 n^{-\frac{1}{2(2m-1)}},$$

where C_1, C_2, C_3 are absolute positive constants. The resulting inequalities and (9) finally imply that for $n \geq 2^{3(2m-1)}$,

$$\mathbf{P}\left(\sup_{A\in\mathscr{B}} |\tilde{Q}_n(A) - G(A)| > C_0 \max(C_2, C_3)\, n^{-\frac{1}{2(2m-1)}} \log n \right) \leqslant Cn^{-2}. \tag{11}$$

Relation (11) strengthens one of the results in [6] as well as the result obtained by this author in [5] (in the sense of an explicit dependence of the constants on the dimension of R^m).

As a corollary of Theorem 2 one can obtain power estimates of the rate of convergence in (4), in essence, for any class $\mathcal{B}$ which is embedded in a finite-dimensional Euclidean space (for example, for the set of all balls or ellipsoids in R^m, and so on).

Let us consider one more example -- to elaborate, in particular, the arguments of Section 3 concerning the unimprovability of the obtained estimates of the convergence rate in (4). Let $P = \{p_i\}$ be an arbitrary discrete distribution in $(\mathcal{X}, \mathcal{A})$. One can assume that $\mathcal{X} = \{a_1, a_2, \ldots\}$, where the $\{a_i\}$ are the atoms of the distribution P, and $\mathcal{A}$ is all possible subsets of $\mathcal{X}$. Set $\mathcal{B} = \mathcal{A}$ and let $\varepsilon_m = \sum_{i>m} p_i$. Let the atoms $\{a_i\}$ be numbered in such a way that the sequence $\{p_i\}$ monotonically decreases. Then it is easy to see that the set $(a_0 \overset{\text{def}}{=} \emptyset)$

$$\{\{a_{i_1}, \ldots, a_{i_m}\}, \{a_{i_1}, \ldots, a_{i_m}\} \cup \{a_i;\ \ i > m\};\ \ 0 \leqslant i_1 \leqslant \ldots \leqslant i_m \leqslant m\}$$

is an ε_m-net with embedding for $\mathcal{A}$, i.e., $H_1(\varepsilon_m) \leq \hat{H}_1(\varepsilon_m) \equiv m+1$. It is shown in [3] that in our case there exists an absolute constant c_0 such that for any natural m,

$$\int_0^{\sqrt{\varepsilon_m}} H_1^{1/2}(x^2)\, dx \leqslant c_0 \sum_{i>m} \sqrt{p_i}.$$

Note that this particular relation implies that condition (5) is necessary for the central limit theorem to be satisfied uniformly for $\mathcal{B} = \mathcal{A}$

for the indicated class of distributions, since in this case the convergence of the series $\sum\sqrt{p_i}$ is a necessary and sufficient condition for (4) to be satisfied (see [3]).

Next, noting the remark after the proof of Lemma 3 in Section 2, we conclude that $\gamma^*(\varepsilon_m, A, P) \leq m$. Let $M(m) = [-\log_2 \varepsilon_m] - 1$. We assume that $\varepsilon_m \leq \frac{1}{2}$. Since $\varepsilon_m \leq 2^{-M(m)-1}$, then obviously

$$\Delta_n(\widetilde{M}(n)) \leqslant \widetilde{C} \min_{m:m+1\leqslant 4n\varepsilon_m} \widehat{\Delta}_n(m)$$

$$\equiv \widetilde{C} \min_{m:m+1\leqslant 4n\varepsilon_m} \max\left\{\left(1+m^{-1}\log n\right)\sum_{i>m}\sqrt{p_i},\ mn^{-1/2}\log n\right\},$$

where $\tilde{C}$ is an absolute constant. Denoting the nonnegative integer minimizing the function $\hat{\Delta}_n(m)$ on the set $\{m \geq 1:\ m+1 \leq 4n\varepsilon_m\}$ by $m_0(n)$, we obtain from (9) that

$$\mathbf{P}\left(\sup_{A\in\mathcal{A}} \left|\widetilde{Q}_n(A) - G(A)\right| > \widetilde{C}C_0\widehat{\Delta}_n(m_0(n))\right) \leqslant Cn^{-2}.$$

2. Proof of the theorems

The proof runs as follows. First we separate in (A, d) a finite 2^{-M}-net $\{A_i\}$ with embedding for $\mathcal{B}$. Next, on a common probability space we construct the sets of random variables $\{Q_n(A_i)\}$ and $\{G(A_i)\}$. The problem of constructing these sets reduces to a similar problem of constructing sets of $\{\bar{Q}_n(A_i)\}$ and $\{\bar{G}(A_i)\}$, where $\bar{Q}_n$ and $\bar{G}$ are defined in (1) and (2) for some discrete distribution $\bar{P}$ (Lemmas 2, 3). In turn, special representations (Lemma 1) hold for $\bar{Q}_n$ and $\bar{G}$ -- they enable us to construct easily random processes $\bar{Q}_n$ and $\bar{G}$ with sufficiently close trajectories on a common probability space. Upon assigning the sets of random variables $\{Q_n(A_i)\}$ and $\{G(A_i)\}$ on a common probability space, we can define more precisely the values of the $Q_n(A)$ and $G(A)$ at other points $A \in A$ as well. To complete the proof, we need to estimate the freak values (the outliers) of the increments of the random processes Q_n and G in neighborhoods of the nodes of the 2^{-M}-net (Lemmas 4, 5).

Thus, we proceed to prove basic auxiliary assertions.

Lemma 1. Let $\bar{P} = \{p_k;\ k = 1,2,\ldots,m\}$ be a discrete distribution in (X,A). Then we have the following relations:

$$\bar{Q}_n(\cdot) \overset{\mathscr{D}}{=} \sum_{i=1}^{m} I_{a_i}(\cdot)\,[V_n(F(i+1)) - V_n(F(i))], \tag{12}$$

$$\bar{G}(.) \overset{\mathscr{D}}{=} \sum_{i=1}^{m} I_{a_i}(\cdot)\,[\overset{\circ}{w}(F(i+1)) - \overset{\circ}{w}(F(i))], \tag{13}$$

where $\{a_i;\ i = 1,2,..,m\}$ are atoms of the distribution $\bar{P}$; $V_n(\cdot)$ is the empirical process on $[0,1]$ constructed from a sample of size n from the uniform distribution on $[0,1]$; $\overset{\circ}{W}(\cdot)$ is the "Brownian bridge" on $[0,1]$, $F(t) = \sum_{i<t} p_i$; the processes $\bar{Q}_n$ and $\bar{G}$ are defined in (1) and (2) for the distribution $\bar{P}$.

Proof. Is essentially given in [5] and therefore we omit it.

We specify random processes $V_n(\cdot)$ and $\overset{\circ}{w}(\cdot)$ on the common probability space using the method of [7]. Then it follows from (12) and (13) that for any m-point distribution in (X,A), uniformly for any subclass $B \subseteq A$ as $n \to \infty$,

$$|\tilde{Q}_n(\cdot) - G(\cdot)| = O(mn^{-1/2}\log n) \tag{14}$$

with probability 1, the constant in the definition of $O(\cdot)$ being absolute. It is not hard to see that the factor m in (14) can substantially be decreased, depending on the properties of class B and the order of numbering the atoms $\{a_i\}$. In the sequel, we call a set of the form $\{a_i:\ k \leq i \leq \ell\}$ a chain of atoms. Clearly, for any fixed $A \in B$ the closeness of $\tilde{Q}_n(A)$ and $G(A)$ defined on the common probability space by the method of [7] does not depend on the number of atoms m, but rather on the number of chains of atoms (denote it by $\gamma(A)$) of the discrete distribution which are contained in A. Obviously, the $\gamma(A)$ essentially depends on the manner of numbering the atoms of the distribution. In view of this remark, we introduce the quantity

$$\gamma_0(\mathscr{B}, \bar{P}) = \min \max_{A \in \mathscr{B}} \gamma(A),$$

where the minimum is taken over all possible enumerations of the atoms $\{a_i\}$. Now we can sharpen estimate (14) by replacing the factor m in

(14) by $\gamma_0(\cdot)$. For example, let $\mathcal{X} = R^k$, let $\mathcal{A}$ be the Borel σ-algebra and let $\mathcal{B} = \{\{z \in R^k: z < t\};\ t \in R^k\}$ be the set of open "nodes" in R^k. Suppose that the atoms $\{a_i\}$ of the discrete distributions $\bar{P}$ are numbered so that $a_i \leqq a_{i+1}$ for any $i < m$. Then, obviously, $\gamma(\cdot) = 1$, and the method suggested makes it possible to get the estimate

$$\sup_{A \in \mathcal{B}} |\bar{Q}_n(A) - \bar{G}(A)| = O(n^{-1/2} \log n)$$

with probability 1.

Here is an example that is more meaningful and important for the application. Let $\bar{P}$ be an arbitrary discrete distribution in R^k, concentrated on the lattice:

$$R(N) = \{(t_{i_1}^{(1)}, \ldots, t_{i_k}^{(k)});\quad i_s = 1, 2, \ldots, N,\ s \leqslant k\},$$

where the $t_i^{(\ell)}$ are arbitrary scalars. Let $\mathcal{B}$ denote the set of open "corners" considered in the previous example. Putting $R(N)$ in lexicographic order, we can show (see [5]) that

$$\gamma_0(\mathcal{B}, \bar{P}) \leqslant N^{k-1}.$$

L e m m a 2. Let $\mathcal{B}_N = \{A_1, \ldots, A_N\}$ be an arbitrary finite set from $\mathcal{A}$. Then there exists a discrete distribution $\bar{P}$ having at most 2^N atoms, and a sequence of independent random variables $\{\bar{X}_k\}$ distributed by the law $\bar{P}$ and given on the same probability space as $\{X_k\}$, for which $\bar{Q}_n(A_i) = Q_n(A_i)$ for any $A_i \in \mathcal{B}_N$, where $\bar{Q}_n(\cdot)$ is the empirical process constructed from the sample $\bar{X}_1, \ldots, \bar{X}_n$.

Proof. Introduce the following notation:

$$B_0 = \mathfrak{X} \setminus \bigcup_{i \leqslant N} A_i, \qquad B_k = A_k \setminus \bigcup_{\substack{i \leqslant N,\\ i \neq k}} A_i, \quad k = 1, 2, \ldots, N,$$

$$B_{k_1 k_2} = A_{k_1} \cap A_{k_2} \setminus \bigcup_{\substack{i \leqslant N,\\ i \neq k_1, k_2}} A_i, \ k_1 < k_2,\ k_l \leqslant N,$$

$$B_{k_1 k_2 k_3} = A_{k_1} \cap A_{k_2} \cap A_{k_3} \setminus \bigcup_{\substack{i \leqslant N,\\ i \neq k_1, k_2, k_3}} A_i, \ k_1 < k_2 < k_3,\ k_l \leqslant N,$$

. . . .

$$B_{k_1 \ldots k_{N-1}} = \bigcap_{l \leqslant N-1} A_{k_l} \setminus \bigcup_{\substack{i \leqslant N,\\ i \neq k_1, \ldots, k_{N-1}}} A_i, \ k_1 < \ldots < k_{N-1},\ k_l \leqslant N, \tag{15}$$

$$B_{k_1 \ldots k_N} = \bigcap_{i \leqslant N} A_i.$$

It is not hard to see that this family $\{B_{.}\}$ decomposes X into disjoint subsets. We assume without loss of generality that $\mathcal{B}_N \neq \{X\}$. Therefore the family (15) contains at least two nonempty sets.

Now, from each nonempty set $B_{k_1,\ldots,k_\ell}$ we take a point $b_{k_1,\ldots,k_\ell}$. We define the sequence $\{\bar{X}_k\}$ as follows: if $X_k \in B_{k_1,\ldots,k_\ell}$, then $\bar{X}_k = b_{k_1,\ldots,k_\ell}$. It follows from (15) that the number of atoms of this discrete distribution does not exceed 2^N.

Let us show that any set $A_i \in \mathcal{B}_N$ can be represented as a union of sets from $\{B_{.}\}$. Indeed, from the system of sets (15) we take all the sets in which one of the multiindex coordinates is equal to i. Obviously, the union of such sets is contained in A_i. On the other hand, for any $a \in A_i$, we take from $\mathcal{B}_N \setminus \{A_i\}$ all the sets $A_{k_1}, A_{k_2}, \ldots, A_{k_s}$ containing the point a. Clearly, then

$$a \in A_i \bigcap_{j \leq s} A_{k_j} \setminus \bigcup_{\substack{l \leq N,\\ l \neq i, k_1, \ldots, k_s}} A_l \equiv B_{\cdot i}\,.$$

Hence, the union of sets from $\{B_{.}\}$ in which one of the multiindex coordinates is equal to i coincides with the set A_i.

Next, using the additivity property of probability measures, we conclude that for the distribuiton $\bar{P}$ of the random variables $\bar{X}_1$ and the associated empirical distribution, we have the equalities

$$\bar{P}(A_i) = P(A_i), \quad \bar{P}_n(A_i) = P_n(A_i), \quad A_i \in \mathcal{B}_N.$$

The lemma is proved.

L e m m a 3. Let $\mathcal{B}_N = \{A_1, \ldots, A_N\}$ and let $\bar{P}$ be the discrete distribution constructed in Lemma 2. Then

$$\gamma_0(\mathcal{B}_N, \bar{P}) \leq N.$$

Proof. First of all, we decompose the class of sets $\{B_{\bullet}\}$ defined in (15) into $N+1$ disjoint classes:

$$\{B_{\cdot}\} = \bigcup_{k=1}^{N} \{B_{k\cdot}\} \cup \{B_0\},$$

where $B_{k\bullet}$ is the set of the form (15) in which the first coordinate of the multiindex is equal to k. For simplicity of arguing, we assume that all the sets $B_{\bullet}$ are nonempty.

Let us consider separately class $\{B_{1\cdot}\}$. It can be represented as follows:

$$\{B_{1\cdot}\} = \{B_1\} \cup \bigcup_{l=2}^{N} \{B_{1l\cdot}\}.$$

Now we assign the index "1" to the set B_1 (more precisely, we number the atom $b_1 \in B_1$ of the distribution $\bar{P}$). Next, we number the elements of the class $\{B_{12\cdot}\}$ in arbitrary order by numbers from 2 to $m_1 = 2^{N-2}$ (without skipping), the elements of class $\{B_{13\cdot}\}$ by numbers from $m_1 + 1$ to $m_1 + m_2 = 2^{N-2} + 2^{N-3}$, and so on, until we reach class $\{B_{1N}\}$ of the singleton B_{1N}, which we number

$$\sum_{i=1}^{N-1} m_i + 1 = \sum_{i=0}^{N-2} 2^i + 1.$$

Thus we have numbered all the elements of the subclass $\{B_{1\cdot}\}$ which is a partition of the set A_1. There is now one chain of atoms inside A_1. At the same time, the consecutive numbers of the elements of the subclass $\{B_{1k\cdot}\}$ also make one chain of atoms inside A_k.

The next stage of the enumeration involves the same procedure for class $\{B_{2\cdot}\}$. That is,

$$\{B_{2\cdot}\} = \{B_2\} \cup \bigcup_{l=3}^{N} \{B_{2l\cdot}\}.$$

We number the set B_2 with $\sum_{i=1}^{N-1} m_i + 2$. Then, as at the previous stage, we number consecutively subclasses $\{B_{23\cdot}\}$, $\{B_{24\cdot}\}$, ..., $\{B_{2N}\}$ (the numbering of the elements inside each subclass $\{B_{2k\cdot}\}$ is without skipping!). We have thus formed a second chain of atoms inside A_2 and all of the elements of the partition of A_2 have been numbered. Moreover, we have constructed the second chain of atoms inside each of the sets $A_3, A_4, \dots, A_N$.

Continuing this procedure, one can construct at most k chains of atoms inside each set A_k.

The lemma is proved.

REMARK. In engineering problems the quantity $\gamma_0(\mathcal{B}_n, \bar{P})$ is, as a rule, substantially smaller than N (in spite of the fact that the estimate in Lemma 3 is unimprovable). For example, let $A_1, \dots, A_N$ be all possible unions (including the empty set) of disjoint sets $B_1, \dots, B_m$. Obviously, in this case $N = 2^m$. At the same time, the family $\{B_i;\ i \leq m\}$ coincides with the system of sets $\{B_\bullet\}$ considered in Lemmas 2, 3. In other words, $\{B_i;\ i \leq m\}$ is a minimal system of generators for $\{A_i;\ i \leq N\}$. Hence for any distribution P one has the estimate $Y_0(\mathcal{B}_N, \bar{P}) < m = \log_2 N$, where the distribution $\bar{P}$ corresponding to P and class $\{A_i;\ i \leq N\}$ is constructed in Lemma 2.

In what follows we shall use the symbol $A_{M,i}$ to denote an element of the ε-net with embedding for $\varepsilon = 2^{-M}$. Also, let

$$\mathscr{B}_{(M,i)} = \{A \in \mathscr{B}:\ d(A, A_{M,i}) \leqslant 2^{-M}\},\quad i = 1, 2, \ldots, N_1(2^{-M}).$$

In the next two lemmas, the random processes $Q_n(\cdot)$ and $G(\cdot)$ are assumed to be d-separable. We shall make a remark in regard to this assumption while proving Theorem 1.

L e m m a 4. For any $r \geq 0$, $M = 0,1,\ldots$ and $y \geq 2r+5$, we have the inequality

$$\mathbf{P}\Big(\sup_{A\in\mathscr{B}(M,i)} |Q_n(A) - Q_n(A_{M,i})| \geqslant (4y+1)\Big[\sum_{k>M}[2^{-k}H_1(2^{-k})]^{1/2} + 5\cdot 2^{-M/2}\Big]\Big) \tag{16}$$
$$\leqslant I(n, M, H_1)\, N_1(2^{-M})^{-r} R_1(y, r) + (1 - I(n, M, H_1))\, N_1(2^{-M})^2 \times \exp\Big\{-yn^{1/2}\sum_{k>M}[2^{-k}H_1(2^{-k})]^{1/2}\Big\},$$

where the functions $R_1(y,r)$ and $I(n,M,H_1)$ are defined in Theorem 1.

Proof. Let $\ell^- = \ell^-(k,A)$ and $\ell^+ = \ell^+(k,A)$ denote respectively the indices of the inner and outer neighborhoods of the 2^{-k}-net for the set $A \in \mathcal{B}$, i.e., $A_{k,\ell^-} \subseteq A \subseteq A_{k,\ell^+}$ and $d(A_{k,\ell^-}, A_{k,\ell^+}) \leq 2^{-k}$. Note that the $\ell^{\pm}(\cdot)$ are, generally speaking, not uniquely determined. Therefore $\ell^{\pm}(\cdot)$ stands for any pair of natural numbers with this property.

We introduce next the following notation:

$$N^* = \min\{k = 0, 1, \ldots : \max(k,\ H_1(2^{-M-k})) \geqslant 4n2^{-M-k}\} - 1. \tag{17}$$

We assume first that $N^* \geq 1$. We decompose class $\mathcal{B}(M,i)$ into disjoint subclasses for each of which we can find a pair $\ell^{\pm}(M+N^*, \cdot)$ common for all elements of the given subclass. Obviously, the number of such subclasses is not larger than

$$N_1(2^{-M-N^*})(N_1(2^{-M-N^*}) - 1)/2. \tag{18}$$

It is easily seen that

$$\sup_{A\in\mathscr{B}(M,i)} |Q_n(A)-Q_n(A_{M,i})| \leqslant \max_{m:l^-(M+N^*,A)=m, A\in\mathscr{B}(M,i)} |Q_n(A_{M+N^*,m})$$
$$-Q_n(A_{M,i})| + \max_{l\pm(M+N^*,\cdot)} \sup_{A\in\mathscr{B}(M,i):A_{M+N^*,l-}\subseteq A\subseteq A_{M+N^*,l+}} |Q_n(A)$$
$$-Q_n\left(A_{M+N^*,l-}\right)|. \tag{19}$$

Let $A_{M+N^*,m}$ be an arbitrary subset with respect to which the maximum is taken in the first summand on the right side of (19). Obviously, we can find a set $A\in B(M,i)$ such that

$$d(A_{M+N^*,m}, A)\leqslant 2^{-M-N^*}.$$

Now, for the given A we take sets

$$\{A^{(m)}_{M+j,l(j)};\ j=1,2,\ldots,N^*-1\}$$

such that

$$d\left(A^{(m)}_{M+j,l(j)}, A\right)\leqslant 2^{-M-j}.$$

Also set

$$A^{(m)}_{M+N^*,\,l(0)}=A_{M+N^*,m},\quad A^{(m)}_{M,\,l(N^*)}=A_{M,\,i}.$$

It follows from the triangle inequality for $d(\cdot)$, that for any $j=1,2,\ldots,N^*$

$$d\left(A^{(m)}_{M+j,l(j)}, A^{(m)}_{M+j-1,l(j-1)}\right)\leqslant 2^{-M-j+2}. \tag{20}$$

Let us estimate from above the first and second summands of the right side of (19). We have

$$\max_{m:l^-(M+N^*,A)=m, A\in\mathscr{B}(M,i)} |Q_n(A_{M+N^*,m})-Q_n(A_{M,i})|$$
$$\leqslant \sum_{j=1}^{N^*} \max_{\substack{m:l^-(\cdot)=m,\\ A\in\mathscr{B}(M,i)}} \left|Q_n\left(A^{(m)}_{M+j,l(j)}\right)-Q_n\left(A^{(m)}_{M+j-1,l(j-1)}\right)\right|$$
$$\leqslant \sum_{k=1}^{N^*} \max_{m,l:d(A_{M+k,m},A_{M+k-1,l})\leqslant 2^{-M-k+2}} |Q_n(A_{M+k,m})-Q_n(A_{M+k-1,l})| \equiv J_1. \tag{21}$$

The estimate for the second summand on the right side of (19), is given by

$$\max_{l\pm(M+N^*,\cdot)} \sup_{A\in\mathscr{B}(M,i):A_{M+N^*,l-}\subseteq A\subseteq A_{M+N^*,l+}} \left|Q_n(A)-Q_n\left(A_{M+N^*,l-}\right)\right|$$

$$= \max_{l \pm (M+N^*, \cdot)} \sup_{\cdots} \max \{Q_n(A) - Q_n(A_{M+N^*,l-}), Q_n(A_{M+N^*,l-}) - Q_n(A)\}$$

$$\leqslant \max_{\substack{k,j: A_{M+N^*,k} \subseteq A_{M+N^*,j} \\ d(A_{M+N^*,k}, A_{M+N^*,j}) \leqslant 2^{-M-N^*}}} \max \{\sqrt{n}[P_n(A_{M+N^*,j}) - P(A_{M+N^*,k})]$$

$$- Q_n(A_{M+N^*,k}), \sqrt{n}[P(A_{M+N^*,j}) - P(A_{M+N^*,k})]\}$$

$$\leqslant \max_{\substack{k,j: A_{M+N^*,k} \subseteq A_{M+N^*,j} \\ d(A_{M+N^*,k}, A_{M+N^*,j}) \leqslant 2^{-M-N^*}}} |Q_n(A_{M+N^*,j}) - Q_n(A_{M+N^*,k})| \tag{22}$$

$$+ \sqrt{n}\, 2^{-M-N^*} \equiv J_2 + \sqrt{n}\, 2^{-M-N^*}.$$

Now we estimate the probabilities of large deviations of the random variables J_1 and J_2. For any positive $y_1, \ldots, y_{N^*}$ we have from (21) that

$$\mathbf{P}\left(J_1 > \sum_{k=1}^{N^*} y_k 2^{-(M+k-2)/2}\right) \leqslant \sum_{k=1}^{N^*} N_1(2^{-M-k})^2 \tag{23}$$

$$\times \sup_{A,B: d(A,B) \leqslant 2^{-M-k+2}} \mathbf{P}(|Q_n(A) - Q_n(B)| > y_k 2^{-(M+k-2)/2}).$$

The quantity $Q_n(A) - Q_N(B)$ is the normalized sum of independent identically distributed random variables with zero mean. Therefore, to estimate the probabilities on the right side of (23), one can use Bernshtein's inequality, which is the following in our case. Let $\xi_1, \ldots, \xi_n$ be independent identically distributed random variables with zero mean. In addition, let $|\xi_i| \leqq 1$ with probability 1 and $D\xi_i \leqq \sigma$. Then for any $y \geqq 0$ (see [8]) one has

$$\mathbf{P}\left(n^{-1/2}\left|\sum_{i=1}^{n} \xi_i\right| > y\sqrt{\sigma}\right) \leqslant 2\exp\left\{-\frac{y^2}{2(1 + y/\sqrt{n\sigma})}\right\}. \tag{24}$$

Note now that for any $A, B \in \mathcal{A}$,

$$\mathbf{D}(Q_n(A) - Q_n(B)) \leqslant d(A, B). \tag{25}$$

Hence from (20), (23) - (25), for any $y_k \geqq 0$ one has the inequality

$$\mathbf{P}\left(J_1 > \sum_{k=1}^{N^*} y_k 2^{-(M+k-2)/2}\right) \leqslant 2\sum_{k=1}^{N^*} N_1(2^{-M-k})$$

$$\times \exp\left\{-\frac{y_k^2}{2(1 + y_k/(n2^{-M-k+2})^{1/2})}\right\} = 2\sum_{k=1}^{N^*} \exp\left\{2H_1(2^{-M-k})\right.$$

$$- \frac{y_k^2}{2\left(1 + y_k/(n2^{-M-k+2})^{1/2}\right)}\Bigg\}.$$

Set

$$y_k = y\left[\max\left(H_1\left(2^{-M-k}\right), k\right)\right]^{1/2}. \tag{26}$$

Then for any $k \leqq N^*$,

$$y_k \leqslant y(n2^{-M-k+2})^{1/2}$$

and consequently

$$\mathbf{P}\left(J_1 > \sum_{k=1}^{N^*} y_k 2^{-(M+k-2)/2}\right)$$
$$\leqslant 2\sum_{k=1}^{N^*} \exp\left\{-\left[\frac{y^2}{2(1+y)} - 2\right]\max\left(H_1\left(2^{-M-k}\right), k\right)\right\}$$
$$= 2\sum_{k=1}^{N^*} \exp\left\{-\left[\frac{(y-r-2)^2 - (r+3)^2 + 1}{2(1+y)} + r\right]\max\left(H_1\left(2^{-M-k}\right), k\right)\right\}$$
$$\leqslant 2N_1\left(2^{-M}\right)^{-r} \sum_{k=1}^{N^*} \exp\left\{-\frac{(y-r-2)^2 - (r+3)^2 + 1}{2(1+y)} \max\left(H_1\left(2^{-M-k}\right), k\right)\right\}, \tag{27}$$

where r is an arbitrary nonnegative scalar. Obviously, for $y \geqq 2r+5$ the arguments of the exponentials under the sum sign on the right side of (27) become negative. Hence for $y \geqq 2r+5$, the quantities $\max\left(H_1\left(2^{-M-k}\right), k\right)$ in (27) can be replaced by k. Thus,

$$\mathbf{P}\left(J_1 > \sum_{k=1}^{N^*} y_k 2^{-(M+k-2)/2}\right)$$
$$\leqslant 2N_1\left(2^{-M}\right)^{-r}\left[\exp\left\{\frac{(y-r-2)^2 - (r+3)^2 + 1}{2(1+y)}\right\} - 1\right]^{-1}. \tag{28}$$

Next we obtain from (26) that

$$\sum_{k=1}^{N^*} y_k 2^{-(M+k-2)/2} \leqslant 2y \sum_{k=M+1}^{N^*} \left[2^{-k} H_1\left(2^{-k}\right)\right]^{1/2}$$
$$+ y2^{1-M/2} \sum_{k=1}^{N^*} \left(k2^{-k}\right)^{1/2}. \tag{29}$$

Consider the right side of (22). Noting relations (18) and (25) and using Bernshtein's inequality (24) we obtain

$$\mathbf{P}(J_2 > z\sqrt{n}2^{-M-N^*}) \leqslant \exp\left\{2H_1(2^{-M-N^*}) - \frac{z^2 n 2^{-M-N^*}}{2(1+z)}\right\}.$$

It follows from (17) that

$$n2^{-M-N^*} \geqslant 4^{-1}\max(H_1(2^{-M-N^*}), N^*),$$

and therefore

$$\mathbf{P}(J_2 > z\sqrt{n}2^{-M-N^*}) \leqslant \exp\left\{-\left[\frac{z^2}{8(1+z)} - 2\right]\max(H_1(2^{-M-N^*}), N^*)\right\}.$$

Now set $z = 4y$. Then it is not hard to check that for $y \geqq 5$,

$$\frac{z^2}{8(1+z)} > \frac{y^2}{2(1+y)}\ .$$

Hence (see the derivation of (27)) for $y \geqq 2r+5$,

$$\begin{aligned} &\mathbf{P}(J_2 > z\sqrt{n}2^{-M-N^*}) \\ &\leqslant \exp\left\{-\left[\frac{y^2}{2(1+y)} - 2\right]\max(H_1(2^{-M-N^*}), N^*)\right\} \\ &\leqslant N_1(2^{-M})\exp\left\{-\frac{(y-r-2)^2-(r+3)^2+1}{2(1+y)}\cdot N^*\right\}, \end{aligned} \tag{30}$$

where r is an arbitrary nonnegative scalar.

Next, from (17) we get

$$\begin{aligned} \sqrt{n}2^{-M-N^*} &\leqslant [\max(H_1(2^{-M-N^*-1}), N^*+1)]^{1/2}\,2^{-(M+N^*+1)/2} \\ &\leqslant [H_1(2^{-M-N^*-1})\,2^{-M-N^*-1}]^{1/2} + [(N^*+1)\,2^{-N^*-1}]^{1/2}\,2^{-M/2}. \end{aligned} \tag{31}$$

Thus, it follows from (22), (30) and (31) that for $y \geqq 2r+5$

$$\begin{aligned} &\mathbf{P}\Bigg(\max_{l\pm(M+N^*,\cdot)} \sup_{\substack{A\in\mathscr{B}(M,i):\\ A_{M+N^*,l-}\subseteq A\subseteq A_{M+N^*,l+}}} |Q_n(A) - Q_n(A_{M+N^*,l-})| \\ &\qquad > (4y+1)\{[H_1(2^{-M-N^*-1})\,2^{-M-N^*-1}]^{1/2} \\ &\qquad\qquad + [(N^*+1)\,2^{-N^*-1}]^{1/2}\,2^{-M/2}\}\Bigg) \\ &\leqslant N_1(2^{-M})^{-r}\exp\left\{-\frac{(y-r-2)^2-(r+3)^2+1}{2(1+y)}\cdot N^*\right\}. \end{aligned} \tag{32}$$

The assertion of the Lemma follows from (19), (21), (28), (29), (32) and the inequality $\sum_{k\geq 1}(k2^{-k})^{\frac{1}{2}} < 5$.

Now let $N^* < 1$ (i.e., $n2^{-M+1} \leq \max(H_1(2^{-M-1}), 1)$). Then, as in (19) and (22), we obtain

$$\sup_{A\in\mathscr{B}(M,i)} |Q_n(A) - Q_n(A_{M,i})|$$
$$\leq \max_{k,j:A_{M,k}\subseteq A_{M,j}, d(A_{M,k},A_{M,j})<2^{-M}} |Q_n(A_{M,k}) - Q_n(A_{M,j})| \tag{33}$$
$$+ \sqrt{n}2^{-M} \equiv J_3 + \sqrt{n}2^{-M}.$$

For any $z \geq 1$, it then follows from Bernshtein's inequality (24) that

$$P\left(J_3 > z\sum_{k>M}[2^{-k}H_1(2^{-k})]^{1/2}\right) \leq$$
$$\leq N_1(2^{-M})^2 \exp\left\{-\frac{z^2 2^M\left(\sum_{k>M}[2^{-k}H_1(2^{-k})]^{1/2}\right)^2}{2\left(1 + z2^M n^{-1/2}\sum_{k>M}[2^{-k}H_1(2^{-k})]^{1/2}\right)}\right\} \tag{34}$$
$$\leq N_1(2^{-M})^2 \exp\left\{-4^{-1}zn^{1/2}\sum_{k>M}[2^{-k}H_1(2^{-k})]^{1/2}\right\}.$$

Moreover, if $N^* < 1$, then

$$\sqrt{n}2^{-M} < [H_1(2^{-M-1})2^{-M-1}]^{1/2} + 2^{-M/2}. \tag{35}$$

Setting $z = 4y$, from (33) - (35) we obtain the required.

The lemma is proved.

REMARK. The specific features of metric entropy with embedding are used, in essence, only in proving Lemma 4. One can use as well other considerations in order to obtain inequalities of the form (16). However, this goes beyond the scope of our investigation.

Lemma 5. If condition (5) is satisfied for any $r \geq 0$, $M = 0,1,...$ and $y \geq (2r+5)^{\frac{1}{2}}$, then

$$P\left(\sup_{A\in\mathscr{B}(M,i)} |G(A) - G(A_{M,i})|\right.$$
$$\left.> 2y\left(\sum_{k>M}[2^{-k}H_1(2^{-k})]^{1/2} + 5\cdot 2^{-M/2}\right)\right)$$
$$\leq N_1(2^{-M})^{-r} R_2(y,r),$$

where the function $R_2(y,r)$ is defined in Theorem 1.

Proof. The proof of the Lemma is based on the well-known methods for estimating the maximal freak value (outlier) of trajectories of a separable Gaussian random process with a finite-dimensional parameter set (see, for example, [4]). Our arguments are, to a certain extent, analogous to those used in proving Lemma 4.

Thus, for any d-separable random process on A, the inequality

$$\sup_{A\in\mathscr{B}(M,i)} |G(A)-G(A_{M,i})|$$

$$\leqslant \sup_{A\in\mathscr{B}(M,i)}\ \sup_{k=1,2,\ldots} \left|G\left(A_{M+k,l^{-}(M+k,A)}-G(A_{M,i})\right|\equiv J_4, \right. \tag{36}$$

holds with probability 1, where, as before $\ell^{-}(M+k, A)$ is defined by the condition

$$d\left(A_{M+k,l^{-}(\cdot)}, A\right)\leqslant 2^{-M-k}.$$

Obviously,

$$J_4\leqslant \sum_{k=1}^{\infty}\sup_{A\in\mathscr{B}(M,i)}\left|G(A_{M+k,l(M+k,A)})-G\left(A_{M+k-1,l^{-}(M+k-1,A)}\right)\right|,$$

where we assume $A_{M,\ell^{-}(M,A)} = A_{M,i}$ for notational convenience. Since the subclass $\{A_{M+k,\ell^{-}(M+k,A)};\ A\in\mathcal{B}(M,i)\}$ is finite, then noting (20) we can again strengthen the preceding inequality:

$$J_4\leqslant \sum_{k=1}^{\infty}\ \max_{k,j:d(A_{M+k,l},A_{M+k-1,j})\leqslant 2^{-M-k+2}} |G(A_{M+k,l})-G(A_{M+k-1,j})|. \tag{37}$$

Hence, by (24), (25) and (37), for any $y_k \geqq 0$ one has

$$\mathbf{P}\left(J_4>\sum_{k=1}^{\infty} y_k 2^{-(M+k-2)/2}\right)\leqslant \sum_{k=1}^{\infty} N_1(2^{-M-k})^2\,\overline{\Phi}(y_k), \tag{38}$$

where

$$\overline{\Phi}(y)=(2\pi)^{-1/2}\int_y^{\infty} e^{-t^2/2}dt.$$

Now, as in proving Lemma 4, in (38) we set

$$y_k=y[\max(H_1(2^{-M-k}),\ k)]^{1/2}.$$

Then for any $r \geqq 0$,

$$\mathbf{P}\left(J_4>\sum_{k=1}^{\infty} y_k 2^{-(M+k-2)/2}\right)\leqslant \sum_{k=1}^{\infty}[\max(H_1(2^{-M-k}),\ k)/2\pi]^{1/2}$$

$$\times\int_y^{\infty}\exp\{-(t^2/2-2)\max\left(H_1(2^{-M-k}),\ k\right)\}dt\leqslant N_1(2^{-M})^{-r}(2\pi)^{-1/2}$$

$$\times \sum_{k=1}^{\infty} \int_{y}^{\infty} \exp\{-((t^2-2r-5)/2)\max\left(H_1(2^{-M-k}),\, k\right) dt. \tag{39}$$

If $y \geq (2r+5)^{\frac{1}{2}}$, then replacing $\max\,(H_1(2^{-M-k}),k)$ by k on the right side of (39), we finally get that

$$\mathbf{P}\left(J_4 > \sum_{k=1}^{\infty} y_k 2^{-(M+k-2)/2}\right)$$
$$\leqslant N_1(2^{-M})^{-r}(2\pi)^{-1/2}\int_{y}^{\infty}[\exp\{(t^2-2r-5)/2\}-1]^{-1}\,dt. \tag{40}$$

The assertion of the Lemma follows from (36), (40) and the inequality

$$\sum_{k=1}^{\infty} y_k 2^{-(M+k-2)/2} \leqslant 2y\left(\sum_{k>M}[2^{-k}H_1(2^{-k})]^{1/2}+5\cdot 2^{-M/2}\right).$$

Proof of Theorem 1. Let M be an arbitrary nonnegative integer. Using Lemmas 1-3, by the method of quantile transformations (see [7]) one can construct on the trajectory of the Brownian bridge $\mathring{w}(\cdot)$ the families of random variables $\{\tilde{Q}_n(A_{M,i});\ i \leq N_1(2^{-M})\}$, $n = 1,2,\ldots,$ and $\{G(A_{M,i});\ i \leq N_1(2^{-M})\}$, such that for any n,

$$\mathbf{P}\left(\max_{i\leqslant N_1(2^{-M})} |\tilde{Q}_n(A_{M,i}) - G(A_{M,i})|\right.$$
$$\left.\geqslant C\gamma^*\left(2^{-M}, \mathscr{B}, P\right)(\log n + x)\, n^{-1/2}\right) \leqslant Ke^{-\lambda x}, \tag{41}$$

where C, k, λ are absolute positive constants, $\gamma^*(2^{-M},\mathcal{B},P) = \gamma_0(\mathcal{B}_N,\bar{P})$, and $\bar{P}$ is defined in Lemma 2 for the family of sets $\mathcal{B}_N = \{A_{M,i};\ i \leq N_1(2^{-M})\}$.

It follows from Lemma 3 that $\gamma^*(\varepsilon,\mathcal{B},P)$ satisfies inequality (7).

Now, using Kolmogorov's theorem for adapted conditional distribution and extending appropriately the probability space constructed--on which $\mathring{w}(\cdot)$ is specified, we can define more precisely the values $\tilde{Q}_n(A)$, $n = 1,2,\ldots,$ and $G(A)$ at all remaining "points" $A \in \mathcal{A}$. In this case $\tilde{Q}_1(\cdot)$, $\tilde{Q}_2(\cdot)$, ..., $G(\cdot)$ are constructed as random processes conditionally independent with respect to the σ-algebra generated by $\mathring{w}(\cdot)$. One can assume without loss of generality in the arguments that these random processes are d-separable (see [4]). We need to note, however, that in deriving estimates (22) and (33) in Lemma 4, the explicit form of the process $Q_n(\cdot)$ was used, which was assumed to be d-separable. On the other hand, if we take into account the procedure of constructing a d-separable modification of the process $\tilde{Q}_n(\cdot)$ (see [4]), it will easily be seen that the corresponding inequalities hold for the separabilized random process $\tilde{Q}_n(\cdot)$ as well.

For the random processes $\tilde{Q}_n(\cdot)$ and $G(\cdot)$ constructed on a common probability space, we have the inequality

$$\begin{aligned}\sup_{A\in\mathscr{B}} |\tilde{Q}_n(A) - G(A)| \leqslant & \max_{i\leqslant N_1(2^{-M})} |\tilde{Q}_n(A_{M,i}) - G(A_{M,i})| \\ + & \max_{i\leqslant N_1(2^{-M})} \sup_{A\in\mathscr{B}(M,i)} |\tilde{Q}_n(A) - \tilde{Q}_n(A_{M,i})| \\ & + \max_{i\leqslant N_1(2^{-M})} \sup_{A\in\mathscr{B}(M,i)} |G(A) - G(A_{M,i})|. \end{aligned} \tag{42}$$

The assertion of the Theorem follows from (41), (42) and Lemmas 4, 5.

Proof of Theorem 2. The only difference between this proof and the preceding one is in the construction of the random processes $\tilde{Q}_n(\cdot)$.

Let $\nu_n(A)$ denote the number of points of the sample $X_1, X_2, \ldots, X_n$, falling into the set $A \in \mathcal{A}$. Now, for each n, we define on some probability space the family of independent random vectors with the multi-index $(n_1, \ldots, n_s)$:

$$\left\{(X_{1n}(n_1, \ldots, n_s), \ldots, X_{nn}(n_1, \ldots, n_s));\ n_j \geqslant 0,\ \sum_{j=1}^{s} n_j = n\right\}, \tag{43}$$

where $s = N_1(2^{-M})$, and the distribution of a vector for a fixed multi-index $(n_1, \ldots, n_s)$ coincides with the conditional distribution of the sample $X_1, \ldots, X_n$ under the condition $\bigcap_{i\leq s} \{\nu_n(A_{M,i}) = n_i\}$. If the event $\bigcap_{i\leq s} \{\nu_n(A_{M,i}) = n_i\}$ has zero probability, then the distribution of the corresponding vector $(X_{1n}(n_1, \ldots, n_s), X_{2n}(n_1, \ldots, n_s), \ldots, X_{nn}(n_1, \ldots, n_s))$ can be assumed arbitrary. For different n, the families of random vectors (43) are assumed to be uniformly independent in the aggregate.

Furthermore, on a trajectory of the Brownian bridge $\mathring{w}(\cdot)$ not depending on the above-mentioned set of random vectors, we construct, as in Theorem 1, the families of random variables $\{\tilde{Q}_n(A_{M,i});\ i \leq s\}$, $n = 1, 2, \ldots$, and $\{G(A_{M,i});\ i \leq s\}$. Let

$$X_{kn} = X_{kn}(\tilde{\nu}_n(A_{M,1}), \ldots, \tilde{\nu}_n(A_{M,s})) ,$$

where

$$\tilde{\nu}(A_{M,i}) = n^{\frac{1}{2}} \tilde{Q}_n(A_{M,i}) + nP(A_{M,i})$$

is a replica of the random variable $\nu_n(A_{M,i})$. Obviously, the random vector $(X_{1n},\ldots,X_{nn})$ coincides in distribution with the sample $X_1,\ldots,X_n$ and the number of points of the sample $X_{1n},\ldots,X_{nn}$ falling into the set $A_{M,i}$ is equal to $\tilde{\nu}_n(A_{M,i})$ with probability 1.

Let us construct next an empirical measure $\tilde{Q}_n(\cdot)$ from the sample $X_{1n},\ldots,X_{nn}$. Obviously, the distributions of any measurable functionals of $Q_n(\cdot)$ and $\tilde{Q}_n(\cdot)$ coincide. Therefore, the notation "$\mathcal{D}$" can be interpreted in this case in a much stronger sense than in (3).

Note also that the random process $G(\cdot)$ is constructed in this case exactly as in the preceding theorem, i.e., using Kolmogorov's theorem on adapted distributions. One can assume the process $G(\cdot)$ to be conditionally (with respect to the σ-algebra generated by the $\{G(A_{M,i});\ i \leq s\}$) independent of $\{\tilde{Q}_n(\cdot);\ n=1,2,\ldots\}$.

The existence of a measurable estimate Λ_n follows from inequality (42) and inequalities obtained in proving Lemma 4.

Theorem 2 is proved.

3. Lower bounds for the rate of convergence

In this section we shall examine the accuracy of estimates of the form (9). We shall show that the estimates of the convergence rate obtained in the central limit theorem for empirical measures are, in some sense, unimprovable.

We first consider the case $\mathcal{X} = R$, $\mathcal{B} = \{(-\infty,t);\ t\in R\}$. It follows from (11) for $m = 1$ that this estimate of the convergence rate has order of smallness $O(n^{-\frac{1}{2}}\log n)$ with probability 1, which, as has been shown in [7], is the largest possible (i.e., for any method of constructing random processes Q_n and G on a common probability space, the order of closeness of their trajectories cannot be greater than $O(n^{-\frac{1}{2}}\log n)$). Note that in this case, $H_1(\varepsilon) \sim |\log \varepsilon|$ as $\varepsilon \to 0$, i.e., in our example the metric entropy with embedding increases sufficiently slowly as $\varepsilon \to 0$. At the same time, if $H_1(\varepsilon)$ increases, say, exponentially as $\varepsilon \to 0$, then the quantity $\Delta_n(\tilde{M}(n))$ in (9) will decrease considerably more slowly than $n^{-1/2} \log n$. The rate of this decrease with respect to $H_1(\cdot)$ can be arbitrarily small. The question arises whether such a deterioration is essentially in the considered generality? A positive answer to this question is provided by

T h e o r e m 3. There exist a distribution P and a class $\mathcal{B} \subseteq \mathcal{A}$ for which:

1) $H_1(\varepsilon) \sim \psi(\varepsilon)\varepsilon^{-1}$

as $\varepsilon \to 0$, where $\psi(\cdot)$ is a positive function satisfying the relations $\lim_{\varepsilon\to 0} \psi(\varepsilon) = 0$, $\lim_{\varepsilon\to 0} \psi(\varepsilon)\varepsilon^{-\alpha} = \infty$ for every $\alpha > 0$, $\sum_{k\geq 1} \psi(2^{-k}) < \infty$;

2) for any method of specifying random processes $\{Q_n(\cdot)\}$ and $G(\cdot)$ on a common probability space, one has

$$\sup_{A\in\mathcal{B}} |Q_n(A) - G(A)| \geqslant c\Delta_n(\widetilde{M}(n)) \tag{44}$$

with probability 1 for all sufficiently large n, where c is an absolute positive constant.

Proof. Let $\bar{P} = \{p_i\}$ be an arbitrary discrete distribution in $\mathcal{X}$ with a countable number of atoms $a_1, a_2, \ldots$, where the $\{a_i\}$ are indexed so that the sequence of the corresponding masses $\{p_i\}$ is monotonically decreasing. One can assume that the processes $\bar{Q}_n(\cdot)$ and $\bar{G}(\cdot)$ corresponding to the given $\bar{P}$ are defined by formulas (12) and (13) for $m = \infty$ (see [3] for more detail). Suppose that the random processes $V_n(\cdot)$ and $\mathring{w}(\cdot)$ in formulas (12) and (13) and thereby the processes $\bar{Q}_n(\cdot)$ and $\bar{G}(\cdot)$ have already been specified on a common probability space.

We set $\mathcal{B} = \mathcal{A}$ and assume that

$$\sum_{i\geqslant 1} \sqrt{p_i} < \infty. \tag{45}$$

In this case, the quantity $\sup_{A\in\mathcal{A}} |\bar{Q}_n(A) - \bar{G}(A)|$ is nothing but the distance in variation between the charges $\bar{Q}_n$ and $\bar{G}$, being absolutely continuous with respect to the counting measure at the points $\{a_i\}$, i.e.,

$$\sup_{A\in\mathcal{A}} |\bar{Q}_n(A) - \bar{G}(A)| \geqslant \frac{1}{2}\sum_{i>m} |\Delta_i V_n - \Delta_i \mathring{w}|$$
$$\geqslant \frac{1}{2}\sum_{i>m} |\Delta_i \mathring{w}| - \frac{1}{2}\sum_{i>m} |\Delta_i V_n| \equiv \zeta, \tag{46}$$

where

$$\Delta_i g = g(F(i+1)) - g(F(i)), \quad F(i) = \sum_{k<i} p_k;$$

m being an arbitrary natural number which will be chosen later on. Note that when (45) is satisfied the random variable ζ is nonsingular (see [3]).

We shall need upper estimates for the probabilities of deviations of each of the two sums on the right side of (46).

L e m m a 6. Let condition (45) be satisfied. Then for any positive $x \leqq \sqrt{n} \sum_{i>m} p_i$ and z, one has the inequalities

$$\mathbf{P}\Big(\sum_{i>m} |\Delta_i V_n| > x + 2\sqrt{n}\sum_{i>m} p_i\Big) \leqslant \exp\Big\{-\frac{x^2}{4\sum_{i>m} p_i}\Big\}, \tag{47}$$

$$\mathbf{P}\Big(\Big|\sum_{i>m} |\Delta_i \overset{\circ}{w}| - \sqrt{\frac{2}{\pi}}\sum_{i>m}\sqrt{p_i}\Big| > z\Big) \leqslant 4\exp\Big\{-\frac{z^2}{8\sum_{i>m} p_i}\Big\}. \tag{48}$$

Proof. Relation (47) follows immediately from Bernshtein's inequality (24), since

$$\sum_{i>m} |\Delta_i V_n| \leqslant \frac{1}{\sqrt{n}} \sum_{j=1}^{n} I_{\xi_j}([F(m+1), 1]) + \sqrt{n}\sum_{i>m} p_i,$$

where $\xi_1,\ldots,\xi_n$ are independent random variables with uniform distribution on $[0,1]$.

To prove (48), we use the well-known representation $\overset{\circ}{w}(t) = w(t) - tw(1)$, where $w(t)$ is the standard Wiener process. Then

$$\mathbf{P}\Big(\Big|\sum_{i>m} |\Delta_i \overset{\circ}{w}| - \sqrt{\frac{2}{\pi}}\sum_{i>m}\sqrt{p_i}\Big| > z\Big)$$

$$\leqslant \mathbf{P}\Big(\Big|\sum_{i>m} |\Delta_i w| - \sqrt{\frac{2}{\pi}}\sum_{i>m}\sqrt{p_i}\Big| > z/2\Big) + \mathbf{P}\Big(|w(1)|\sum_{i>m} p_i > z/2\Big)$$

$$\leqslant \mathbf{P}\Big(\sum_{i>m} |\Delta_i w| < \sqrt{\frac{2}{\pi}}\sum_{i>m}\sqrt{p_i} - z/2\Big) \tag{49}$$

$$+ \mathbf{P}\Big(\sum_{i>m} |\Delta_i w| > \sqrt{\frac{2}{\pi}}\sum_{i>m}\sqrt{p_i} + z/2\Big) + 2\exp\Big\{-\frac{z^2}{8\big(\sum_{i>m} p_i\big)^2}\Big\}.$$

Let us estimate the first summand on the right side of (49). Applying

Chebyshev's exponential inequality and noting the independence of the random variable $\{\Delta_i w\}$, we obtain for any $t > 0$ that

$$\mathbf{P}\left(\sum_{i>m}|\Delta_i w| < \sqrt{\frac{2}{\pi}}\sum_{i>m}\sqrt{p_i} - z/2\right)$$
$$\leqslant \exp\left\{t\sqrt{\frac{2}{\pi}}\sum_{i>m}\sqrt{p_i} - tz/2\right\}\prod_{i>m}\mathbf{E}\exp\{-t|\Delta_i w|\}. \tag{50}$$

Note that

$$\mathbf{E}|\Delta_i w| = \sqrt{2p_i/\pi}, \quad \mathbf{E}(\Delta_i w)^2 = p_i.$$

Using elementary inequalities

$$e^{-x} \leqslant 1 - x + x^2/2, \quad x \geqslant 0,$$
$$1 + y \leqslant e^y, \quad y \in R,$$

we easily establish that

$$\prod_{i>m}\mathbf{E}\exp\{-t|\Delta_i w|\} \leqslant \exp\left\{-t\sqrt{\frac{2}{\pi}}\sum_{i>m}\sqrt{p_i} + \frac{t^2}{2}\sum_{i>m}p_i\right\}.$$

Substituting the resulting estimate into (50) and setting $t = z(2\sum_{i>m}p_i)^{-1}$, we finally have

$$\mathbf{P}\left(\sum_{i>m}|\Delta_i w| < \sqrt{\frac{2}{\pi}}\sum_{i>m}\sqrt{p_i} - z/2\right) \leqslant \exp\left\{-\frac{z^2}{8\sum_{i>m}p_i}\right\}.$$

One can show similarly that an analogous estimate holds also for the second summand on the right side of (49).

The lemma is proved.

Next, set $x = z = \sqrt{n}\sum_{i>m}p_i$ in (47) and (48). Then it follows from (46) that on the set of elementary outcomes, the probability of which is not smaller than $1 - 5\exp\{-n\sum_{i>m}p_i/8\}$, we have the inequality

$$\zeta \geqslant \frac{1}{\sqrt{2\pi}}\sum_{i>m}\sqrt{p_i} - 2\sqrt{n}\sum_{i>m}p_i. \tag{51}$$

We shall show that for a sufficiently broad class of discrete distributions, for an appropriate choice of m, the right side of (51), coincides with respect to the order of smallness with the upper estimate obtained in Section 1. At the same time, we shall establish that the convergence rate in the central limit theorem can be arbitrarily small.

From condition (45) and the monotonicity of $\{p_i\}$, one has the representation

$$p_i = \phi^2(i)i^{-2} , \tag{52}$$

where $\lim_{i\to\infty} \phi(i) = 0$. One can assume without loss of generality that $\phi(x)$ is positive and continuously differentiable on $(0,\infty)$. Consider the class of discrete distributions for which the function $\phi(x)$ in (52) for all $x > 0$ satisfies the inequality

$$|\varphi'(x)| \leqslant \gamma(x)\varphi(x)x^{-1}, \tag{53}$$

where $\lim_{x\to\infty} \gamma(x) = 0$. For any $\varepsilon \in (0,\frac{1}{2})$, we denote by $m(\varepsilon,\gamma)$ the smallest natural number defined by the relation $\gamma(x) \leqq \varepsilon$ for all $x \geqq m(\varepsilon,\gamma)$. Since for any natural m

$$\int_{m+1}^{\infty} x^{-2}\varphi^2(x)\,dx \leqslant \sum_{i>m} p_i \leqslant \int_{m}^{\infty} x^{-2}\varphi^2(x)\,dx,$$

then, noting the equality

$$\int_{m}^{\infty} x^{-2}\varphi^2(x)\,dx = m^{-1}\varphi^2(m) + \int_{m}^{\infty} x^{-1}\varphi'(x)\,dx$$

and taking (53) into account, we obtain that for all $m \geqq m(\varepsilon,\gamma)$,

$$\frac{\varphi^2(m+1)}{(1+\varepsilon)(m+1)} \leqslant \sum_{i>m} p_i \leqslant \frac{\varphi^2(m)}{(1-\varepsilon)m} \tag{54}$$

yielding, in turn, the relations

$$\frac{\varphi^2(m)}{m} \leqslant \frac{m(1+\varepsilon)}{m-1-\varepsilon} \sum_{i>m} p_i, \tag{55}$$

$$\sum_{i>m} p_i \leqslant \frac{(1-\varepsilon)(m+1)+1}{(1-\varepsilon)(m+1)} \cdot \frac{\varphi^2(m+1)}{m+1}. \tag{56}$$

Set

$$m(n) = \max\{i : np_i \geqslant 1\},$$
$$m^*(n) = \max\{m(n),\ m(\varepsilon,\gamma)+1\}.$$

In (51) set $m = m^*(n)$. Then from (56) and the definition of $m^*(n)$, we easily derive

$$\sqrt{n} \sum_{i>m^*(n)} p_i \leqslant C\varphi(m^*(n)+1).$$

Next, by using l'Hôpital's rule and (53), we see that as $m \to \infty$,

$$\varphi(m)\left[\int_m^\infty x^{-1}\varphi(x)\,dx\right]^{-1}\to 0.$$

Since $\lim_{n\to\infty} m^*(n) = \infty$ and

$$\sum_{i>m}\sqrt{p_k}\geqslant\int_{m+1}^\infty x^{-1}\varphi(x)\,dx, \tag{57}$$

then for any arbitrarily small $\alpha > 0$ and all $n \geqq n_0(\alpha,\phi)$, we have the inequality

$$\sqrt{n}\sum_{i>m^*(n)} p_i\leqslant\alpha\sum_{i>m^*(n)}\sqrt{p_i}. \tag{58}$$

Thus, it follows from (51), (57), (58) and the definition of $m^*(n)$, that, with probability not smaller than $1 - 5\exp\{-m^*(n)/8\}$, for $n \geqq n_0(\alpha,\phi)$ we have the inequality

$$\zeta\geqslant\left[(2\pi)^{-1/2}-2\alpha\right]\int_{m^*(n)+1}^\infty x^{-1}\varphi(x)\,dx, \tag{59}$$

where, of course, it is assumed that $\alpha < (2\sqrt{2}\,\pi)^{-1}$.

Now let us consider an upper bound for the convergence rate. We showed in Section 1 that for any discrete distribution (with monotonic decrease of the masses $\{p_i\}$), one has

$$\Delta_n(\widetilde{M}(n))\leqslant\delta(n)$$
$$\equiv\widetilde{C}\min_{m:\,m<4n\sum_{i>m}p_i}\max\left(\left[1+m^{-1}\log n\right]\sum_{i>m}\sqrt{p_i},\ mn^{-1/2}\log n\right).$$

Note that (55) and the definition of $m^*(n)$ yield the estimates

$$m^*(n)n^{-1/2}\leqslant\varphi(m^*(n)),$$
$$m^*(n)\leqslant\frac{nm^*(n)(1+\varepsilon)}{m^*(n)-1-\varepsilon}\sum_{i>m^*(n)}p_i<3n\sum_{i>m^*(n)}p_i.$$

In other words,

$$\delta(n)\leqslant\widetilde{C}\max\left(\left[1+\frac{\log n}{m^*(n)}\right]\int_{m^*(n)}^\infty x^{-1}\varphi(x)\,dx,\quad\varphi(m^*(n))\log n\right). \tag{60}$$

Now it is not hard to construct an example of a distribution for which the orders of smallness of the upper and lower bounds coincide. However, estimate (60) can be determined even more specifically by narrowing the class of discrete distributions defined by (53). That is, in addition to

(53) we require that the function $\phi(x)$ in (52) be monotonically decreasing as $x \to \infty$. In this case, we have the representation

$$\varphi(x)=\beta(x)(\log x)^{-1}, \quad x>1, \tag{61}$$

where $\lim_{x\to\infty} \beta(x) = 0$ and, furthermore, condition (53) implies that $\phi(x)$ and $\beta(x)$ are slowly varying functions as $x \to \infty$, i.e., for any $t > 0$,

$$\lim_{n\to\infty} m^*(n)\, n^{-(1/2-t)} = \infty. \tag{62}$$

Obviously, for $n \geqq n_0(\cdot)$ from (60), noting (61) and (62), one has the estimate

$$\delta(n) \leqslant 2\widetilde{C} \max\left(\int_{m^*(n)}^{\infty} \beta(x)(x\log x)^{-1}dx,\ \beta(m^*(n))\right), \tag{63}$$

which essentially is the same as the estimate obtained in (9). At the same time, the right side of (59) is a lower bound, below which the variable $\sup_{A\in\mathcal{A}} |\bar{Q}_n(A) - P(A)|$, starting with some n, does not go with probability 1 (this follows from the Borel-Cantelli lemma, since $\exp\{-m^*(n)/8\} \leqq \exp\{-n^{1/4}/8\}$ for all sufficiently large n).

As an example we consider the discrete distribution with masses

$$p_k = C\left[[\log^{(r)}(k+d)]^{\varepsilon} \prod_{1\leqslant i\leqslant r} \log^{(i)}(k+d)\right]^{-2}, \tag{64}$$

where $\log^{(i)}(\cdot)$ is the i-fold logarithm, $\log^{(r)} d > 0$, $\varepsilon > 0$, and C is a normalizing constant. It is easily seen that in this case condition (45) is satisfied and, moreover,

$$\begin{gathered} m^*(n) \underset{n\to\infty}{\sim} c_1 \sqrt{n}(\log^{(r)} n)^{\varepsilon} \prod_{1\leqslant i\leqslant r} \log^{(i)} n, \\ \varphi(x)\log x \equiv \beta(x) = c_2\left[[\log^{(r)}(x+d)]^{\varepsilon} \prod_{2\leqslant i\leqslant r} \log^{(i)}(x+d)\right]^{-1}, \end{gathered} \tag{65}$$

$$\int_m^{\infty} \frac{\beta(x)}{x\log x}\,dx \underset{m\to\infty}{\sim} c_3(\log^{(r)} m)^{-\varepsilon}, \tag{66}$$

where, by definition, $\prod_{k\leqq i\leqq m} = 1$ for $k > m$ (case $r = 1$).

Since the order of smallness in (65) as $n \to \infty$ is not smaller than in (66), the upper and lower bounds of the distribution (65) have the same order $(\log^{(n)} n)^{-\varepsilon}$. Furthermore, it follows from the results of Section 1 and from relation (54) that under these conditions

$$H_1(\varepsilon) \underset{\varepsilon \to 0}{\sim} \psi(\varepsilon)\varepsilon^{-1},$$

where $\psi(\varepsilon) = c_4\phi(1/\varepsilon)$. Inequality (44) follows from (59), (63) - (66). The Theorem is proved.

REMARK. It follows from the definition of $m^*(n)$, (59) and (63) that if the function $\beta(x)$ in (61) is slowly varying as $x \to \infty$ (i.e., the entropy $H_1(\varepsilon)$ is close to critical), then the estimate of the convergence rate in (4) for $\mathcal{B} = \mathcal{A}$ for any discrete distribution is essentially determined by the value of the integral $\int_{\sqrt{n}}^{\infty} \beta(x)(x \log x)^{-1}dx$, of which we demand no more than its existence. Therefore the rate of convergence in (4) can be arbitrarily small in the considered generality.

REFERENCES

[1] Dudley, R.M. "Limit Theorems for Empirical Measures." In *Vtoraya Vil'nyusskaya konferentsiya po teorii veroyatnostej i matematicheskoj statistike*, Tezisy dokladov (The Second Vilnius Conference on Probability Theory and Mathematical Statistics, Abstracts), pp. 40-41 (vol.3). Vilnius, 1977.

[2] ______. "Central Limit Theorems for Empirical Measures." *The Annals of Probability*, vol.6, no.6 (1978): 899-929.

[3] Borisov, I.S. "Problem of Accuracy of Approximation in the Central Limit Theorem for Empirical Measures." *Siberian Mathematical Journal*, vol.24, no.6 (1983): 833-43. (English transl.)

[4] Fernique, Xavier. "Regularitédes trajectoires des fonctions aléatoires gaussiennes." *Lecture Notes in Mathematics*, 480, pp. 1-96. New York Berlin Heidelberg Tokyo: Springer Verlag, 1975.

[5] Borisov, I.S. "Approximation of Empirical Fields Constructed With Respect to Vector Observations With Dependent Components." *Siberian Mathematical Journal*, vol.23, no.5 (1982): 615-24. (English transl.)

[6] Philipp, Walter, and Pinzur, L. "Almost Sure Approximation Theorems for the Multivariate Empirical Process." *Z. Wahrscheinlichkeitstheorie verw. Gebiete*, B.54, H.1 (1980): 1-13.

[7] Komlós, J., Major, P., and Tusnády, G. "An Approximation of Partial Sums of Independent RV'-s, and the Sample DF. I." *Z. Wahrscheinlichkeitstheorie verw. Gebiete*, B.32, H.1/2 (1975): 111-33.

[8] Yurinskij (Yurinskii), V.V. "Exponential Inequalities for Sums of Random Variables." *J. Multivariate Analysis*, vol.6, no.4 (1976): 473-99.

V.S. Lugavov

ON THE DISTRIBUTION OF THE SOJOURN TIME ON A HALF-AXIS AND THE FINAL POSITION OF A PROCESS WITH INDEPENDENT INCREMENTS, CONTROLLED BY A MARKOV CHAIN

1. Introduction

We shall consider the two-dimensional Markov process

$$\mathscr{L}_t = \{\xi_t, \varkappa_t;\ t \geqslant 0\},\ \ \xi_0 = 0\,,$$

which is time-homogeneous and is additive in the first component. The second component $\{\kappa_t;\ t \geqq 0\}$ is a homogeneous Markov process with values $1, 2, \ldots, N$, and the first coordinate $\{\xi_t;\ t \geqq 0\}$ is a process with conditionally independent increments for fixed κ_t (see [1]). The evolution of the process is given by the matrix

$$\Psi(s,\ t) = \|\mathbf{M}(\exp\{-s(\xi_{t+u} - \xi_u)\};\ \varkappa_{t+u} = j \mid \varkappa_u = i)\|,\ \ u \geqslant 0.$$

The matrix $\Psi(s,t)$ (see [2]) has the form $\Psi(s,t) = \exp(tA(s))$, where

$$A(s) = \begin{Vmatrix} -g_1 + K_1(s) & g_{12}\Psi_{12}(s) \ldots g_{1N}\Psi_{1N}(s) \\ g_{21}\Psi_{21}(s) & -g_2 + K_2(s) \ldots g_{2N}\Psi_{2N}(s) \\ g_{N1}\Psi_{N1}(s) & g_{N2}\Psi_{N2}(s) \ldots -g_N + K_N(s) \end{Vmatrix}, \tag{1.1}$$

where

$$g_{ij} \geqslant 0;\ \ g_i = \sum_{j=1}^{N} g_{ij};\ \ \Psi_{ij}(s) = \int_{-\infty}^{\infty} e^{-sx}\, dF_{ij}(x);$$

F_{ij} is the distribution of the jump of the process ξ_t when the process κ_t makes the transition from state i to state j; the $K(s)$ are the cumulants corresponding to homogeneous processes with independent increments:

$$K_l(s) = -sb_l + \frac{s^2\sigma_l^2}{2} + \int_{-\infty}^{\infty} \left(e^{-sx} - 1 + \frac{sx}{1+x^2}\right) S_l(dx). \tag{1.2}$$

If $\int_{|x|<1} S_\ell(dx) < \infty$, then we can rewrite (1.2) in the form

$$K_l(s) = -sa_l + \frac{s^2\sigma_l^2}{2} + \int_{-\infty}^{\infty} (e^{-sx} - 1)\, S_l(dx),$$

where a_ℓ is called the drift coefficient. If $\sigma_\ell^2 = 0$ and S_ℓ is a finite measure, then $K_\ell(s)$ is called the cumulant of a generalized Poisson process.

Set

$$\phi(x) = \begin{cases} 1 & \text{for } x > 0, \\ 0 & \text{for } x \leqq 0, \end{cases}$$

and introduce the functional

$$L_t = \int_0^t \varphi(\xi_u)\, du,$$

where L_t is the sojourn time of the process $\{\xi_u, 0 \leqq u \leqq t\}$ on the positive half-axis; one may similarly define L_t^x as the sojourn time of this process on the half-axis, (x,∞).

The boundary value functionals L_t, L_t^x are studied in [3] and [4] where integral transformations of the generating functions of L_t and L_t^x are obtained under additional conditions on L_t. The main result of this article is the transformation of the joint distribution of ξ_t, κ_t, L_t^x without additional constraints on L_t. In this article, a scheme of discrete approximation is used for processes with continuous time (see [5], [6], [7]). In [5], a transformation of the joint distribution of the random variables (ξ_t, L_t) is obtained, in particular, for $N = 1$. We have also used the results of [8] which contain integral transformations of the joint distribution of the sojourn time on the half-axis (x,∞) and of the final position of a random walk controlled by a Markov chain.

2. Factorization identities for sums of random variables given on a finite Markov chain

Consider a homogeneous two-dimensional Markov chain $(S_n, \kappa_n)_{n=0}^\infty$, with the second coordinate $\{\kappa_n\}_{n=0}^\infty$ being a homogeneous Markov chain with state space $\{1,2,..,N\}$. The matrix defining the evolution of the process $\{S_n, \kappa_n\}_{n=0}^\infty$ for any $y \in R$ has the form

$$\Psi(s) = \|\mathbf{M}(\exp\{-s(S_{n+1} - S_n)\};\ \varkappa_{n+1} = j \mid S_n = y,\ \varkappa_n = i)\|.$$

The structure of a matrix $\Psi(s)$ has been examined in [10].

Suppose $S_0 = 0$, Δ_n^x is the number of elements of the sequence $S_1, \dots, S_n$, greater than x ($\Delta_0^x = 0$), I is the identity matrix of order N, and

$$U_{ij}^x(s, \omega, n) = \mathbf{M}\left(e^{-s(S_n - x)}\omega^{\Delta_n^x}; \ \varkappa_n = j \mid \varkappa_0 = i\right),$$
$$U_n^x(s, \omega) = \left\| U_{ij}^x(s, \omega, n) \right\|_{i,j=\overline{1,N}}, \quad U_0^x(s, \omega) = e^{sx} I,$$
$$U^x(s, \omega, \rho) = \sum_{n=0}^{\infty} \rho^n U_n^x(s, \omega).$$

Introduce also the projection operator

$$T_{\{\alpha,\beta\}}\left(\int_{-\infty}^{\infty} e^{-sx}\mu(dx)\right) = \int_{\{\alpha,\beta\}} e^{-sx}\mu(dx), \ T_{(-\infty,0]} \equiv T_0^*, \ T_{(0,\infty)} \equiv T_0.$$

With this notation, from Theorem 5 of [8] one has

T h e o r e m 1. For Re (s) = 0, $|\rho| < 1$, $|\rho\omega| < 1$

$$U^x(s, \omega, \rho) = \begin{cases} \left[T_0 \{U_0^x(s, \omega)\left[\Psi^-(s, \omega, \rho)\right]^{-1}\}\right] \times \left[\Psi^+(s, \omega, \rho)\right]^{-1} \\ \qquad \times [I - \rho\Psi(s)]^{-1} \quad \text{for} \quad x < 0, \\ \left[T_0^* \{U_0^x(s, \omega)\Psi^+(s, \omega, \rho)\}\right] \times \left[\Psi^+(s, \omega, \rho)\right]^{-1} \\ \qquad \times [I - \rho\Psi(s)]^{-1} \quad \text{for} \quad x \geqslant 0, \end{cases} \tag{2.1}$$

where for Re (s) = 0,

$$[I - \rho\Psi(s)]^{-1}[I - \rho\omega\Psi(s)] = \Psi^+(s, \omega, \rho)\Psi^-(s, \omega, \rho) \tag{2.2}$$

and $\Psi^+(s,\omega,\rho)$, $\Psi^-(s,\omega,\rho)$ satisfy the following conditions:

A_1. The matrix $\Psi^+(s,\omega,\rho)$ has inverse $[\Psi^+(s,\omega,\rho)]^{-1}$ for Re (s) ≧ 0; $\Psi^+(s,\omega,\rho)$ and $[\Psi^+(s,\omega,\rho)]^{-1}$ are bounded and continuous for Re (s) ≧ 0; they are regular for Re (s) > 0; $\Psi^+(\infty,\omega,\rho) = I$.

A_2. The matrix $\Psi^-(s,\omega,\rho)$ has inverse $[\Psi^-(s,\omega,\rho)]^{-1}$ for Re (s) ≦ 0; $\Psi^-(s,\omega,\rho)$ and $[\Psi^-(s,\omega,\rho)]^{-1}$ are bounded and continuous for Re (s) ≦ 0; they are regular for Re (s) < 0.

Representation (2.2) is the so-called matrix factorization of $[I - \rho\Psi(s)]^{-1}[I - \rho\omega\Psi(s)]$ (see also [10], [11]). It follows from Liouville's theorem that conditions A_1, A_2 and (2.2) uniquely determine $\Psi^+(s,\omega,\rho)$ and $\Psi^-(s,\omega,\rho)$.

Theorem 1 is proved in [9] for x = 0. Note that in [8], assertions (133), (134), which could be used for deriving Theorem 1, are erroneous: they contradict Theorem 5 thereof.

We give a probabilistic interpretation to the components of the factorization $\Psi^{+}(s,\omega,\rho)$, $\Psi^{-}(s,\omega,\rho)$.

L e m m a 1. The following relations hold:

$$\Psi^{+}(s,\omega,\rho) = I + (1-\omega)\sum_{k=1}^{\infty}\rho^{k}\left\| \mathbf{M}\left(e^{-sS_k}\delta(S_k>0)\,\omega^{\sum_{i=1}^{k-1}\delta(S_k<S_i)};\ \varkappa_k=j\mid\varkappa_0=i\right)\right\|, \tag{2.3}$$

$$\Psi^{-}(s,\omega,\rho) = I + (1-\omega)\sum_{k=1}^{\infty}\rho^{k}\left\| \mathbf{M}\left(e^{-sS_k}\delta(S_k\leqslant 0)\,\omega^{\sum_{i=1}^{k-1}\delta(S_i>0)};\ \varkappa_k=j\mid\varkappa_0=i\right)\right\|, \tag{2.4}$$

$$[\Psi^{+}(s,\omega,\rho)]^{-1} = I - (1-\omega)\sum_{k=1}^{\infty}\rho^{k}\left\| \mathbf{M}\left(e^{-sS_k}\delta(S_k>0)\,\omega^{\sum_{i=1}^{k-1}\delta(S_i>0)};\ \varkappa_k=j\mid\varkappa_0=i\right)\right\|, \tag{2.5}$$

$$[\Psi^{-}(s,\omega,\rho)]^{-1} = I - (1-\omega)\sum_{k=1}^{\infty}\rho^{k}\left\| \mathbf{M}\left(e^{-sS_k}\delta(S_k\leqslant 0)\,\omega^{\sum_{i=1}^{k-1}\delta(S_k<S_i)};\ \varkappa_k=j\mid\varkappa_0=i\right)\right\|, \tag{2.6}$$

where $\delta(X)$ is the indicator of the event X.

Proof. Let A, B, C, D denote the matrices on the right sides of (2.3), (2.4), (2.5), (2.6), respectively.

Let us verify the equality

$$[I-\rho\Psi(s)]^{-1}[I-\rho\omega\Psi(s)] = A\times B. \tag{2.7}$$

For the left side of relation (2.7), we have

$$[I-\rho\Psi(s)]^{-1}[I-\rho\omega\Psi(s)] = [I-\rho\Psi(s)]^{-1}[I-\rho\Psi(s)+\rho(1-\omega)\Psi(s)]$$

$$= I + (1-\omega)\sum_{k=1}^{\infty}\rho^{k}\left\|\mathbf{M}\left(e^{-sS_k};\ \varkappa_k=j\mid\varkappa_0=i\right)\right\|.$$

The right side of (2.7) becomes

$$A\times B = I + (1-\omega)\sum_{k=1}^{\infty}\rho^{k}\left\|\mathbf{M}\left[e^{-sS_k}\left(\omega^{\sum_{i=1}^{k-1}\delta(S_i>0)}\delta(S_k\leqslant 0) + \omega^{\sum_{i=1}^{k-1}\delta(S_k<S_i)}\delta(S_k>0)\right);\ \varkappa_k=j\mid\varkappa_0=i\right]\right\|$$

$$+ (1-\omega)^2 \sum_{k=2}^{\infty} \rho^k \sum_{\substack{m,n\geqslant 1\\ m+n=k}} \left\| \mathbf{M}\left(e^{-sS_m}\omega^{\sum_{i=1}^{m-1}\delta(S_m<S_i)}\delta(S_m>0);\quad \varkappa_m=j\,|\,\varkappa_0=i\right)\right\|$$

$$\times \left\| \mathbf{M}\left(e^{-sS_n}\delta(S_n\leqslant 0)\,\omega^{\sum_{i=1}^{n-1}\delta(S_i>0)};\ \varkappa_n=j\,|\,\varkappa_0=j\right)\right\|.$$

The last summand of this equality is equal to

$$(1-\omega)^2 \sum_{k=2}^{\infty} \rho^k \sum_{m=1}^{k-1} \left\| \mathbf{M}\left(e^{-sS_k}\omega^{\sum_{i=1}^{m-1}\delta(S_m<S_i)}\delta(S_m>0)\right.\right.$$

$$\left.\left.\times\ \omega^{\sum_{i=m+1}^{k-1}\delta(S_i>S_m)}\delta(S_k\leqslant S_m);\ \varkappa_k=j\,|\,\varkappa_0=i\right)\right\|.$$

Thus we need to verify

$$I+(1-\omega)\sum_{k=1}^{\infty}\rho^k\left\|\mathbf{M}\left(e^{-sS_k};\ \varkappa_k=j\,|\,\varkappa_0=i\right)\right\|$$

$$=I+(1-\omega)\sum_{k=1}^{\infty}\rho^k\left\|\mathbf{M}\left[e^{-sS_k}\left(\delta(S_k\leqslant 0)\,\omega^{\sum_{i=1}^{k-1}\delta(S_i>0)}\right.\right.\right.$$

$$\left.\left.\left.+\ \omega^{\sum_{i=1}^{k-1}\delta(S_k<S_i)}\delta(S_k>0)\right);\ \varkappa_k=j\,|\,\varkappa_0=i\right]\right\|$$

$$+(1-\omega)^2\sum_{k=2}^{\infty}\rho^k\sum_{m=1}^{k-1}\left\|\mathbf{M}\left(e^{-sS_k}\omega^{\sum_{i=1}^{m-1}\delta(S_m<S_i)+\sum_{i=m+1}^{k-1}\delta(S_i>S_m)}\delta(S_m>0)\right.\right.$$

$$\left.\left.\times\ \delta(S_k\leqslant S_m);\ \varkappa_k=j\,|\,\varkappa_0=i\right)\right\|.$$

For this, it suffices to see that for $k \geqq 2$ one has the equality

$$1=\omega^{\sum_{i=1}^{k-1}\delta(S_i>0)}\delta(S_k\leqslant 0)+\omega^{\sum_{i=1}^{k-1}\delta(S_k<S_i)}\delta(S_k>0)$$

$$+(1-\omega)\sum_{m=1}^{k-1}\omega^{\sum_{i=1}^{m-1}\delta(S_m<S_i)+\sum_{i=m+1}^{k-1}\delta(S_i>S_m)}\delta(S_m>0)\,\delta(S_k\leqslant S_m). \tag{2.8}$$

Let $S_k > 0$. In this case, relation (2.8) takes on the form

$$1=\omega^{\sum_{i=1}^{k-1}\delta(S_k<S_i)}+(1-\omega)\sum_{m=1}^{k-1}\omega^{\sum_{i=1}^{m-1}\delta(S_m<S_i)+\sum_{i=m+1}^{k-1}\delta(S_i>S_m)}\delta(S_k\leqslant S_m).$$

Consider all the indices $\ell_j < k$ for $S_{\ell_j} \geqq S_k$, $j = 1,\dots,r$. It suffices to prove that

$$1 = \omega^r + (1-\omega)\sum_{j=1}^{r} \omega^{\sum_{i=1}^{l_j-1}\delta(S_{l_j}<S_i)+\sum_{i=l_j+1}^{k-1}\delta(S_i>S_{l_j})}. \tag{2.9}$$

Let $(p_1,\dots,p_r)$ be a permutation of the indices $\ell_1,\dots,\ell_r$: $S_{p_1} \leqq \dots \leqq S_{p_r}$; $p_i > p_{i+1}$ if $S_{p_i} = S_{p_{i+1}}$. Then (2.9) follows from the equality

$$\sum_{j=1}^{r} \omega^{\sum_{i=1}^{p_j-1}\delta(S_{p_j}<S_i)+\sum_{i=p_j+1}^{k-1}\delta(S_i>S_{p_j})} = \sum_{j=1}^{r}\omega^{r-j} = \frac{1-\omega^r}{1-\omega}.$$

Let $S_k \leqq 0$ and the ℓ_j be such that $\ell_j < k$, $S_{\ell_j} > 0$, $j = 1,\dots,r$. Then relation (2.8) again becomes (2.9). This completes the proof of (2.7).

To verify that $C \times A = I$, it suffices to show that

$$\begin{aligned}
&\sum_{k=1}^{\infty}\rho^k\left\|\mathbf{M}\left(e^{-sS_k}\omega^{\sum_{i=1}^{k-1}\delta(S_k<S_i)}\delta(S_k>0);\ \varkappa_k=j\,|\,\varkappa_0=i\right)\right\| \\
&= \sum_{k=1}^{\infty}\rho^k\left\|\mathbf{M}\left(e^{-sS_k}\omega^{\sum_{i=1}^{k-1}\delta(S_i>0)}\delta(S_k>0);\ \varkappa_k=j\,|\,\varkappa_0=i\right)\right\| \\
&+ (1-\omega)\sum_{k=1}^{\infty}\rho^k\left\|\mathbf{M}\left(e^{-sS_k}\omega^{\sum_{i=1}^{k-1}\delta(S_i>0)}\delta(S_k>0);\ \varkappa_k=j\,|\,\varkappa_0=i\right)\right\| \\
&\times \sum_{k=1}^{\infty}\rho^k\left\|\mathbf{M}\left(e^{-sS_k}\omega^{\sum_{i=1}^{k-1}\delta(S_k<S_i)}\delta(S_k>0);\ \varkappa_k=j\,|\,\varkappa_0=i\right)\right\|
\end{aligned}$$

or, after elementary transformations,

$$\begin{aligned}
&\sum_{k=1}^{\infty}\rho^k\left\|\mathbf{M}\left(e^{-sS_k}\omega^{\sum_{i=1}^{k-1}\delta(S_k<S_i)}\delta(S_k>0);\ \varkappa_k=j\,|\,\varkappa_0=i\right)\right\| \\
&= \sum_{k=1}^{\infty}\rho^k\left\|\mathbf{M}\left(e^{-sS_k}\omega^{\sum_{i=1}^{k-1}\delta(S_i>0)}\delta(S_k>0);\ \varkappa_k=j\,|\,\varkappa_0=i\right)\right\| \\
&+ (1-\omega)\sum_{k=2}^{\infty}\rho^k\sum_{m=1}^{k-1}\left\|\mathbf{M}\left(e^{-sS_k}\omega^{\sum_{i=1}^{m-1}\delta(S_i>0)+\sum_{i=m+1}^{k-1}\delta(S_k<S_i)}\delta(S_m>0)\,\delta(S_k \right.\right. \\
&\qquad \left.\left. >S_m);\ \varkappa_k=j\,|\,\varkappa_0=i\right)\right\|.
\end{aligned}$$

For the last relation to hold, it suffices that for $k \geqq 2$ the equality

$$\omega^{\sum_{i=1}^{k-1}\delta(S_k<S_i)}\delta(S_k>0)=\omega^{\sum_{i=1}^{k-1}\delta(S_i>0)}\delta(S_k>0)$$
$$+(1-\omega)\sum_{m=1}^{k-1}\omega^{\sum_{i=1}^{m-1}\delta(S_i>0)}\delta(S_m>0)\,\delta(S_k>S_m)\,\omega^{\sum_{i=m+1}^{k-1}\delta(S_k<S_i)} \tag{2.10}$$

hold. Obviously, for $S_k \leqq 0$ equality (2.10) holds. Let $S_k > 0$. Let ℓ_i $(i = 1,\dots,r;\ \ell_1 < \cdots < \ell_r)$ denote the indices of the sums S_{ℓ_i} for which $\delta(S_{\ell_i} > 0)\,\delta(S_k > S_{\ell_i}) = 1$. Set

$$p_1=\sum_{j=1}^{l_1-1}\delta(S_j>0),\quad p_2=\sum_{j=l_1+1}^{l_2-1}\delta(S_j>0),\ \dots$$
$$\dots,\ p_r=\sum_{j=l_{r-1}+1}^{l_r-1}\delta(S_j>0),\quad p_{r+1}=\sum_{j=l_r+1}^{k-1}\delta(S_j>0).$$

Then the relation (2.10) being verified can be transformed as follows:

$$\omega^{\sum_{j=1}^{r+1}p_j}=\omega^{r+\sum_{j=1}^{r+1}p_j}+(1-\omega)\sum_{i=1}^{r}\omega^{\left(i-1+\sum_{j=1}^{i}p_j\right)+\sum_{j=i+1}^{r+1}p_j}.$$

As is easy to see, the last relation holds true. Thus the equality $C\times A = I$ is proved.

We next show that $B\times D = I$ or, equivalently,

$$\sum_{k=1}^{\infty}\rho^k\mathrm{M}\left(e^{-sS_k}\delta(S_k\leqslant 0)\,\omega^{\sum_{i=1}^{k-1}\delta(S_i>0)};\ \varkappa_k=j\mid\varkappa_0=i\right)$$
$$=\sum_{k=1}^{\infty}\rho^k\mathrm{M}\left(e^{-sS_k}\delta(S_k\leqslant 0)\,\omega^{\sum_{i=1}^{k-1}\delta(S_k<S_i)};\ \varkappa_k=j\mid\varkappa_0=i\right)$$
$$+(1-\omega)\sum_{k=2}^{\infty}\rho^k\sum_{m=1}^{k-1}\mathrm{M}\left(e^{-sS_k}\omega^{\sum_{i=1}^{m-1}\delta(S_i>0)}\delta(S_m\leqslant 0)\right.$$
$$\left.\times\ \omega^{\sum_{i=m+1}^{k-1}\delta(S_k<S_i)}\delta(S_k\leqslant S_m);\ \varkappa_k=j\mid\varkappa_0=i\right).$$

To prove the latter, it suffices to consider the case $S_k \leqq 0$ and show that for $k \geqq 2$ we have the equality

$$\omega^{\sum_{i=1}^{k-1}\delta(S_i>0)} = \omega^{\sum_{i=1}^{k-1}\delta(S_k\leqslant S_i)}$$

$$+ (1-\omega)\sum_{m=1}^{k-1}\omega^{\sum_{i=1}^{m-1}\delta(S_i>0)+\sum_{i=m+1}^{k-1}\delta(S_k\leqslant S_i)}\delta(S_m\leqslant 0)\,\delta(S_k\leqslant S_m). \qquad (2.11)$$

Let ℓ_i $(i = 1,..,r;\ \ell_1 < \cdots < \ell_r)$ denote the indices of the sums S_{ℓ_i} such that $\delta(S_{\ell_i} \leqq 0)\,\delta(S_k \leqq S_{\ell_i}) = 1$. Set

$$p_1 = \sum_{j=1}^{l_1-1}\delta(S_j>0), \qquad p_2 = \sum_{j=l_1+1}^{l_2-1}\delta(S_j>0), \ldots$$

$$\ldots,\ p_r = \sum_{j=l_{r-1}+1}^{l_r-1}\delta(S_j>0), \quad p_{r+1} = \sum_{j=l_r+1}^{k-1}\delta(S_j>0).$$

Then (2.11) becomes

$$\omega^{\sum_{j=1}^{r+1}p_j} = \omega^{r+\sum_{j=1}^{r+1}p_j} + (1-\omega)\sum_{i=1}^{r}\omega^{\sum_{j=1}^{i}p_j+\left(r-i+\sum_{j=i+1}^{r+1}p_j\right)},$$

which is easily verified. Then, $B \times D = I$.

Furthermore, as is easy to show, the matrices A, C and B, D satisfy conditions A_1 and A_2 respectively.

This completes the proof of the Lemma.

3. Auxiliary results

We classify a homogeneous process wtih independent increments as A_+, A_-, A_0, if it is a generalized Poisson process with positive, negative, or zero drift, respectively, and all the remaining homogeneous processes with independent increments as B. We assign the value i ($i \in \{1,\ldots,N\}$) of the coordinate κ_t to the set a_+, if in the representation (1.1) $K_i(s)$ is the cumulant of a process of type A_+. We analogously define the sets a_-, a_0, b. Thus the set of values of the coordinate κ_t is decomposed into four disjoint subsets a_+, a_-, a_0, b.

L e m m a 2. Let $u > 0$, $k \in b$. Then for arbitrary x,

$$\mathbf{P}\{\xi_u = x \mid \varkappa_0 = k\} = 0.$$

Proof. We have

$$\|\mathbf{M}(e^{i\lambda\xi_u};\ \varkappa_u = j \mid \varkappa_0 = i)\| = \|\mathbf{M}(e^{i\lambda\xi_t};\ \varkappa_t = j \mid \varkappa_0 = i)\|$$
$$\times \|\mathbf{M}(e^{i\lambda\xi_{u-t}};\ \varkappa_{u-t} = j \mid \varkappa_0 = i)\|$$

yielding

$$\mathbf{M}\left(e^{i\lambda\xi_u} \mid \varkappa_0 = k\right)$$
$$= \sum_{j=1}^{N}\sum_{i=1}^{N} \mathbf{M}\left(e^{i\lambda\xi_t};\ \varkappa_t = i \mid \varkappa_0 = k\right)\mathbf{M}\left(e^{i\lambda\xi_{u-t}};\ \varkappa_{u-t} = j \mid \varkappa_0 = i\right)$$

implying, in turn,

$$\left|\mathbf{M}\left(e^{i\lambda\xi_u} \mid \varkappa_0 = k\right)\right| \leqslant N\sum_{j=1}^{N} \left|\mathbf{M}\left(e^{i\lambda\xi_t};\ \varkappa_t = j \mid \varkappa_0 = k\right)\right|.$$

Let t be sufficiently small. Then

$$\mathbf{M}\left(e^{i\lambda\xi_u} \mid \varkappa_0 = k\right)| \leqslant N\left\{\left|\mathbf{M}\left(e^{i\lambda\xi_t};\ \varkappa_t = k \mid \varkappa_0 = k\right)\right| + \varepsilon/2\right\}$$
$$\leqslant N\left\{\left|\mathbf{M}\left(e^{i\lambda\xi_t};\ \forall s \in [0, t]\ \varkappa_s = k \mid \varkappa_0 = k\right)\right| + \varepsilon\right\}.$$

We assume that $P\{\kappa_0 = k\} \neq 0$. Then

$$N\left\{\left|\mathbf{M}\left(e^{i\lambda\xi_t};\ \forall s \in [0, t]\ \varkappa_s = k \mid \varkappa_0 = k\right)\right| + \varepsilon\right\}$$
$$\leqslant N\left\{\left|\mathbf{M}\left(e^{i\lambda\xi_t} \mid \forall s \in [0, t]\ \varkappa_s = k\right)\right| + \varepsilon\right\}.$$

Set

$$f(\lambda) = \mathbf{M}\left(e^{i\lambda\xi_u} \mid \varkappa_0 = k\right),\ f_t(\lambda) = \mathbf{M}\left(e^{i\lambda\xi_t} \mid \forall s \in [0, t]\ \varkappa_s = k\right).$$

We have thus shown that

$$|f(\lambda)| \leqslant N\{|f_t(\lambda)| + \varepsilon\}$$

yielding the inequality

$$\lim_{T\to\infty} \frac{1}{2T}\int_{-T}^{T} |f(\lambda)|^2\,d\lambda \leqslant N^2 \lim_{T\to\infty} \frac{1}{2T}\int_{-T}^{T} |f_t(\lambda)|^2 d\lambda + N^2(2\varepsilon + \varepsilon^2).$$

Since $k \in b$, $f_t(\lambda)$ is the characteristic function of continuous distribution function and consequently (see [12])

$$\lim_{T\to\infty} \frac{1}{2T}\int_{-T}^{T} |f_t(\lambda)|^2 d\lambda = 0.$$

Thus,

$$\lim_{T\to\infty} \frac{1}{2T}\int_{-T}^{T} |f(\lambda)|^2 d\lambda \leqslant N^2(2\varepsilon + \varepsilon^2).$$

Since ε was arbitrary, the lemma is proved.

Lemma 3. Let $0 \leqq s \leqq t < u$, $i \in \{1,\ldots,N\}$. Then $P\{\xi_i = \xi_u,\ \kappa_\alpha \in b$ for some $\alpha \in [t,u] \mid \kappa_s = i\} = 0$.

Proof. Let θ be the set of rationals in the interval (t,u). Then

$$\mathbf{P}\{\xi_t=\xi_u,\ \varkappa_\alpha\in b \quad \text{for} \quad \alpha\in[t,u]\,|\,\varkappa_s=i\}\leqslant\sum_{k\in b}\sum_{l\in\theta}\mathbf{P}\{\xi_t=\xi_u,\ \varkappa_l=k\,|\,\varkappa_s=i\}.$$

For the right side of this inequality to be zero, it suffices that $P\{\xi_i=\xi_u,\ \kappa_\ell=k\mid\kappa_t=j\}=0$. Which follows from Lemma 2, as well as the time-homogeneity of the process $\{\xi_t,\kappa_t\}$ considered and its additivity in the first component.

We use $P_i\{\cdot\}$ to denote $P\{\cdot\mid\kappa_0=i\}$.

L e m m a 4. For almost all $s>0$, $0\leqq\alpha\leqq s$, $i\in\{1,\dots,N\}$,

$$\mathbf{P}_i\{\xi_s=0,\ \varkappa_\alpha\in a_+\cup a_-\}=0.$$

To prove the Lemma, we need the following notation:

W is the union of the set on which the discrete components of the spectral measures S_ℓ are concentrated for $\ell\in a_+\cup a_-\cup a_0$, with the set of discontinuity points of the distribution functions $F_{ij}(i,j\in\{1,\dots,N\})$ of the jumps of the process ξ_t at the transition instants of the controlling chain from one state to another;

W^+ is the additive group generated by the set W;

R_ℓ^s is the event that ℓ transitions of the controlling chain occur on the segment $[0,s]$;

K_b^s is the event that for $\forall u\in[0,s]$, $\kappa_u\notin b$;

α_i are the drift coefficients for the processes in $A_+\cup A_-\cup A_0$ (the i indicates that the process corresponding to the state i is being considered).

Proof. By Lemma 2, 3 it suffices to consider the case $i\notin b$ and prove that

$$\mathbf{P}_i\{\xi_s=0,\ \varkappa_\alpha\in a_+\cup a_-,\ K_b^s\}=0.$$

We have

$$\mathbf{P}_i\{\xi_s=0,\ \varkappa_\alpha\in a_+\cup a_-,\ K_b^s\}=\sum_{l=0}^{\infty}\mathbf{P}_i\{\xi_s=0,\ \varkappa_\alpha\in a_+\cup a_-,\ K_b^s,\ R_l^s\}.$$

Consider the first summand on the right side. If $i\in a_0$, it is equal to zero, since the events $\{\kappa_\alpha\in a_+\cup a_-\}$ and R_0^s are incompatible in this case. In order that it be zero for $i\in a_+\cup a_-$ it suffices that $-\alpha_i s\notin W^+$.

Consider the other summands. For $\ell\geqq 1$, we have

$$\mathbf{P}_i\{\xi_s = 0,\ \varkappa_\alpha \in a_+ \cup a_-,\ R_l^s,\ K_b^s\}$$
$$\leqslant \mathbf{P}_i\Big\{\exists p_1, \ldots, p_l \in a_0 \cup a_- \cup a_+ : \{i, p_1, \ldots$$
$$\ldots, p_l\} \cap (a_- \cup a_+) \neq \varnothing,\ \alpha_i\tau_i + S(0, \tau_i) + \alpha_{p_1}\tau_{p_1} + S(\tau_i, \tau_i + \tau_{p_1}) + \ldots$$
$$+ \alpha_{p_{l-1}}\tau_{p_{l-1}} + S(\tau_i + \ldots + \tau_{p_{l-2}}, \tau_i + \ldots + \tau_{p_{l-1}})$$
$$+ \alpha_{p_l}\Big(s - \tau_i - \sum_{j=1}^{l-1} \times \tau_{p_j}\Big) + S(\tau_i + \ldots + \tau_{p_{l-1}}, s) + \gamma_{ip_1} + \gamma_{p_1p_2}$$
$$+ \ldots + \gamma_{p_{l-1}p_l} = 0\Big\},$$

where $\tau_i, \ldots, \tau_i+\tau_{p_1}+\cdots+\tau_{p_{\ell-1}}$ are the transition instants of the controlling chain κ_t, $S(\alpha,\beta)$ is the sum of jumps of ξ_t over the time (α,β), γ_{ij} is the size of the jump of ξ_t at the transition instant of the controlling chain from state i to state j.

If there are distinct numbers among the $\alpha_i, \alpha_{p_1}, \ldots, \alpha_{p_{\ell-1}}, \alpha_{p_\ell}$, the right side of the last inequality is equal to zero, since the random variables

$$\tau_i(\alpha_i-\alpha_{p_\ell}) + S(0,\tau_i),\ \tau_{p_{\ell-1}}(\alpha_{p_{\ell-1}}-\alpha_{p_\ell}) + S(\tau_i+\ldots+\tau_{p_{\ell-2}},$$
$$\tau_i+\ldots+\tau_{p_{\ell-1}}),\ S(\tau_i+\ldots+\tau_{p_{\ell-1}}, s),\quad \gamma_{ip_1},\ldots,\gamma_{p_{\ell-1}p_\ell})$$

are independent and at least one of the first ℓ random variables has continuous distribution. It is easy to see that the right side of the last inequality is equal to zero also for $\alpha_i = \alpha_{p_1} = \cdots = \alpha_{p_\ell} \neq 0$ if $\alpha_{p_\ell}s \notin W^+$. Since W^+ is countable, the lemma is proved.

It is not hard to derive from Lemma 4 the

L e m m a 5. For almost all $s-t$, $0 \leqq u \leqq t < s$, $i \in \{1,\ldots,N\}$,

$$\mathbf{P}\{\xi_t = \xi_s,\ \exists\alpha \in [t, s]:\ \varkappa_\alpha \in \alpha_+ \cup \alpha_- \mid \varkappa_u = i\} = 0.$$

Next, along with the process $L_t = \{\xi_t,\kappa_t;\ t \geqq 0\}$ we consider the processes $L_t^{(n)} = \{\xi_t^n,\kappa_t^n;\ t \geqq 0\}$, where $\xi_t^n = \xi_{k2^{-n}}$, $\kappa_t^n = \kappa_{k2^{-n}}$ for $k2^{-n} \leqq t < (k+1)2^{-n}$. We use $M_t = \int_0^t \phi(\xi_t-\xi_u)\,du$ to denote the sojourn time of the process on the half-axis $(-\infty,\xi_t)$. Also, let

$$L^n(t) = \int_0^t \varphi(\xi_u^n)\,du, \quad M^n(t) = \int_0^t \varphi(\xi_t^n - \xi_u^n)\,du,$$

where the function $\phi(x)$ is defined in the Introduction.

In addition, we consider the function

$$\phi_m(x) = \begin{cases} 1 & \text{for } x > 1/m, \\ mx & \text{for } 0 < x \leqq 1/m, \\ 0 & \text{for } x \leqq 0 \end{cases}$$

and define the following functionals:

$$L_m(t) = \int_0^t \varphi_m(\xi_u)\,du, \quad L_m^n(t) = \int_0^t \varphi_m(\xi_u^n)\,du,$$

$$M_m^n(t) = \int_0^t \varphi_m(\xi_t^n - \xi_u^n)\,du, \quad M_m(t) = \int_0^t \varphi_m(\xi_t - \xi_u)\,du.$$

L e m m a 6.

$$\forall \varepsilon > 0 \lim_{m\to\infty} \lim_{n\to\infty} \mathbf{P}_i\{|L_m^n(u) - L^n(u)| > \varepsilon\} = 0.$$

Introduce the notation

$$\mathbf{M}_i(\cdot) = \mathbf{M}(\cdot|\varkappa_0 = i), \quad \eta(x, n, m) = \varphi_m(\xi_x^n) - \varphi(\xi_x^n),$$

$S(x,y)$ is the event that κ_t has a jump on the interval (x,y),

$\bar{S}(s,y)$ is the complement of $S(x,y)$.

Proof. Introduce the random variable

$$\gamma_u = \min\{u, \min\{t\colon \varkappa_t \neq a_0\}\}.$$

For p such that $u/p < \varepsilon/2$, we have

$$\mathbf{P}_i\left\{\left|\int_0^u \eta(x, n, m)\,dx\right| > \varepsilon\right\}$$

$$\leqslant \sum_{k=0}^{p-1} \mathbf{P}_i\left\{\left|\int_0^u \eta(x, n, m)\,dx\right| > \varepsilon,\ \gamma_u \in [ku/p,\ (k+1)u/p]\right\}$$

$$\leqslant \sum_{k=0}^{p-1} \mathbf{P}_i\left\{\int_0^{ku/p} |\eta(x, n, m)|\,dx + \int_{ku/p}^{(k+1)u/p} |\eta(x, n, m)|\,dx\right.$$

$$\left. + \int_{(k+1)u/p}^{u} |\eta(x, n, m)|\,dx > \varepsilon,\quad \gamma_u \in [ku/p, (k+1)u/p]\right\}$$

$$\leqslant \sum_{k=1}^{p-1} \mathbf{P}_i\left\{\int_0^{ku/p} |\eta(x, n, m)|\,dx > \varepsilon/4,\ \gamma_u \in [ku/p,\ (k+1)\,u/p]\right\}$$

$$+ \sum_{k=0}^{p-2} \mathbf{P}_i\left\{\int_{(k+1)u/p}^{u} |\eta(x, n, m)|\,dx > \varepsilon/4,\ \gamma_u \in [ku/p,\ (k+1)\,u/p]\right\} \equiv A + B.$$

To estimate B:

$$B \leqslant 4\varepsilon^{-1} \sum_{k=0}^{p-2} \mathbf{M}_i\left(\int_{(k+1)u/p}^{u} |\eta(x, n, m)|\,dx;\quad \gamma_u \in [ku/p, (k+1)\,u/p]\right)$$

$$\leqslant 4\varepsilon^{-1} \sum_{k=0}^{p-2} \int_{(k+1)u/p}^{u} \mathbf{P}_i\{0 < \xi_x^n \leqslant 1/m,\quad \gamma_u \in [ku/p, (k+1)\,u/p]\}\,dx.$$

Passing to the double limit in the last inequality, first as $n \to \infty$, then as $m \to \infty$, we obtain

$$\lim_{m\to\infty} \lim_{n\to\infty} \sum_{k=0}^{p-2} \mathbf{P}_i\left\{\int_{(k+1)u/p}^{u} |\eta(x, n, m)|\,dx > \varepsilon/4,\ \gamma_u \in [ku/p,\ (k+1)\,u/p]\right\} \leqslant 4\varepsilon^{-1} \sum_{k=0}^{p-2} \int_{(k+1)u/p}^{u} \mathbf{P}_i\{\xi_x = 0,$$

$$\gamma_u \in [ku/p,\ (k+1)\,u/p]\}\,dx = 0.$$

The last equality holds by Lemmas 3, 5, since

$$\{\gamma_u \in [ku/p,\ (k+1)u/p\} \subseteq \{\exists t \leqslant x:\ \varkappa_t \in a_- \cup a_+ \cup b\}$$

for $x \geqslant (k+1)u/p$.

To estimate A:

$$A \leqslant 4\varepsilon^{-1} \sum_{k=1}^{p-1} \int_0^{ku/p} \mathbf{P}_i\{0 < \xi_x^n \leqslant 1/m,\quad \gamma_u \in [ku/p,\ (k+1)\,u/p]\}\,dx.$$

To estimate the integrand in the last inequality:

$$\mathbf{P}_i\{0 < \xi_x^n \leqslant 1/m,\quad \gamma_u \in [ku/p,\ (k+1)\,u/p]\}$$

$$\leqslant \mathbf{P}_i\{0 < \xi_x^n \leqslant 1/m,\quad \forall t \leqslant x\ \ \varkappa_t \in a_0\} \leqslant \mathbf{P}_i\{S([x2^n]/2^n, x)\}$$

$$+ \mathbf{P}_i\{0 < \xi_x \leqslant 1/m\} + \mathbf{P}_i\{\xi_x \neq \xi_x^n,\quad \forall_t < x\ \ \varkappa_t \in a_0,\ \overline{S}([x2^n]/2^n, x)\},$$

Passing to the limit in the extreme terms of the inequality first as

$n \to \infty$, then $m \to \infty$, we obtain that the double limit of the integrand at hand is equal to zero a.e. Therefore,

$$\lim_{m\to\infty}\lim_{n\to\infty}\sum_{k=1}^{p-1} \mathbf{P}_i\left\{\int_0^{ku/p} |\eta(x, n, m)|\,dx > \varepsilon/4, \quad \gamma_u \in [ku/p, (k+1)u/p]\right\} = 0.$$

The lemma is proved.

Lemma 7.
$$L_m^n(u) \underset{\mathbf{P}_i}{\longrightarrow} L_m(u).$$

Proof.

$$\forall\varepsilon > 0 \quad \mathbf{P}_i\{|L_m^n(u) - L_m(u)| > \varepsilon\}$$
$$\leqslant \varepsilon^{-1}\int_0^u \mathbf{M}_i|\varphi_m(\xi_t^n) - \varphi_m(\xi_t)|\,dt \leqslant \varepsilon^{-1}\int_0^u \mathbf{M}_i(|\varphi_m(\xi_t^n) - \varphi_m(\xi_t)|;$$
$$|\xi_t^n - \xi_t| < \delta)\,dt + \varepsilon^{-1}\int_0^u \mathbf{M}_i(|\varphi_m(\xi_t^n) - \varphi_m(\xi_t)|;\ |\xi_t^n - \xi_t| \geqslant \delta)\,dt$$
$$\leqslant \varepsilon^{-1}\left(m\delta u + \int_0^u \mathbf{P}_i\{|\xi_t^n - \xi_t| \geqslant \delta\}\,dt\right).$$

Passing to the limit as $n \to \infty$, then as $\delta \to 0$, we get the assertion of the lemma.

One can similarly prove the next assertions.

Lemma 8. $L_m(u) \underset{P_i}{\to} L_u$.

Lemma 9. $L^n(u) \underset{P_i}{\to} L_u$.

The proof follows from Lemmas 6, 7, 8 and the inequality

$$\mathbf{P}_i\{|L^n(u) - L_u| > \varepsilon\} \leqslant \mathbf{P}_i\{|L^n(u) - L_m^n(u)| > \varepsilon/3\}$$
$$+ \mathbf{P}_i\{|L_m^n(u) - L_m(u)| > \varepsilon/3\} + \mathbf{P}_i\{|L_m(u) - L_u| > \varepsilon/3\}.$$

Lemma 10.

$$\forall\varepsilon > 0 \ \lim_{m\to\infty}\lim_{n\to\infty} \mathbf{P}_i\{|\mathbf{M}_m^n(u) - M^n(u)| > \varepsilon\} = 0.$$

Proof. Introduce the notation:

$$r(u, t, m, n) = \varphi(\xi_u^n - \xi_t^n) - \varphi_m(\xi_u^n - \xi_t^n),$$
$$\gamma_u^* = \min\{u, \max\{s: \varkappa_s \in a_- \cup a_+ \cup b\}\}.$$

For p such that $u/p < \varepsilon/2$, we have

$$\mathbf{P}_i\left\{\left|\int_0^u r(u, t, m, n)\,dt\right| > \varepsilon\right\}$$

$$\leqslant \sum_{k=1}^{p-1} \mathbf{P}_i \left\{ \left| \int_0^{ku/p} r(u, t, m, n)\, dt \right| > \varepsilon/4, \quad \gamma_u^* \in [ku/p, (k+1)u/p] \right\}$$

$$+ \sum_{k=0}^{p-2} \mathbf{P}_i \left\{ \left| \int_{(k+1)u/p}^{u} r(u, t, m, n)\, dt \right| > \varepsilon/4, \right.$$

$$\left. \gamma_u^* \in [ku/p, (k+1)u/p] \right\} \equiv C + D.$$

Let us estimate D:

$$\sum_{k=0}^{p-2} \mathbf{P}_i \left\{ \left| \int_{(k+1)u/p}^{u} r(u, t, m, n)\, dt \right| > \varepsilon/4, \quad \gamma_u^* \in [ku/p, (k+1)u/p] \right\}$$

$$\leqslant 4\varepsilon^{-1} \sum_{k=0}^{p-2} \int_{(k+1)u/p}^{u} \mathbf{P}_i \{ 0 < \xi_u^n - \xi_t^n \leqslant 1/m, \quad \forall s \in [t, u] \ \varkappa_s \in a_0 \}\, dt.$$

Consider the integrand on the right side of the last inequality. We have

$$\mathbf{P}_i \{ 0 < \xi_u^n - \xi_t^n \leqslant 1/m, \quad \forall s \in [t, u] \ \varkappa_s \in a_0 \} \leqslant \mathbf{P}_i \{ S[[t2^n]/2^n, t] \}$$

$$+ \mathbf{P}_i \{ S[[u2^n]/2^n, u] \} + \mathbf{P}_i \{ 0 < \xi_u^n - \xi_t^n \leqslant 1/m, \quad \forall s \in [t, u] \ \varkappa_s \in a_0,$$

$$\bar{S}[[t2^n]/2^n, t], \bar{S}[[u2^n]/2^n, u] \}.$$

Recall that here, just as in Lemma 6, S[a,b] is the event that κ_t has a jump in [a,b]; $\bar{S}$[a,b] is the complement of S[a,b].

Just as we did in estimating A in Lemma 6, we can show that the right side of the last inequality is not greater than the sum

$$\mathbf{P}_i \{ S[[t2^n]/2^n, t] \} + \mathbf{P}_i \{ S[[u2^n]/2^n, u] \} + \mathbf{P}_i \{ 0 < \xi_u - \xi_t \leqslant 1/m \}$$

$$+ \mathbf{P}_i \{ \xi_t^n \neq \xi_t, \ \varkappa_s \in a_0 \quad \text{for} \quad s \in [[t2^n]/2^n, t], \ \bar{S}[[t2^n]/2^n, t] \}$$

$$+ \mathbf{P}_i \{ \xi_u^n \neq \xi_u, \ \varkappa_s \in a_0 \quad \text{for} \quad s \in [[u2^n]/2^n, u], \ \bar{S}[[u2^n]/2^n, u] \}.$$

This implies that

$$\lim_{m\to\infty} \lim_{n\to\infty} \mathbf{P}_i \{ 0 < \xi_u^n - \xi_t^n \leqslant 1/m, \quad \forall s \in [t, u] \ \varkappa_s \in a_0 \} = 0,$$

and therefore

$$\lim_{m\to\infty} \lim_{n\to\infty} \sum_{k=0}^{p-2} \mathbf{P}_i \left\{ \left| \int_{(k+1)u/p}^{u} r(u, t, m, n)\, dt \right| > \varepsilon/4, \right.$$

$$\left. \gamma_u^* \in [ku/p, (k+1)u/p] \right\} = 0.$$

Let us estimate C. We have the following chain of inequalities:

$$C = \sum_{k=1}^{p-2} \mathbf{P}_i \left\{ \left| \int_0^{ku/p} r(u, t, m, n)\, dt \right| > \varepsilon/4, \quad \gamma_u^* \in [ku/p, (k+1)u/p] \right\}$$

$$+ \mathbf{P}_i \left\{ \left| \int_0^{(p-1)u/p} r(u, t, m, n)\, dt \right| > \varepsilon/4, \quad \gamma_u^* \in [(p-1)u/p, u] \right\}$$

$$\leqslant 4\varepsilon^{-1} \sum_{k=1}^{p-2} \int_0^{ku/p} \mathbf{P}_i \{ 0 < \xi_u^n - \xi_t^n \leqslant 1/m, \quad \gamma_u^* \in [ku/p, (k+1)u/p] \}$$

$$+ \mathbf{P}_i \left\{ \left| \int_0^{(p-1)u/p} r(u, t, m, n)\, dt \right| > \varepsilon/4, \quad \exists s \in [(p-1)u/p, u] \ \varkappa_s \notin a_0 \right\}$$

$$+ \mathbf{P}_i \left\{ \left| \int_0^{(p-1)u/p} r(u, t, m, n)\, dt \right| > \varepsilon/4, \quad \forall s \in [0, u] \ \varkappa_s \in a_0 \right\}$$

$$\leqslant 4\varepsilon^{-1} \sum_{k=1}^{p-2} \int_0^{ku/p} \mathbf{P}_i \{ 0 < \xi_u^n - \xi_t^n \leqslant 1/m, \quad \exists s \in]t, u] \ \varkappa_s \notin a_0 \}\, dt$$

$$+ \int_0^{(p-1)u/p} \mathbf{P}_i \{ 0 < \xi_u^n - \xi_t^n \leqslant 1/m, \quad \exists s \in [t, u] \ \varkappa_s \notin a_0 \}\, dt$$

$$+ \int_0^{(p-1)u/p} \mathbf{P}_i \{ 0 < \xi_u^n - \xi_t^n \leqslant 1/m, \quad \forall s \in [0, u] \ \varkappa_s \in a_0 \}\, dt.$$

Passing to the double limit in the extreme terms of this chain of inequalities, we get

$$\lim_{m\to\infty} \lim_{n\to\infty} \sum_{k=1}^{p-1} \mathbf{P}_i \left\{ \left| \int_0^{ku/p} r(u, t, m, n)\, dt \right| > \varepsilon/4, \quad \gamma_u^* \in [ku/p, (k+1)u/p] \right\}$$

$$\leqslant 4\varepsilon^{-1} \sum_{k=1}^{p-2} \int_0^{ku/p} \mathbf{P}_i \{ \xi_u - \xi_t, \quad \exists s \in [t, u] \ \varkappa_s \notin a_0 \}\, dt$$

$$+ \int_0^{(p-1)u/p} \mathbf{P}_i \{ \xi_u = \xi_t, \quad \exists s \in [t, u] \ \varkappa_s \notin a_0 \}\, dt$$

$$+ \lim_{m\to\infty} \lim_{n\to\infty} \int_0^{(p-1)u/p} \mathbf{P}_i \{ 0 < \xi_u^n - \xi_t^n \leqslant 1/m, \quad \forall s \in [0, u] \ \varkappa_s \in a_0 \}\, dt.$$

The first two summands on the right side of the last inequality are equal to zero by Lemmas 3, 5. The third summand is equal to zero from the inequalities used above to estimate D.

In the same way as Lemmas 7, 9 we prove

Lemma 11.

$$M_m^n(u) \underset{\mathbf{P}_i}{\rightarrow} M_m(u).$$

Lemma 12.

$$M^n(u) \underset{\mathbf{P}_i}{\rightarrow} M_u.$$

4. Factorization identities for processes with independent increments defined on a finite Markov chain

Theorem 2. For $\mu,\nu > 0$ there exist matrices $Q^+(s,\mu,\nu)$ and $Q^-(s,\mu,\nu)$ such that

$$[\mu I - A(s)]^{-1}[(\mu+\nu)I - A(s)] = Q^+(s,\mu,\nu)Q^-(s,\mu,\nu) \qquad (4.1)$$

for $\mathrm{Re}(s) = 0$, where $Q^+(s,\mu,\nu)$ and $Q^-(s,\mu,\nu)$ satisfy the conditions:

D_1. The matrix $Q^+(s,\mu,\nu)$ has inverse $[Q^+(s,\mu,\nu)]^{-1}$ for $\mathrm{Re}(s) \geqq 0$; $Q^+(s,\mu,\nu)$ and $[Q^+(s,\mu,\nu)]^{-1}$ are bounded and continuous for $\mathrm{Re}(s) \geqq 0$; they are regular for $\mathrm{Re}(s) > 0$; $Q^+(\infty,\mu,\nu) = I$.

D_2. The matrix $Q^-(s,\mu,\nu)$ has inverse $[Q^-(s,\mu,\nu)]^{-1}$ for $\mathrm{Re}(s) \leqq 0$; $Q^-(s,\mu,\nu)$ and $[Q^-(s,\mu,\nu)]^{-1}$ are bounded and continuous for $\mathrm{Re}(s) \leqq 0$; they are regular for $\mathrm{Re}(s) < 0$.

$Q^+(s,\mu,\nu)$ and $Q^-(s,\mu,\nu)$ are uniquely determined by the conditions D_1, D_2 and (4.1), and

$$Q^+(s,\mu,\nu) = I + \nu\int_0^\infty e^{-(\mu+\nu)t}\mathbf{M}\left(e^{-s\xi_t+\nu M_t}\delta(\xi_t>0);\quad \varkappa_t = j \mid \varkappa_0 = i\right)dt,$$

$$Q^-(s,\mu,\nu) = I + \nu\int_0^\infty e^{-\mu t}\mathbf{M}\left(e^{-s\xi_t-\nu L_t}\delta(\xi_t\leqslant 0);\quad \varkappa_t = j \mid \varkappa_0 = i\right)dt,$$

$$[Q^+(s,\mu,\nu)]^{-1} = I - \nu\int_0^\infty e^{-\mu t}\mathbf{M}\left(e^{-s\xi_t-\nu L_t}\delta(\xi_t>0);\quad \varkappa_t = j \mid \varkappa_0 = i\right)dt,$$

$$[Q^-(s,\mu,\nu)]^{-1} = I - \nu\int_0^\infty e^{-(\mu+\nu)t}\mathbf{M}\left(e^{-s\xi_t+\nu M_t}\delta(\xi_t\leqslant 0);\quad \varkappa_t = j \mid \varkappa_0 = i\right)dt.$$

Proof. Consider the processes

$$\mathscr{L}_t^{(n)} = \{\xi_t^n,\ \varkappa_t^n;\ t \geqslant 0\}$$

and for $n = 1,2,\ldots$ set

$$\Psi^{(n)}(s) \equiv \Psi(s,\ 2^{-n}) = \exp(2^{-n}A(s)).$$

Denote by $\Psi^{+(n)}(s,\rho,\omega)$, $\Psi^{-(n)}(s,\rho,\omega)$ the components of the left factorization of the matrix $[I-\rho\Psi^{(n)}(s)]^{-1}[I-\rho\omega\Psi^{(n)}(s)]$, i.e.,

$$[I-\rho\Psi^{(n)}(s)]^{-1}[I-\rho\omega\Psi^{(n)}(s)] = \Psi^{+(n)}(s, \rho, \omega)\Psi^{-(n)}(s, \rho, \omega) \tag{4.2}$$

for $\operatorname{Re}(s) = 0$, where $\Psi^{+(n)}(s,\omega,\rho)$ and $\Psi^{-(n)}(s,\omega,\rho)$ satisfy conditions A_1, A_2 of Theorem 1, respectively. The expressions for the factorization components are given in Lemma 1. Multiplying both sides of equality (4.2) by $(1-\rho)(1-\omega\rho)^{-1}$, setting

$$\omega = \exp(-\nu 2^{-n}), \quad \rho = \exp(-\mu 2^{-n}) \quad (\nu, \mu > 0),$$
$$S_k^{(n)} = \xi_{k2^{-n}}, \quad \varkappa_k^{(n)} = \varkappa_{2k^{-n}}$$

and passing to the limit as $n \to \infty$ (the validity of passing to the limit will be proved below), we get:

$$\begin{aligned}
&[I-\rho\Psi^{(n)}(s)]^{-1}[I-\rho\omega\Psi^{(n)}(s)](1-\rho)(1-\omega\rho)^{-1} \\
&= (1-e^{-\mu 2^{-n}})2^n \times \sum_{k=0}^{\infty} 2^n e^{-\mu k 2^{-n}}\Psi(s, k2^{-n})\Big[(1-e^{-(\mu+\nu)2^{-n}})2^n \\
&\times \sum_{k=0}^{\infty} 2^{-n} e^{-(\mu+\nu)2^{-n}}\Psi(s, k2^{-n})\Big]^{-1} \xrightarrow[n\to\infty]{} \mu(\mu+\nu)^{-1}\int_0^\infty e^{-\mu t}\Psi(s, t)\,dt \\
&\times \left[\int_0^\infty e^{-(\mu+\nu)t}\Psi(s, t)\,dt\right]^{-1} = \mu(\mu+\nu)^{-1}[\mu I - A(s)]^{-1}[(\mu+\nu)I - A(s)]
\end{aligned}$$

(We recall that $\Psi(s, t) = \exp(tA(s))$),

$$\begin{aligned}
\Psi^{+(n)}(s, \rho, \omega) &= I + (1-e^{-\nu 2^{-n}})\sum_{k=1}^{\infty} e^{-\mu k 2^{-n}} \mathbf{M}\Big(e^{-sS_k^{(n)}}\delta(S_k^{(n)} > 0) \\
&\times \exp\Big(-\nu 2^{-n}\sum_{i=1}^{k-1}\delta(S_k^{(n)} \leqslant S_i^{(n)})\Big);\ \varkappa_k^{(n)} = j \mid \varkappa_0^{(n)} = i\Big) \xrightarrow[n\to\infty]{} I + \nu\int_0^\infty e^{-(\mu+\nu)t} \\
&\times \mathbf{M}(e^{-s\xi_t + \nu M_t}\delta(\xi_t > 0);\ \varkappa_t = j \mid \varkappa_0 = i)\,dt \equiv Q^+(s, \mu, \nu),
\end{aligned}$$

$$\begin{aligned}
\Psi^{-(n)}(s, \rho, \omega) &= I + (1-e^{-\nu 2^{-n}})\sum_{k=1}^{\infty} e^{-\mu k 2^{-n}} \mathbf{M}\Big(e^{-sS_k^{(n)}}\delta(S_k^{(n)} \leqslant 0) \\
&\times \exp\Big(-\nu 2^{-n} \times \sum_{i=1}^{k-1}\delta(S_i^{(n)} > 0)\Big);\ \varkappa_k^{(n)} = j \mid \varkappa_0^{(n)} = i\Big) \xrightarrow[n\to\infty]{} I + \nu\int_0^\infty e^{-\mu t} \\
&\times \mathbf{M}(e^{-s\xi_t - \nu L_t}\delta(\xi_t \leqslant 0);\ \varkappa_t = j \mid \varkappa_0 = i)\,dt \equiv Q^-(s, \mu, \nu).
\end{aligned}$$

Similarly

$$[Q^{+}(s,\mu,\nu)]^{-1}\equiv\lim_{n\to\infty}\left[\Psi^{+(n)}\left(s,e^{-\mu 2^{-n}},e^{-\nu 2^{-n}}\right)\right]^{-1}$$

$$=\lim_{n\to\infty}\left[I-\left(1-e^{-\nu 2^{-n}}\right)\sum_{k=1}^{\infty}e^{-\mu k 2^{-n}}\mathbf{M}\left(e^{-sS_k^{(n)}}\exp\left(-\nu 2^{-n}\sum_{i=1}^{k-1}\delta\left(S_i^{(n)}>0\right)\right)\right.\right.$$

$$\left.\left.\times\delta\left(S_k^{(n)}>0\right);\ \varkappa_k^{(n)}=j\,|\,\varkappa_0^{(n)}=i\right)\right]=I-\nu\int_0^{\infty}e^{-\mu t}\mathbf{M}\left(e^{-s\xi_t-\nu L_t}\delta(\xi_t>0);\ \varkappa_t\right.$$

$$\left.=j\,|\,\varkappa_0=i\right)dt,\quad [Q^{-}(s,\mu,\nu)]^{-1}\equiv\lim_{n\to\infty}\left[\Psi^{-(n)}\left(s,e^{-\mu 2^{-n}},e^{-\nu 2^{-n}}\right)\right]^{-1}$$

$$=\lim_{n\to\infty}\left[I-\left(1-e^{-\nu 2^{-n}}\right)\sum_{k=1}^{\infty}e^{-\mu k 2^{-n}}\mathbf{M}\left(e^{-sS_k^{(n)}}\delta\left(S_k^{(n)}\leqslant 0\right)\right.\right.$$

$$\left.\left.\times\exp\left(-\nu 2^{-n}\sum_{i=1}^{k-1}\delta\left(S_k^{(n)}\leqslant S_i^{(n)}\right)\right);\ \varkappa_k^{(n)}=j\,|\,\varkappa_0^{(n)}=i\right)\right]$$

$$=I-\nu\int_0^{\infty}e^{-(\mu+\nu)t}\mathbf{M}\left(e^{-s\xi_t+\nu M_t}\delta(\xi_t\leqslant 0);\ \varkappa_t=j\,|\,\varkappa_0=i\right)dt.$$

Now let us discuss the possibility of passing to the limit. To be able to pass to the limit,

$$\Psi^{-(n)}(s,\rho,\omega)\to Q^{-}(s,\mu,\nu)\quad\text{as}\quad n\to\infty$$

it is sufficient that

$$\begin{aligned}&\mathbf{M}_i\left(\exp\left(-s\xi_u^n-\nu L^n(u)\right)\delta\left(\xi_u^n\leqslant 0\right)\delta\left(\varkappa_u^n=j\right)\right)\to\\&\to\mathbf{M}_i\left(\exp\left(-s\xi_u-\nu L_u\right)\delta(\xi_u\leqslant 0)\,\delta(\varkappa_u=j)\right)\end{aligned} \qquad 3)$$

as $n\to\infty$ for the random variable u.

Let

$$s=s_1+is_2(\operatorname{Im}s_1=\operatorname{Im}s_2=0,\ s_1\leqslant 0)\ .$$

Then

$$e^{-s\xi_u^n}=e^{-s_1\xi_u^n}\cos\left(s_2\xi_u^n\right)-ie^{-s_1\xi_u^n}\sin\left(s_2\xi_u^n\right).$$

We have:

$$\mathbf{P}_i\left\{\left|\operatorname{Re}\left(e^{-s\xi_u^n}\delta\left(\xi_u^n\leqslant 0\right)\right)-\operatorname{Re}\left(e^{-s\xi_u}\delta(\xi_u\leqslant 0)\right)\right|>\varepsilon\right\}$$

$$=\mathbf{P}_i\left\{\left|\operatorname{Re}\left(e^{-s\xi_u^n}\delta\left(\xi_u^n\leqslant 0\right)\right)-\operatorname{Re}\left(e^{-s\xi_u}\delta(\xi_u\leqslant 0)\right)\right|>\varepsilon,\right.$$

$$\left.\delta\left(\xi_u^n\leqslant 0\right)=\delta(\xi_u\leqslant 0)\right\}+\mathbf{P}_i\left\{\left|\delta\left(\xi_u^n\leqslant 0\right)-\delta(\xi_u\leqslant 0)\right|=1\right\}$$

$$\leqslant \mathbf{P}_i\left\{\left|\mathrm{Re}\left(e^{-s\xi_u^n}\right)-\mathrm{Re}\left(e^{-s\xi_u}\right)\right|>\varepsilon,\ \xi_u^n\leqslant 0,\ \xi_u\leqslant 0\right\}$$
$$+\ \mathbf{P}_i\{|\delta(\xi_u^n\leqslant 0)-\delta(\xi_u\leqslant 0)|=1\}\leqslant \mathbf{P}_i\{|\xi_u^n-\xi_u|>\gamma(\varepsilon)>0\}$$
$$+\ \mathbf{P}_i\{|\delta(\xi_u^n\leqslant 0)-\delta(\xi_u\leqslant 0)|=1\}.$$

Here we have used the fact that for $x \leqq 0$ $(s_1 \leqq 0)$, the function $e^{-s_1 x}\cos(s_2 x)$ is uniformly continuous. This implies

$$\mathrm{Re}\left(e^{-s\xi_u^n}\delta(\xi_u^n\leqslant 0)\right)\underset{\mathbf{P}_i}{\rightarrow}\mathrm{Re}\left(e^{-s\xi_u}\delta(\xi_u\leqslant 0)\right).$$

Further, by the uniform continuity of the function $e^{-\nu x}$ for $x \in [0,u]$, we have

$$\mathbf{P}_i\left\{\left|e^{-\nu L^n(u)}-e^{-\nu L_u}\right|>\varepsilon\right\}\leqslant \mathbf{P}_i\{|L^n(u)-L_u|>r(\varepsilon)>0\},$$

and by Lemma 9

$$e^{-\nu L^n(u)}\underset{\mathbf{P}_i}{\rightarrow}e^{-\nu L_u}.$$

Finally, since the random variables

$$\mathrm{Re}\left(e^{-s\xi_u^n}\delta(\xi_u^n\leqslant 0)\right),\qquad \exp(-\nu L^n(u)),\ \delta(\varkappa_u^n=j)$$

are bounded, we have

$$\mathrm{Re}\left(\exp\left(-s\xi_u^n-\nu L^n(u)\right)\delta(\xi_u^n\leqslant 0)\,\delta(\varkappa_u^n=j)\right)$$
$$\underset{\mathbf{P}_i}{\rightarrow}\mathrm{Re}\left(\exp\left(-s\xi_u-\nu L_u\right)\delta(\xi_u\leqslant 0)\,\delta(\varkappa_u=j)\right)$$

and, by the uniform integrability in n of

$$\mathrm{Re}\left(e^{-s\xi_u^n-\nu L^n(u)}\delta(\xi_u^n\leqslant 0)\,\delta(\varkappa_u^n=j)\right)$$

(see [13]) that

$$\mathbf{M}_i\left(\mathrm{Re}\left(\exp\left(-s\xi_u^n-\nu L^n(u)\right)\delta(\xi_u^n\leqslant 0)\,\delta(\varkappa_u^n=j)\right)\right)$$
$$\underset{n\to\infty}{\rightarrow}\mathbf{M}_i\left(\mathrm{Re}\left(\exp\left(-s\xi_u-\nu L_u\right)\delta(\xi_u\leqslant 0)\,\delta(\varkappa_u=j)\right)\right).$$

Analogously,

$$\mathbf{M}_i\left(\mathrm{Im}\left(\exp\left(-s\xi_u^n-\nu L^n(u)\right)\delta(\xi_u^n\leqslant 0)\,\delta(\varkappa_u^n=j)\right)\right)\rightarrow$$
$$\rightarrow\mathbf{M}_i\left(\mathrm{Im}\left(\exp\left(-s\xi_u-\nu L_u\right)\delta(\xi_u\leqslant 0)\,\delta(\varkappa_u=j)\right)\right)\quad\text{as } n\rightarrow\infty.$$

Thus the passage to the limit

$$\Psi^{-(n)}(s,\rho,\omega)\rightarrow Q^{-}(s,\mu,\nu)\quad\text{as } n\rightarrow\infty$$

is proved. To prove the passage to the limit

$$\Psi^{+(n)}(s,\rho,\omega)\rightarrow Q^{+}(s,\mu,\nu)\quad\text{as } n\rightarrow\infty$$

it suffices that for almost all u,

$$\begin{aligned} &\mathbf{M}_i\left(\exp\left(-s\xi_u^n+\nu M^n(u)\right)\delta\left(\xi_u^n>0\right)\delta\left(\varkappa_u^n=j\right)\right) \\ &\to \mathbf{M}_i\left(\exp\left(-s\xi_u+\nu M_u\right)\delta\left(\xi_u>0\right)\delta\left(\varkappa_u=j\right)\right) \quad \text{as} \quad n\to\infty. \end{aligned} \tag{4.4}$$

The validity of this convergence follows from Lemma 12 and arguments analogous to those in proving (4.3). The validity of the limit transitions

$$[\Psi^{\pm(n)}(s,\rho,\omega)]^{-1}\to[Q^{\pm}(s,\mu,\nu)]^{-1} \quad \text{as} \quad n\to\infty$$

follows from (4.3) and (4.4).

The matrices $Q^+(s,\mu,\nu)$ and $Q^-(s,\mu,\nu)$ satisfy, obviously, conditions D_1, D_2 and (4.1). It follows from Liouville's theorem that these conditions determine the matrices uniquely.

From Theorems 1, 2 we have

T h e o r e m 3. For $\mu,\nu>0$, and $\operatorname{Re}(s)=0$,

$$\left\|\int_0^\infty e^{-\mu t}\mathbf{M}\left(e^{-s\xi_t-\nu L_t^x};\ \varkappa_t=j\mid\varkappa_0=i\right)dt\right\|$$

$$=\begin{cases}[T(x,0]\{[Q^-(s,\mu,\nu)]^{-1}\}]\times[Q^+(s,\mu,\nu)]^{-1}[\mu I-A(s)]^{-1} & \text{as } x<0,\\ [T[0,x]\{Q^+(s,\mu,\nu)\}]\times[Q^+(s,\mu,\nu)]^{-1}[\mu I-A(s)]^{-1} & \text{as } x\geqslant 0,\end{cases}$$

where $Q^{\pm}(s,\mu,\nu)$ are defined in Theorem 2.

Let $\bar{L}_t^x$ be the sojourn time of the process $\{\xi_u,\ 0\leqq u\leqq t\}$ on the half-axis $[x,\infty)$. It is not hard to derive from Theorem 3 the following

COROLLARY. For $\mu,\nu>0$, $\operatorname{Re}(s)=0$,

$$\left\|\int_0^\infty e^{-\mu t}\mathbf{M}\left(e^{-s\xi_t-\nu\bar{L}_t^x};\ \varkappa_t=j\mid\varkappa_0=i\right)dt\right\|$$

$$=\begin{cases}[T[x,0]\{[Q^-(s,\mu,\nu)]^{-1}\}]\times[Q^+(s,\mu,\nu)]^{-1}[\mu I-A(s)]^{-1} & \text{as } x\leqslant 0,\\ [T[0,x)\{Q^+(s,\mu,\nu)\}]\times[Q^+(s,\mu,\nu)]^{-1}[\mu I-A(s)]^{-1} & \text{as } x>0.\end{cases}$$

Next, let $N=1$, $x=0$. Consider the infinitely divisible factorization of the function $u/(u-A(s))$ (see [6]):

$$u/(u-A(s))=\Psi_{u+}(s)\Psi_{u-}(s).$$

Expressions for the factorization components $\Psi_{u+}(s)$, $\Psi_{u-}(s)$ are given in [5]. We have the representation

$$[\mu-A(s)]^{-1}[\mu+\nu-A(s)]=\frac{\mu+\nu}{\mu}\times\frac{\Psi_{\mu+}(s)\Psi_{\mu-}(s)}{\Psi_{(\mu+\nu)+}(s)\Psi_{(\mu-\nu)-}(s)}.$$

Let

$$H(\mu, \nu) = \lim_{s\to\infty} \frac{\Psi_{\mu+}(s)}{\Psi_{(\mu+\nu)+}(s)} = \exp\left\{\int_0^\infty \frac{e^{-\mu u}}{u}\left(e^{-\nu u} - 1\right)\mathbf{P}\{\xi_u > 0\}\,du\right\}.$$

It follows from Theorem 3 that

$$\int_0^\infty e^{-\mu t}\mathbf{M}\left(e^{-s\xi_t - \nu L_t}\right)dt = \left[Q^+(s, \mu, \nu)\right]^{-1}[\mu - A(s)]^{-1}$$
$$= \mu^{-1}\Psi_{(\mu+\nu)+}(s)\times H(\mu, \nu)\,\Psi_{\mu-}(s).$$

The last relation coincides with identity (2.7) of [5].

The author is familiar with D.V. Gusak's work [14] in which expressions are found for

$$s\int_0^\infty e^{-st}\mathbf{M}\left(e^{-\mu L_t^x}\right)dt \qquad \text{and} \qquad \sum_{n=1}^\infty u^n \mathbf{M} z^{\Delta_n^x}$$

(Theorem 1 and Theorem 2, respectively). Theorem 1 of [14] follows from Theorem 3 (for $s = 0$, $N = 1$) and Theorem 2 (for $N = 1$) in our paper. Likewise, Theorem 2 of [14] follows from Theorem 1 and Lemma 1 herein.

REFERENCES

[1] Ezhov, I.I., and Skorokhod, A.V. "Markov Processes with Homogeneous Second Component, I." *Theory Probab. Applications*, vol.14, no.1 (1969): 1-13. (English transl.)

[2] ______. "Markov Processes with Homogeneous Second Component, II." *Theory Probab. Applications*, vol.14, no.4 (1969): 652-67. (English transl.)

[3] Peresypkina, S.I. "On the Sojourn Time Above the Zero Level of a Homogeneous Process with Independent Increments, Given on a Markov Chain" (in Russian). In *Issledovaniya po teorii sluchajnykh protsessov* (Investigations in the Theory of Random Processes), pp. 130-34. Kiev: Izd-vo Instituta Matematiki Akademii Nauk SSSR, 1976.

[4] Gusak, D.V., and Peresypkina, S.I. "The Time Spent Above a Fixed Level by a Class of Controlled Random Processes." *Ukrainian Mathematical Journal*, vol.30, no.3 (1978): 270-74. (English transl.)

[5] Pecherskij (Pecherskii), E.A., and Rogozin, B.A. "On Joint Distributions of Random Variables Associated with Fluctuations of a Process with Independent Increments." *Theory Probab. Applications*, vol.14, no.3 (1969): 410-23. (English transl.)

[6] Rogozin, B.A. "On the Distributions of Functionals Related to Boundary Problems for Processes with Independent Increments." *Theory Probab. Applications*, vol.11, no.4 (1966): 580-91. (English transl.)

[7] Mogul'skij (Mogul'skii), A.A. "Factorization Identities for Processes with Independent Increments, Given on a Finite Markov Chain." *Theory of Probab. and Mathemat. Statistics* 11 (1976): 87-98. (English transl.)

[8] Takács, Lajos. "On Fluctuations of Sums of Random Variables." Gian-Carlo Rota, ed. *Studies in Probability and Ergodic Theory. Advances in Mathematics. Supplementary Studies*, vol.2, pp. 45-93. New York San Francisco London: Academic Press, 1978.

[9] Semenov, A.T. "Sojourn Time of a Semi-Markov Random Walk on the Semi-axis." *Lithuanian Mathemat. Journal*, vol.15, no.4 (1975): 659-64. (English transl.)

[10] Presman, È.L. "Factorization Methods and Boundary Problems for Sums of Random Variables Given on Markov Chains." *Math. USSSR Izvestiya*, vol.3, no.4 (1969): 815-52. (English transl.)

[11] Borovkov, A.A. *Stochastic Processes in Queueing Theory*. New York Berlin Heidelberg Tokyo: Springer-Verlag, 1976. (English transl.)

[12] Lukacs, Eugene. *Characteristic Functions*. 2nd ed. rev. London: Griffin, 1970.

[13] Billingsley, Patrick. *Convergence of Probability Measures*. New York: Wiley & Sons Publishers, 1968.

[14] Gusak, D.V. "On the Sojourn Time Above the Level of a Class of Random Processes." *Mezhdunarodnaya konferentsiya po teorii veroyatnostej i matematicheskoj statistike. Tezisy dokladov* (The International Conference on Probability Theory and Mathematical Statistics. Abstracts), vol.1. Vilnius, 1981.

A.L. Miroshnikov

INEQUALITIES FOR THE INTEGRAL CONCENTRATION FUNCTION

1. Introduction

Let R^d be d-dimensional Euclidean space, $E \subset R^d$, $\bar{0} \in E$, where E is a convex set in R^d for which Minkowski's functional

$$p_E(x) = \inf\{\alpha > 0;\ x \in \alpha E\},\quad x \in R^d,\quad E = \{x \in R^d;\ p_E(x) \leqslant 1\}$$

is defined.

Let ξ be a random vector in R^d. If ξ' and ξ'' are independent and distributed identically with ξ, we set $\tilde{\xi} = \xi' - \xi''$. If $P(\cdot)$ is the distribution of ξ, then $\tilde{P}(\cdot)$ is the distribution of $\tilde{\xi}$.

Introduce the characteristic $U(\xi,E)$ being the integral concentration function of the random vector ξ given by

$$U(\xi, E) = \int_{R^d} \exp\{-p_E(x)\}\,\tilde{\mathbf{P}}(dx),$$

the definition of which is due to Anan'evskij [1]. Now let us note results for the integral concentration function.

T h e o r e m A (Anan'evskij). The following estimates are valid:

1. $0 < U(\xi, E) \leqslant 1$;
2. If $\lambda_1 \geqslant \lambda_2 > 0$, then $U(\xi, \lambda_1 E) \geqslant U(\xi, \lambda_2 E)$;
3. $M(p_E(\tilde{\xi}) \wedge 1^2 \geqslant (1 - U(\xi, E))^2$;
4. $U(\xi, E) \leqslant (1 + \sum_{k=1}^{\infty} e^{-k}(kd^2 + 1)^d)Q(\xi, E)$,

where E is a convex, closed, bounded, symmetric set with nonempty interior; for scalars a, b, by definition, $a \wedge b = \min(a,b)$; $Q(\xi,E)$ is the (Lévy) concentration function of the random vector ξ on R^d and

$$Q(\xi,E) = \sup_{y \in R^d} \mathbf{P}(\xi \in y + E).$$

In this paper we study the behavior of the integral concentration function for $S_n = \sum_{k=1}^{n} \xi_k$, where $\xi_1, \ldots, \xi_n$ are independent random vectors in R^d. Such estimates were first obtained by Enger (see [2]). It should, however, be pointed out that Enger does not introduce it explicitly but uses it as an auxiliary variable to estimate the Lévy concentration function.

Let us define a class Ω of convex sets in R^d. Let

$$x \in R^d,\ x = (x_1, \ldots, x_d), (x, y) = \sum_{i=1}^{d} x_i y_i,\ \ y \in R^d,\ |x| = (x, x)^{1/2}.$$

Assume that A is the class of convex sets in R^d, such that

$$p_E(x) = \sum_{i=1}^{m} |A_i x|,$$

where $A_1, \ldots, A_m$ are arbitrary linear operators in R^d. Then Ω is the class of convex sets in R^d, being limiting for a set in class A in the sense that there exists a sequence $\{E_k\}_{k \in N}$ of sets in A such that $p_{E_k}(x) \to p_E(x)$ for all $x \in R^d$ as $k \to \infty$. The set

$$\left\{x \in R^d;\ \sum_{i=1}^{d} |x_i| \leqslant 1\right\} \in \mathbf{\Omega}.$$

The set

$$\left\{x \in R^d;\ \max_{1 \leqslant i \leqslant d} |x_i| \leqslant 1\right\} \notin \mathbf{\Omega} \quad \text{as } d \geqslant 3.$$

From Enger's results, one has

T h e o r e m B. Let $\xi_1,\dots,\xi_n$ be independent random vectors in R^d. Let

$$E \in \Omega, \quad \lambda \geqslant \max_{1\leqslant k\leqslant n} \lambda_k, \quad S_n = \sum_{k=1}^{n} \xi_k,$$

Then

$$U(S_n, \lambda E) \leqslant C\lambda \times \left(\sum_{k=1}^{n} \mathbf{M}(p_E(\widetilde{\xi}_k) \wedge \lambda_k)^2\right)^{-1/2}$$

and

$$U(S_n, \lambda E) \leqslant C\lambda \left(\sum_{k=1}^{n} \lambda_k^2 (1 - U(\xi_k, \lambda_k E))^2\right)^{-1/2}. \qquad (*)$$

Inequality (*) follows from Theorem A.

Here and below, if not stated otherwise, C is an absolute constant. Taking a wider class of convex sets, we obtain a dependence of such estimates on the dimension of the space.

Let $\xi_1,\dots,\xi_n$ be independent random vectors in R^d, $S_n = \sum_{k=1}^{n} \xi_k$. Then we have

T h e o r e m C (Anan'evskij). If $E \subset R^d$ is a convex, closed, bounded, symmetric set with nonempty interior, then

$$U(S_n, E) \leqslant Cd\left(1 + \sum_{k=1}^{\infty} e^{-k}(kd^2 + 1)^d\right)\left(\sum_{k=1}^{n} (1 - U(\xi_k, E))^2\right)^{-1/2}.$$

This result follows from Theorem A and from [2].

T h e o r e m D (Enger). If E is a convex symmetric set in R^d and $\lambda \geqq \max_{1\leqq k\leqq n} \lambda_k$, then

$$Q(S_n, \lambda E) \leqslant Cd\lambda \left(\sum_{k=1}^{n} \mathbf{M}(p_E(\widetilde{\xi}_k) \wedge \lambda_k)^2\right)^{-1/2}.$$

T h e o r e m E (Enger). If E is a bounded, absorbing, symmetric and convex set in R^d, then

$$Q(S_n, \lambda E) \leqslant C(d) \inf_{u>0} \frac{\lambda^d + u^d}{u^d + \left(\sum_{i=1}^{n} \chi_i^E(u)\right)^{d/2}},$$

where

$$\chi_i^E(u) = \inf_{p_{E^*}(t)=1} \int_{p_E(x)<u} (t, x)^2 \widetilde{\mathbf{P}}_i(dx), \quad E^* = \{t \in R^d;\ |(t, x)| \leqslant 1, \forall x \in E\},\ E = \{x \in R^d;\ p_E(x) \leqslant 1\},$$

and $P_i(\cdot)$ is the distribuiton of the vector $\tilde{\xi}_i$.

The integral concentration function is a multidimensional characteristic of one of the one-dimensional generalized concentration functions (see [1]). Hence, the estimates for $U(S_n, E)$ in Theorems A, B, C are multidimensional analogs of Kolmogorov-Rogozin inequalities (see [3]-[8]). Our objective is to obtain local bounds for the integral concentration function (Kesten-type estimates) (see [8]-[14]).

These results entail all the estimates mentioned above (Theorems A, B, C, D, E).

2. The statement of the basic results

Theorem 1. Let $\xi_1, \ldots, \xi_n$ be independent random vectors in R^d and let

$$S_n = \sum_{k=1}^{n} \xi_k, \quad \lambda \geqslant \max_{1 \leqslant k \leqslant n} \lambda_k, \quad E \in \Omega,$$

Then

$$U(S_n, \lambda E) \leqslant C\lambda \left(\sum_{k=1}^{n} \mathbf{M}(p_E(\tilde{\xi}_k) \wedge \lambda_k)^2 U^{-2}(\xi_k, \lambda E) \right)^{-1/2}.$$

COROLLARY 1.1. Under the conditions of Theorem 1, for identically distributed random variables ξ_k, for $\lambda_k = \lambda_0$, $k = \overline{1,n}$, one has

$$U(S_n, \lambda E) \leqslant C \frac{\lambda}{\lambda_0} \frac{1}{\sqrt{n}} \frac{U(\xi_1, \lambda E)}{1 - U(\xi_1, \lambda_0 E)}.$$

Theorem 2. Let E be a convex and symmetric set in R^d, let $\xi_1, \ldots, \xi_n$ be independent random vectors in R^d and let

$$S_n = \sum_{k=1}^{n} \xi_k, \quad \lambda \geqslant \max_{1 \leqslant k \leqslant n} \lambda_k$$

Then

$$U(S_n, \lambda E) \leqslant Cd\lambda \left(\sum_{k=1}^{n} \mathbf{M}(p_E(\tilde{\xi}_k) \wedge \lambda_k)^2 \right)^{-1/2}.$$

Let H be a separable Hilbert space and let $\xi_1, \ldots, \xi_n$ be independent elements in H, $\|x\| = \sqrt{(x,x)}$. The class A is constructed of sets of the form

$$E = \left\{ x \in H;\ p_E(x) = \sum_{i=1}^{m} \|A_i x\| \leqslant 1 \right\},$$

where A_i, $i = \overline{1,m}$, are bounded linear operaors in H.

Class $\mathbf{\Omega}$ is constructed as in the finite-dimensional case (see [2]).

T h e o r e m 3. In the separable Hilbert space H, for $E \in \mathbf{\Omega}$, and $\lambda \geqq \max_{1\leqq k\leqq n} \lambda_k$ we have

$$U(S_n, \lambda E) \leqslant C\lambda \left(\sum_{k=1}^{n} \mathbf{M}(p_E(\tilde{\xi}_k) \wedge \lambda_k)^2 U^{-2}(\xi_k, \lambda E) \right)^{-1/2}.$$

COROLLARY 3.1. Under the conditions of Theorem 3, for identically distributed elements in H for $\lambda_k = \lambda_0$ and $k = \overline{1,n}$, we obtain

$$U(S_n, \lambda E) \leqslant C \frac{\lambda}{\lambda_0} \frac{1}{\sqrt{n}} \frac{U(\xi_1, \lambda E)}{1 - U(\xi_1, \lambda_0 E)}.$$

T h e o r e m 4. If E is a bounded, absorbing, symmetric and convex set in R^d, then

$$U(S_n, \lambda E) \leqslant C(d) \inf_{u>0} \frac{\lambda^d + u^d}{u^d + \left(\sum_{i=1}^{n} \chi_i^E(u) \right)^{d/2}},$$

and if $\xi_1, \ldots, \xi_n$ are identically distributed, one has

$$U(S_n, \lambda E) \leqslant C(d) \inf_{n>0} \frac{\lambda^d + u^d}{(\chi_1^E(u))^{d/2}} n^{-d/2}.$$

3. Proof of the auxiliary results

L e m m a 1. Let ξ_1 and ξ_2 be independent random vectors in R^d and $E \in \mathbf{\Omega}$. Then

$$U(\xi_1 + \xi_2, E) \leqslant U(\xi_1, E) \wedge (U(\xi_2, E).$$

Proof. Let $E \in A$. Then as follows from [2], $\exp\{-p_E(x)\}$ is the characteristic function of some probability measure $\mu(\cdot)$. Hence, if ξ_1 and ξ_2 are two independent random vectors in R^d, $\tilde{P}_1(\cdot)$ is the distribution of $\tilde{\xi}$ and $\tilde{P}(\cdot)$ is the distribution of $\tilde{\xi}_1 + \tilde{\xi}_2$, one has

$$U(\xi_1 + \xi_2, E) = \int_{R^d} \exp\{-p_E(x)\} \tilde{\mathbf{P}}(dx)$$

$$= \int_{R^d}\int_{R^d} \cos(t, x) \tilde{\mathbf{P}}(dx) \mu(dt) = \int_{R^d} |f_1(t)|^2 |f_2(t)|^2 \mu(dt)$$

$$\leqslant \int_{R^d} |f_1(t)|^2 \mu(dt) = \int_{R^d}\int_{R^d} \cos(t, x) \tilde{\mathbf{P}}_1(dx) \mu(dt)$$

$$= \int_{R^d} e^{-p_E(x)} \tilde{\mathbf{P}}_1(dx) = U(\xi_1, E),$$

where $f_1(t)$ is the characteristic function of ξ_1, $f_2(t)$ is the characteristic fucntion of ξ_2. Therefore, for $E \in A$ we have

$$U(\xi_1 + \xi_2, E) \leqq U(\xi_1, E) \quad ,$$

and similarly

$$U(\xi_1 + \xi_2, E) \leqq U(\xi_2, E) \quad .$$

Lemma 1 is proved for $E \in A$. Consider the sequence of sets $\{E_k\}_{k \in N} \in A$ for all k such that $p_{E_k}(x) \to p_E(x)$ as $k \to \infty$ for all $x \in R^d$. Since Lemma 1 is valid for the sets E_k, then, letting $k \to \infty$, we extend the result to class $\mathbf{\Omega}$.

L e m m a 2. Let

$$E \in \Omega,\ p_k > 0,\ k = \overline{1, n},\ \sum_{k=1}^{n} p_k^{-1} = 1,\ \xi_1, \ldots, \xi_n$$

be independent random vectors in R^d, and let

$$S_n = \sum_{k=1}^{n} \xi_k,\ \tilde{S}_m^{(k)} = \sum_{i=1}^{m} \tilde{\xi}_k^{(i)},$$

where $\tilde{\xi}_k^{(i)}$, $i = \overline{1,m}$, be independent random vectors identically distributed with $\tilde{\xi}_k$. Here $\tilde{P}_n(\cdot)$ is the distribution of the vector $\tilde{S}_n$, $\tilde{P}_m^{(k)}$ is the distribution of the vector $\tilde{S}_m^{(k)}$, and $[\cdot]$ is the integer part of a number. Then

$$\int_{R^d} e^{-p_E(x)} \tilde{\mathbf{P}}_n(dx) \leqslant \sum_{k=1}^{n} p_k^{-1} \int_{R^d} e^{-p_E(x)} \tilde{\mathbf{P}}_{[p_k]}^{(k)}(dx)$$

or

$$U(S_n, E) \leqslant \sum_{k=1}^{n} p_k^{-1} U\left(S_{[p_k]}^{(k)}, E\right),\ \sum_{k=1}^{n} p_k^{-1} = 1.$$

Proof. We prove Lemma 2 for sets of class A, because estimates for class $\mathbf{\Omega}$ will follow by a simple passage to the limit, as in the proof of Lemma 1.

Let $E \in A$. Then $\exp\{-p_E(x)\} = \int_d \cos(t, x)\, \mu(dt)$,

where $\mu(\cdot)$ is some probability measure in R^d. We have

$$\int_{R^d} e^{-p_E(x)} \tilde{\mathbf{P}}_n(dx) = \int_{R^d} \prod_{k=1}^{n} |f_k(t)|^2 \mu(dt),$$

where $f_k(t)$ is the characteristic function of the random vector ξ_k. Let $p_k > 0$ and $\sum_{k=1}^{n} p_k^{-1} = 1$. Then by Hölder's inequality,

$$\int\limits_{R^d}\prod_{k=1}^{n}|f_k(t)|^2\mu(dt)\leqslant\prod_{k=1}^{n}\left(\int\limits_{R^d}|f_k(t)|^{2[p_k]}\mu(dt)\right)^{p_k^{-1}}$$

$$\leqslant\sum_{k=1}^{n}p_k^{-1}\int\limits_{R^d}|f_k(t)|^{2[p_k]}\mu(dt)$$

$$=\sum_{k=1}^{n}p_k^{-1}\int\limits_{R^d}\int\limits_{R^d}\cos(t,x)\,\widetilde{\mathbf{P}}_{[p_k]}^{(k)}(dx)\,\mu(dt)$$

$$=\sum_{k=1}^{n}p_k^{-1}\int\limits_{R^d}e^{-p_E(x)}\widetilde{\mathbf{P}}_{[p_k]}^{(k)}(dx)=\sum_{k=1}^{n}p_k^{-1}U\left(S_{[p_k]}^{(k)},E\right),$$

and the Lemma is proved.

Lemma 3. Let ξ be a random vector in R^d, $E\in\boldsymbol{\Omega}$. Then $Q(\tilde{\xi},\lambda E)\leqq eU(\xi,\lambda E)$.

Proof. For simplicity, set $\lambda=1$ and $E\in A$. Then for any $y\in R^d$, one has

$$\mathbf{P}(\tilde{\xi}\in y+E)=\int\limits_E\widetilde{\mathbf{P}}_y(dx),$$

where $\tilde{P}_y(\cdot)$ is the distribution of the random vector $\tilde{\xi}-y$. We have

$$\int\limits_E\widetilde{\mathbf{P}}_y(dx)\leqslant e\int\limits_E e^{-p_E(x)}\widetilde{\mathbf{P}}_y(dx)\leqslant e\int\limits_{R^d}e^{-p_E(x)}\widetilde{\mathbf{P}}(dx)=e\int\limits_{R^d}\int\limits_{R^d}\cos(t,x)$$

$$\times\widetilde{\mathbf{P}}_y(dx)\,\mu(dt)\leqslant e\int\limits_{R^d}\int\limits_{R^d}\cos(t,x)\,\widetilde{\mathbf{P}}_0(dx)\,\mu(dt)=eU(\xi,E).$$

We proceed to class $\boldsymbol{\Omega}$. Take $\delta>0$ and $E_{k,\delta}=(1+\delta)F_k$, where $E_k\in A$, $k=1,2,\ldots$, are such that $p_{E_k}(x)\to p_E(x)$ as $k\to\infty$ for any $x\in R^d$. Then

$$E\subset\liminf_{k\to\infty}E_{k,\delta}\quad\text{and}\quad\lim_{\delta\to0}\lim_{k\to\infty}p_{E_{k,\delta}}(x)=p_E(x),$$

and

$$\mathbf{P}(\tilde{\xi}\in y+E)\leqslant\mathbf{P}\left(\tilde{\xi}\in y+\underline{\lim}_{\delta\to0}\,\underline{\lim}_{k\to\infty}E_{k,\delta}\right)\leqslant\underline{\lim}_{\delta\to0}\,\underline{\lim}_{k\to\infty}\mathbf{P}(\tilde{\xi}\in y+E_{k,\delta})$$

$$\leqslant e\,\underline{\lim}_{\delta\to0}\,\underline{\lim}_{k\to\infty}\int\limits_{R^d}e^{-p_{E_{k,\delta}}(x)}\widetilde{\mathbf{P}}_0(dx)=e\int\limits_{R^d}e^{-p_E(x)}\widetilde{\mathbf{P}}_0(dx)=eU(\xi,E),$$

and the Lemma is proved.

Lemma 4. Let ξ be a random vector in R^d, $E\in\boldsymbol{\Omega}$, $x\in R^d$,

$$U_x(\xi, E) = \int_{R^d} e^{-p_E(y)} \tilde{\mathbf{P}}_x(dy),$$

where $\tilde{P}_x(\cdot)$ is the distribution of the random vector $\tilde{\xi} + x$. Then

$$\sup_{x \in R^d} U_x(\xi, E) = U(\xi, E).$$

Proof. Obviously,

$$U(\xi, E) \leqslant \sup_{x \in R^d} U_x(\xi, E),$$

On the other hand, take any $x \in R^d$, $E \in A$, and let $\exp\{-p_E(y)\}$ be the characteristic function with some probability measure $\mu(\cdot)$, i.e.,

$$e^{-p_E(y)} = \int_{R_d} \cos(t, y)\, \mu(dt),$$

Therefore,

$$U_x(\xi, E) = \int_{R^d} \int_{R^d} \cos(t, y)\, \tilde{\mathbf{P}}_x(dy)\, \mu(dt),$$

where the inner integral is the characteristic function of the random vector $\tilde{\xi}$ multiplied by $e^{i(t,x)}$. Estimating the modulus of this integral yields

$$U_x(\xi, E) \leqslant \int_{R^d} \int_{R^d} \cos(t, y)\, \tilde{\mathbf{P}}_0(dy)\, \mu(dt) = U(\xi, E).$$

It remains to make the transition to Ω by passing to the limit, and the Lemma is proved.

L e m m a 5. Let $\xi_1, \ldots, \xi_n$ be independent random vectors in R^d that are identically distributed wtih the vector ξ, and

$$S_n = \sum_{k=1}^{n} \xi_k, \quad E \in \Omega, \quad \lambda \geqslant \lambda_0 .$$

Then

$$U(S_n, \lambda E) \leqslant C \frac{1}{\sqrt{n}} \lambda \frac{U(\xi, \lambda E)}{\sqrt{\mathbf{M}(p_E(\tilde{\xi}) \wedge \lambda_0)^2}} .$$

Proof. We break up the proof into several parts.

1. If $n = 1$, inequality (1) follows from the estimate

$$U(\xi, \lambda E) \leqslant \frac{\lambda}{\lambda_0} U(\xi, \lambda E) \leqslant \lambda \frac{U(\xi, \lambda E)}{\sqrt{\mathbf{M}(p_E(\tilde{\xi}) \wedge \lambda_0)^2}} .$$

If $Q(\tilde{\xi}, \lambda E) \geqq \frac{1}{2}$, it follows from Theorem B, since for $\lambda_0 = \lambda_k$, $k = 1, 2, \ldots, n$:

$$U(S_n, \lambda E) \leqslant \frac{C}{\sqrt{n}} \lambda (\mathbf{M}(p_E(\tilde{\xi}) \wedge \lambda_0)^2)^{-1/2} \leqslant 2C \frac{\lambda}{\sqrt{n}} Q(\tilde{\xi}, \lambda E)$$
$$\times (\mathbf{M}(p_E(\tilde{\xi}) \wedge \lambda_0)^2)^{-1/2} \leqslant 2eC \frac{\lambda}{\sqrt{n}} \frac{U(\xi, \lambda E)}{\sqrt{\mathbf{M}(p_E(\tilde{\xi}) \wedge \lambda_0)^2}},$$

where the last inequality follows form Lemma 3.

Hence we suppose that $n \geqq 2$ and $Q(\tilde{\xi}, \lambda E) < \frac{1}{2}$, and make the proof for $E \in \mathsf{A}$, since one can get to $\boldsymbol{\Omega}$ by passing to the limit.

2. Set

$$A_k = \left\{x \in R^d;\ k < \frac{p_E(x)}{\lambda} \leqslant k+1\right\},\ B_m = m\lambda E, \quad E = \{x \in R^d;\ p_E(x) \leqslant 1\}.$$

Let the integer $m \geqq 2$ and the random vector ξ be such that $P(\tilde{\xi} \in B_m) \geqq \frac{1}{2}$, but $P(\tilde{\xi} \in B_{m-1}) < \frac{1}{2}$. Letting $D = \{\tilde{\xi} \in B_m\}$, one has $P(D) \geqq \frac{1}{2}$, and if $\bar{D}$ is the complement of the event D, $P(\bar{D}) \leqq \frac{1}{2}$.

Let Q be the distribution of the random vector S which is independent of the random vector $\tilde{\xi}$. Then

$$\mathbf{P}(S + \tilde{\xi} \in A_k) = \int_{R^d} \mathbf{P}(\tilde{\xi} \in A_k - x)\, Q(dx) = \int_{R^d} \mathbf{P}(\{\tilde{\xi} \in A_k - x\} \cap D)\, Q(dx)$$
$$+ \int_{R^d} \mathbf{P}(\{\tilde{\xi} \in A_k - x\}/\bar{D})\, \mathbf{P}(\bar{D})\, Q(dx) \leqslant \int_{R^d} \mathbf{P}(\{\tilde{\xi} \in A_k - x\} \cap D)\, Q(dx)$$
$$+ \frac{1}{2} \mathbf{P}(S + \tilde{\xi}' \in A_k). \tag{2}$$

Here the random vector $\tilde{\xi}'$ does not depend on S and has distribution

$$\mathbf{P}(\tilde{\xi}' \in \beta) = \mathbf{P}(\tilde{\xi} \in \beta/\bar{D}),$$

where β is a Borel set in R^d.

Let $a \vee b = \max(a,b)$. Let us estimate the integral in formula (2):

$$\int_{R^d} \mathbf{P}(\{\tilde{\xi} \in A_k - x\} \cap D)\, Q(dx) \leqslant \int_{2(m \vee (k+1))\lambda E} \mathbf{P}(\{\tilde{\xi} \in A_k - x\}$$
$$\cap \{\tilde{\xi} \in B_m\})\, Q(dx) \leqslant \int_{2(m \vee (k+1))\lambda E} [\mathbf{P}(\tilde{\xi} \in A_k - x) \wedge \mathbf{P}(\xi \in B_m)]\, Q(dx).$$

Indeed,

$$\mathbf{P}(\{\tilde{\xi} \in A_k - x\} \cap \{\tilde{\xi} \in B_m\}) = 0,$$

if $x \notin A_k - B_m = \{y - z;\ y \in A_k,\ z \in B_m\}$. If F is a symmetric convex set such that $F = \{x \in R^d;\ p_F(x) \leq 1\}$, then $F - F \subset 2F$, since for $x \in F - F$ it follows that $x = y - z$, where $y \in F$, $z \in F$, and

$$p_F(x) = p_F(y-z) \leqq p_F(y) + p_F(-z) = p_F(y) + p_F(z) \leqq 2 ,$$

i.e., $p_F(x) \leqq 2$ and therefore $x \in 2F$. It remains to note that $A_k \subset (k+1)\lambda E = B_{k+1}$. Thus, we have the inequality

$$\mathbf{P}(S + \tilde{\xi} \in A_k) \leqslant \int_{2(m \vee (k+1))\lambda E} [\mathbf{P}(\tilde{\xi} \in A_k - x) \wedge \mathbf{P}(\tilde{\xi} \in B_m)]\, Q(dx) + \frac{1}{2}\mathbf{P}(S + \tilde{\xi}' \in A_k). \tag{3}$$

3. Set

$$\tilde{S}_n^{(0)} = \tilde{S}_n, \quad \tilde{S}_n = \sum_{k=1}^{n} \tilde{\xi}_k, \quad \tilde{S}_n^{(j)} = \sum_{k=1}^{n-j} \tilde{\xi}_k + \sum_{k=n-j+1}^{n} \tilde{\xi}'_k, \quad j = 1, \ldots, n-1.$$

The random vectors $\tilde{\xi}_1, \ldots, \tilde{\xi}_n$, $\tilde{\xi}'_1, \ldots, \tilde{\xi}'_n$ are mutually independent. Furthermore, $\tilde{\xi}_k$, $k = \overline{1,n}$, are identically distributed with $\tilde{\xi}$, and the $\tilde{\xi}'_k$, $k = \overline{1,n}$, are identically distributed with $\tilde{\xi}'$. Setting $S = \tilde{S}_{n-1}$ and $\tilde{\xi} = \tilde{\xi}_n$ in (3), we obtain

$$\mathbf{P}(\tilde{S}_{n-1} + \tilde{\xi}_n \in A_k) \leqslant \int_{2(m \vee (k+1))\lambda E} [\mathbf{P}(\tilde{\xi}_n \in A_k - x) \wedge \mathbf{P}(\tilde{\xi}_n \in B_m)] \times \tilde{\mathbf{P}}_{n-1}(dx) + \frac{1}{2}\mathbf{P}(\tilde{S}'_n \in A),$$

where $\tilde{P}_{n-1}(\cdot)$ is the distribution of $\tilde{S}_{n-1}$. We again apply (3) to estimate $P(\tilde{S}_n^{(1)} \in A_k)$, putting

$$S = \tilde{S}_n^{(')} - \tilde{\xi}_{n-1}, \quad \tilde{\xi} = \tilde{\xi}_{n-1}$$

in inequality (3).

If we again apply inequality (3) to estimate $P(\tilde{S}_n^{(2)} \in A_k)$, putting $S = \tilde{S}_n^{(2)} - \tilde{\xi}_{n-2}$, $\tilde{\xi} = \tilde{\xi}_{n-2}$, after the $(n-1)th$ application of this procedure we obtain

$$\mathbf{P}(\tilde{S}_n \in A_k) \leqslant \sum_{j=0}^{n-2} 2^{-j} D_j(n, k) + \frac{1}{2^{n-1}}\mathbf{P}(\tilde{S}_n^{(n-1)} \in A_k).$$

Here

$$D_j(n, k) = \int_{2(m \vee (k+1))\lambda E} [\mathbf{P}(\tilde{\xi} \in A_k - x) \wedge \mathbf{P}(\tilde{\xi} \in B_m)]\, \tilde{L}_n^{(j)}(dx),$$

where $\tilde{L}_n^{(j)}(\cdot)$ is the distribution of

$$\tilde{S}_n^{(j)} - \tilde{\xi}_{n-j} = \sum_{k=1}^{n-j-1} \tilde{\xi}_k + \sum_{k=n-j+1}^{n} \tilde{\xi}'_k, \quad j = 0, \ldots, n-1,$$

$\tilde{L}_n^{(0)}(\cdot) = \tilde{P}_{n-1}(\cdot)$, $\tilde{L}_n^{(n-1)}(\cdot)$ is the distribution of $\sum_{k=2}^{n} \tilde{\xi}'_k$. Therefore

$$\mathbf{P}(\tilde{S}_n \in A_k) \leqslant \sum_{j=0}^{n-2} 2^{-j} D_j(n, k) + \frac{1}{2^{n-1}} \int_{R^d} \mathbf{P}(\tilde{\xi} \in A_k - x) \tilde{L}_n^{(n-1)}(dx). \tag{4}$$

4. From inequality (4) we obtain

$$\sum_{k=0}^{\infty} e^{-k} \mathbf{P}(\tilde{S}_n \in A_k) \leqslant \sum_{k=0}^{m-1} e^{-k} \sum_{j=0}^{n-2} 2^{-j} \int_{2m\lambda E} \mathbf{P}(\tilde{\xi} \in A_k - x) \tilde{L}_n^{(j)}(dx)$$
$$+ \sum_{k=m}^{\infty} e^{-k} \sum_{j=0}^{n-2} 2^{-j} \mathbf{P}(\tilde{\xi} \in B_m) \int_{2(k+1)\lambda E} \tilde{L}_n^{(j)}(dx)$$
$$+ \sum_{k=0}^{\infty} e^{-k} 2^{-n+1} \int_{R^d} \mathbf{P}(\tilde{\xi} \in A_k - x) \tilde{L}_n^{(n-1)}(dx).$$

Next we estimate

$$I_1(j) = \sum_{k=0}^{m-1} e^{-k} \int_{2m\lambda E} \mathbf{P}(\tilde{\xi} \in A_k - x) \tilde{L}_n^{(j)}(dx),$$

$$I_2(j) = \sum_{k=m}^{\infty} e^{-k} \int_{2(k+1)\lambda E} \tilde{L}_n^{(j)}(dx) \mathbf{P}(\tilde{\xi} \in B_m),$$

$$I_3 = \sum_{k=0}^{\infty} e^{-k} 2^{-n+1} \int_{R^d} \mathbf{P}(\tilde{\xi} \in A_k - x) \tilde{L}_n^{(n-1)}(dx).$$

5. We start with an estimate for I_3. We have

$$I_3 = 2^{-n+1} \int_{R^d} \left[\sum_{k=0}^{\infty} e^{-k} \mathbf{P}(\tilde{\xi} \in A_k - x) \right] \tilde{L}_n^{(n-1)}(dx)$$
$$= 2^{-n+1} \int_{R^d} \left[\sum_{k=0}^{\infty} e^{-k} \int_{A_k} \tilde{\mathbf{P}}_x(dy) \right] \tilde{L}_n^{(n-1)}(dx),$$

where $\tilde{P}_x(\cdot)$ is the distribution of $\tilde{\xi} + x$. Therefore

$$I_3 \leqslant 2^{-n+1} \int_{R^d} \left[\sum_{k=0}^{\infty} e \int_{A_k} e^{-\frac{p_E(y)}{\lambda}} \tilde{\mathbf{P}}_x(dy) \right] \tilde{L}_n^{(n-1)}(dx)$$
$$= 2^{-n} 2e \int_{R^d} U_x(\xi, \lambda E) \tilde{L}_n^{(n-1)}(dx) \leqslant \text{ (by Lemma 4)}$$
$$\leqslant 2^{-n} 2e U(\xi, \lambda E) \int_{R^d} \tilde{L}_n^{(n-1)}(dx) = 2^{-n} 2e U(\xi, \lambda E).$$

Consequently $I_3 \leqslant \frac{C_3}{\sqrt{n}} U(\xi, \lambda E)$.

6. Now we estimate $I_1(j)$:

$$I_1(j) = \int_{2m\lambda E} \left[\sum_{k=0}^{m-1} e^{-k} \int_{k < \frac{p_E(y)}{\lambda} \leqslant k+1} \widetilde{\mathbf{P}}_x(dy) \right] \widetilde{L}_n^{(j)}(dx)$$

$$\leqslant e \int_{2m\lambda E} \left[\int_{\frac{p_E(y)}{\lambda} \leqslant m} e^{-\frac{p_E(y)}{\lambda}} \widetilde{P}_x(dy) \right] \widetilde{L}_n^{(j)}(dx) \leqslant e \int_{2m\lambda E} \left[\int_{\mathbf{R}^d} e^{-\frac{p_E(y)}{\lambda}} \widetilde{\mathbf{P}}_x(dy) \right]$$

$$\times \widetilde{L}_n^{(j)}(dx) \leqslant eU(\xi, \lambda E) \int_{2m\lambda E} \widetilde{L}_n^{(j)}(dx),$$

where the last inequality follows form Lemma 4. Therefore

$$I_1(j) \leqslant eU(\xi, \lambda E)\, Q\left(\widetilde{S}_n^{(j)} - \widetilde{\xi}_{n-j}, 2m\lambda E\right)$$

$$\leqslant e^2 U(\xi, \lambda E)\, U\left(\sum_{k=1}^{n-j-1} \xi_k + \sum_{k=n-j+1}^{n} \xi_k', 2m\lambda E \right) \leqslant \quad \text{(by Lemma 4)}$$

$$\leqslant e^2 U(\xi, \lambda E)\, U(S_{n-j-1}, 2m\lambda E).$$

Then we obtain from Theorem B that

$$I_1(j) \leqslant e^2 U(\xi, \lambda E)\, U(S_{n-j-1}, 2m\lambda E)$$

$$\leqslant C \frac{2m}{m-1} \frac{1}{\sqrt{n-j-1}} \frac{U(\xi, \lambda E)}{\sqrt{\mathbf{M}\left(\frac{p_E(\widetilde{\xi})}{(m-1)\lambda} \wedge 1 \right)^2}}.$$

Therefore,

$$I_1(j) \leqslant \frac{C}{\sqrt{n-j-1}} \frac{U(\xi, \lambda E)}{\sqrt{1 - \mathbf{P}(\widetilde{\xi} \in (m-1)\lambda E)}} \leqslant \frac{C_1}{\sqrt{n-j-1}} U(\xi, \lambda E).$$

The last inequality follows from the fact that

$$\mathbf{P}(\widetilde{\xi} \in B_{m-1}) < \frac{1}{2}.$$

Thus

$$I_1(j) \leqslant \frac{C_1}{\sqrt{n-j-1}} U(\xi, \lambda E).$$

7. Now we estimate $I_2(j)$:

$$I_2(j) \leqslant \sum_{k=m}^{\infty} e^{-(k-m)} \int_{B_m} e^{-\frac{p_E(y)}{\lambda}} \widetilde{\mathbf{P}}(dy) \int_{2(k+1)\lambda E} \widetilde{L}_n^{(j)}(dx)$$

$$\leqslant U(\xi, \lambda E) \sum_{k=0}^{\infty} e^{-k} \int_{2(k+m+1)\lambda E} \widetilde{L}_n^{(j)}(dx) \leqslant U(\xi, \lambda E) \sum_{k=0}^{\infty} e^{-k} \int_{2(k+1)(m+1)\lambda E} \widetilde{L}_n^{(j)}(dx)$$

$$\leqslant eU(\xi, \lambda E) \sum_{k=0}^{\infty} e^{-k} \int_{2(k+1)(m+1)\lambda E} e^{-\frac{p_E(x)}{2(k+1)(m+1)\lambda}} \widetilde{L}_n^{(j)}(dx).$$

Therefore,

$$I_2(j) \leqslant eU(\xi, \lambda E) \sum_{i=0}^{\infty} e^{-i} U\left(\sum_{k=1}^{n-j-1} \xi_k + \sum_{k=n-j+1}^{n} \xi_k',\ 2(i+1)(m+1)\lambda E\right).$$

By Lemma 1 we have

$$I_2(j) \leqslant eU(\xi, \lambda E) \sum_{k=0}^{\infty} e^{-k} U(S_{n-j-1},\ 2(k+1)(m+1)\lambda E).$$

It follows from Theorem B that

$$I_2(j) \leqslant eU(\xi, \lambda E) \sum_{k=0}^{\infty} e^{-k} \frac{C}{\sqrt{n-j-1}} \frac{2(k+1)(m+1)\lambda}{(m-1)\lambda}$$

$$\times \frac{1}{\sqrt{\mathbf{M}\left(\frac{p_E(\widetilde{\xi})}{(m-1)\lambda} \wedge 1\right)^2}} \leqslant CU(\xi, \lambda E) \sum_{k=0}^{\infty} e^{-k}(k+1) \frac{1}{\sqrt{n-j-1}}$$

$$\leqslant CU(\xi, \lambda E) \frac{1}{\sqrt{n-j-1}},$$

since

$$\mathbf{M}\left(\frac{p_E(\widetilde{\xi})}{(m-1)\lambda} \wedge 1\right)^2 \geqslant \int_{p_E(x) > \lambda(m-1)} \widetilde{\mathbf{P}}(dx) = 1 - \mathbf{P}(\widetilde{\xi} \in B_{m-1}) \geqslant \frac{1}{2},$$

and

$$\sum_{k=0}^{\infty} e^{-k}(k+1) < C.$$

Thus,

$$I_2(j) \leqslant C_2 \frac{U(\xi, \lambda E)}{\sqrt{n-j-1}}.$$

8. Therefore, we have

$$\sum_{k=0}^{\infty} e^{-k} \mathbf{P}(\widetilde{S}_n \in A_k) \leqslant \sum_{j=0}^{n-2} 2^{-j} \frac{C_1}{\sqrt{n-j-1}} U(\xi, \lambda E)$$

$$+ \sum_{j=0}^{n-2} 2^{-j} \frac{C_2}{\sqrt{n-j-1}} U(\xi, \lambda E) + \frac{C_3}{\sqrt{n}} U(\xi, \lambda E) \leqslant \frac{C}{\sqrt{n}} U(\xi, \lambda E),$$

since

$$\sum_{j=0}^{n-2} 2^{-j} \frac{1}{\sqrt{n-j-1}} \leqslant \frac{C_4}{\sqrt{n}} \cdot$$

Finally, we have

$$U(S_n, \lambda E) = \int_{R^d} \exp\left\{-\frac{p_E(x)}{\lambda}\right\} \widetilde{\mathbf{P}}_n(dx)$$

$$= \sum_{k=0}^{\infty} \int_{A_k} \exp\left\{-\frac{p_E(x)}{\lambda}\right\} \widetilde{\mathbf{P}}_n(dx) \leqslant \sum_{k=0}^{\infty} e^{-k} \int_{A_k} \widetilde{\mathbf{P}}_n(dx) = \sum_{k=0}^{\infty} e^{-k} \mathbf{P}(\widetilde{S}_n \in A_k)$$

$$\leqslant \frac{C}{\sqrt{n}} U(\xi, \lambda E) \leqslant \frac{C}{\sqrt{n}} \frac{\lambda}{\lambda_0} U(\xi, \lambda E) \leqslant C\lambda \frac{1}{\sqrt{n}} \frac{U(\xi, \lambda E)}{\sqrt{\mathbf{M}(p_E(\xi) \wedge \lambda_0)^2}}.$$

The lemma is generalized to class Ω by passing to the limit. The proof is complete.

4. Proof of the main theorems

Proof of Theorem 1. For $k = 1,2,\ldots,n$ we consider the variables

$$p_k = \frac{\sum_{k=1}^{n} \mathbf{M}(p_E(\widetilde{\xi}_k) \wedge \lambda_k)^2 U^{-2}(\xi_k, \lambda E)}{\mathbf{M}(p_E(\widetilde{\xi}_k) \wedge \lambda_k)^2 U^{-2}(\xi_k, \lambda E)}, \quad \lambda \geqslant \max_{1 \leqslant k \leqslant n} \lambda_k.$$

Obviously,

$$p_k \geqslant 1 \quad \text{and} \quad \sum_{k=1}^{n} p_k^{-1} \equiv 1.$$

By Lemma 2 we have

$$U(S_n, \lambda E) \leqslant \sum_{k=1}^{n} p_k^{-1} U\left(S_{[p_k]}^{(k)}, \lambda E\right).$$

Then by Lemma 5 we obtain

$$U(S_n, \lambda E) \leqslant \sum_{k=1}^{n} p_k^{-1} C\lambda \frac{1}{\sqrt{p_k}} \frac{U(\xi_k, \lambda E)}{\sqrt{\mathbf{M}(p_E(\widetilde{\xi}_k) \wedge \lambda_k)^2}}$$

$$= \sum_{k=1}^{n} p_k^{-1} C\lambda \left(\sum_{k=1}^{n} \mathbf{M}(p_E(\widetilde{\xi}_k) \wedge \lambda)^2 U^{-2}(\xi_k, \lambda E)\right)^{-1/2},$$

and since $\sum_{k=1}^{n} p_k^{-1} = 1$, Theorem 1 is proved.

Proof of Corollary 1.1. Follows from Theorem 1 and Theorem A (Section 3).

Proof of Theorem 2. We estimate the integral concentration $U(S_n, \lambda E)$ in terms of the Lévy concentration function:

$$U(S_n, \lambda E) = \int_{R^d} \exp\left\{-\frac{p_E(x)}{\lambda}\right\} \widetilde{\mathbf{P}}_n(dx)$$

$$= \sum_{k=0}^{\infty} \int_{k < \frac{p_E(x)}{\lambda} < k+1} \exp\left\{-\frac{p_E(x)}{\lambda}\right\} \widetilde{\mathbf{P}}_n(dx) \leqslant \sum_{k=0}^{\infty} e^{-k} \int_{k < \frac{p_E(x)}{\lambda} < k+1} \widetilde{\mathbf{P}}_n(dx)$$

$$\leqslant \sum_{k=0}^{\infty} e^{-k} \int_{(k+1)\lambda E} \widetilde{\mathbf{P}}_n(dx) \leqslant \sum_{k=0}^{\infty} e^{-k} Q(\widetilde{S}_n, (k+1)\lambda E)$$

$$\leqslant \sum_{k=0}^{\infty} e^{-k} Q(S_n, (k+1)\lambda E).$$

Consequently, we obtain from Theorem D that

$$U(S_n, \lambda E) \leqslant C \sum_{k=0}^{\infty} (k+1) e^{-k} \lambda d \left(\sum_{j=1}^{n} \mathbf{M} (p_E(\widetilde{\xi}_j) \wedge \lambda_j)^2 \right)^{-1/2}$$

$$\leqslant C \lambda d \left(\sum_{k=1}^{n} \mathbf{M} (p_E(\widetilde{\xi}_k) \wedge \lambda_k)^2 \right)^{-1/2},$$

and the proof is complete.

The result of the theorem, although the proof is simple, substantially improves the estimate in Theorem C.

Now we consider an infinite-dimensional space and prove Theorem 3.

Proof of Theorem 3. Let H be a separable Hilbert space. Let H_d be the subspace with basis $e_1, \ldots, e_d$, and let P_d be the projection operator from H onto H_d.

Consider the following variables:

$$\xi_{d,k} = P_d \xi_k, \quad S_{d,n} = \sum_{k=1}^{n} \xi_{d,k},$$

$A_1, \ldots, A_m$ being bounded linear operators in H, and

$$E_d = \left\{ x \in H; \ \sum_{j=1}^{m} \| P_d A_j P_d x \| \leqslant 1 \right\}.$$

Since the $\xi_{d,k}$ can be viewed as d-dimensional random vectors with values in H_d, we have for H_d:

$$U(S_n, \lambda E_d) = \mathbf{M} \exp\{-p_{E_d}(\widetilde{S}_n)/\lambda\}$$

$$= \mathbf{M} \exp\{-p_{E_d}(\widetilde{S}_{d,n})/\lambda\} = U(S_{d,n}, \lambda E_d),$$

since $\tilde{S}_{d,n} = P_d\tilde{S}_n$,

$$p_{E_d}(x) = \sum_{j=1}^{m} \| P_d A_j P_d x \| = \sum_{j=1}^{m} \| P_d A_j P_d (P_d x) \| = p_{E_d}(P_d(x))$$

for any $x \in H$ and, thus, by Theorem 1 we obtain

$$U(S_n, \lambda E_d) \leqslant C\lambda \left(\sum_{k=1}^{n} \mathbf{M}\left(p_{E_d}(\tilde{\xi}_k) \wedge \lambda_k\right)^2 U^{-2}(\xi_k, \lambda E_d) \right)^{-1/2}.$$

It remains only to note that, as follows from [2],

$$p_{E_d}(x) \to p'_E(x) \quad \text{as} \quad d \to \infty \quad \text{for all} \quad x \in H ,$$

where

$$E = \left\{ x \in H, \ \sum_{j=1}^{m} \| A_j x \| \leqslant 1 \right\}.$$

Passing to class Ω poses no problem, and the proof is complete.

Proof of Corollary 3.1. We carry out the proof analogous to Theorem 3, but instead of Theorem 1, we use the estimate from Corollary 1.1.

Proof of Theorem 4. We use the following inequalities:

$$U(S_n, \lambda E) = \sum_{k=0}^{\infty} \int_{k < \frac{p_E(x)}{\lambda} \leqslant k+1} \exp\left\{ -\frac{p_E(x)}{\lambda} \right\} \tilde{\mathbf{P}}_n(dx)$$

$$\leqslant \sum_{k=0}^{\infty} e^{-k} Q(S_n, \lambda(k+1)E) \leqslant \text{(by Theorem E)} \leqslant C(d) \sum_{k=0}^{\infty} e^{-k}$$

$$\times \inf_{u>0} \frac{(\lambda^d + u^d)(k+1)^d}{u^d + \left(\sum_{i=1}^{n} \chi_i^E(u) \right)^{d/2}} \leqslant C(d) \sum_{k=0}^{\infty} e^{-k}(k+1)^d \inf_{n>0} \frac{\lambda^d + u^d}{u^d + \left(\sum_{i=1}^{n} \chi_i^E(u) \right)^{d/2}},$$

but

$$\sum_{k=0}^{\infty} e^{-k}(k+1)^d \leqslant C(d) < \infty$$

for any fixed d, and if the $\xi_1, \ldots, \xi_n$ are identically distributed, then

$$u^d + \left(\sum_{i=1}^{n} \chi_i^E(u) \right)^{d/2} \geqslant n^{d/2} (\chi_1^E(u))^{d/2},$$

and the Theorem is proved.

The author is indebted to B.A. Rogozin for his continued interest and support.

REFERENCES

[1] Anan'evskij, S.M. *Funktsii kontsentratsii summ nezavisimykh sluchajnykh velichin* (Concentration Functions of Sums of Independent Random Variables. Kandidatskaya dissertatsiya. Leningrad: Izd-vo LGU, 1980.

[2] Enger, J. *Bounds for the Concentration Functions of a Sum of Independent Random Vectors with Values in a Euclidean or a Hilbert Space. Thesis.* Uppsala: Uppsala University, 1975.

[3] Kolmogorov, A.N. "Sur les propriétés des fonctions de concentrations de M.P. Lévy." *Ann. de l'Inst. Henri Poincaré*, XVI, 1 (1958): 27-34.

[4] Rogozin, B.A. "An Estimate for Concentration Functions. *Theory Probab. Applications*, vol.6, no.1 (1961): 94-96. (English transl.)

[5] ______. "On the Increase of Dispersion of Sums of Independent Random Variables." *Theory Probab. Applications*, vol.6, no.1 (1961): 97-99. (English transl.)

[6] Esseen, C.G. "On the Kolmogorov-Rogozin Inequality for Concentration Functions." *Z. Wahrscheinlichkeitstheorie verw. Gebiete*, B.5, H.4 (1966): 210-16.

[7] ______. "On the Concentration Function of a Sum of Independent Random Variables." *Z. Wahrscheinlichkeitstheorie verw. Gebiete*, B.9, H.4 (1968): 290-308.

[8] Hengartner, W., and Theodorescu, R. *Concentration Functions*. New York: Academic Press, 1973.

[9] Kesten, Harry. "A Sharper Form of the Doeblin-Lévy-Kolmogorov-Rogozin Inequality for Concentration Functions." *Math. Scand.*, vol.25, no.1 (1969): 133-44.

[10] ______. "Sums of Independent Random Variables Without Moment Conditions." *Ann. Math. Statistics*, vol.43, no.3 (1972): 701-732.

[11] Postnikova, L.P., and Judin, A.A. "Analytic Method of Estimates of the Concentration Function." *Proceedings of the Steklov Institute of Mathematics of the Academy of Sciences of the USSR*, issue 1, (1980): 153-61. Providence, Rhode Island: American Math. Society. (English transl.)

[12] ______. "A Sharper Form of the Inequality for the Concentration Function." *Theory Probab. Applications*, vol.23, no.2 (1978): 359-62. (English transl.)

[13] Halasz, G. "Estimates for the Concentration Functions of Sums of Independent Random Variables." In *Tret'ya Vil'nyusskaya konferentsiya po teorii veroyatnostej i matematicheskoj statistike. Tezisy dokladov* (The Third Vilnius Conference on Probability Theory and Mathematical Statistics. Abstracts), p. 123. Vilnius, 1981.

[14] Miroshnikov, A.L., and Rogozin, B.A. "Inequalities for the Concentration Functions." *Theory Probab. Applications*, vol.25, no.1 (1980): 176-80. (English transl.)

A.V. Pozhidayev

ASYMPTOTIC NORMALITY OF SOLUTIONS OF PARABOLIC EQUATIONS WITH RANDOM COEFFICIENTS

1. Formulation of the main result and fundamentals of differential equations theory

We investigate the limiting behavior, as $\varepsilon \to 0$, of a parabolic Cauchy problem with random coefficients:

$$\left.\begin{aligned} Qw \equiv \partial w/\partial t - \left(b^{ij}(x, t) + a^{ij}(x, t/\varepsilon, \omega)\right) w_{x_i x_j} = f(x, t), \\ w(0, x) = 0, \end{aligned}\right\} \quad (1)$$

$$x \in R^n, \quad t \in [0, T], \quad T < \infty.$$

The limiting behavior of (1) is analyzed under the following assumptions.

There exist positive constants C_1, C_2 such that for any real vector $\gamma = \{\gamma_1, \ldots, \gamma_n\}$, the inequalities

$$C_1|\gamma|^2 \leqslant (b^{ij}(x, t) + a^{ij}(x, t/\varepsilon, \omega))\gamma_i\gamma_j \leqslant C_2|\gamma|^2 \quad (2)$$

are satisfied wtih probability 1. Here and below, the tensor notation is used to denote summation. The nonrandom coefficients $B^{ij}(x,t)$ are assumed to be smooth, and for all x, t, y, τ:

$$|b^{ij}(x, t) - b^{ij}(y, \tau)| \leqslant C(|x - y|^\alpha + |t - \tau|^{\alpha/2}), \quad (3)$$

where $0 \leqq \alpha \leqq 1$, $C > 0$ is a constant. In the sequel, the letter C designates various constants not depending on ε. Furthermore,

$$\sup_{x,t} \left| b^{ij}_{x_h x_s}(x, t) \right| < C, \quad (4)$$

where

$$u_{x_h} = \partial u/\partial x_h, \ u_{x_h x_s} = \partial^2 u/\partial x_h \partial x_s.$$

The nonrandom right side is finite and differentiable with respect to the spatial arguments. For every t, $f(x,t) \in C_0^1(V)$, where V is a nonrandom bounded region in R^n. Being a function of the pair (x,t), $f(x,t)$ is continuous.

The random fields $a^{ij}(x, t/\varepsilon, \omega)$ belong to class $C_0^2(V)$ with respect to the spatial arguments. For all x, t, y, ε one has

$$|a^{ij}(x, t/\varepsilon, \omega) - a^{ij}(y, t/\varepsilon, \omega)| \leqslant C|x-y|^{\alpha}, \tag{5}$$

with probability 1 for a nonrandom C. Moreover, with probability 1

$$\sup_{x,t} \left| a^{ij}_{x_k x_s}(x, t/\varepsilon, \omega) \right| < C. \tag{6}$$

The matrix-valued random field $a = \|a^{ij}\|$ is assumed to be m-dependent with respect to time: the matrices $a(x,t)$, $a(y,\tau)$ with random coefficients are independent for $|t-\tau| > m$, for any spatial variables x, y. In the sequel, we assume without loss of generality that $m = 1$.

The fields $a^{ij}(x, t/\varepsilon, \omega)$ are centered:

$$\mathbf{M} a^{ij}(x, t/\varepsilon, \omega) = 0. \tag{7}$$

Under the conditions imposed on the analytic properties of the coefficients, problem (1) is uniquely solvable with probability 1 [1].

By a random solution of problem (1) we mean any random field the realizations of which satisfy (1) with probability 1.

Let w_ε be a solution of (1) and let U be a solution of the averaged problem:

$$\left.\begin{aligned} LU \equiv U_t - b^{ij}(x, t)\, U_{x_i x_j} &= f(x, t), \\ U(0, x) &= 0. \end{aligned}\right\} \tag{8}$$

We shall consider later restrictions of solutions of problems (1), (8) to V for fixed t: $0 < t \leqq T$, where $V \subset R^n$ is the support of the right side and the random coefficients, denoting the restrictions by w_ε, U.

Under these conditions, we can assert that for β such that $0 < \beta < \frac{1}{2}$ (see Section 2) we can find a constant $C(\beta) > 0$ such that

$$\mathbf{M} \| w_\varepsilon - U \|^2_{L_2(V)} \leqslant C(\beta)\, \varepsilon^{1/2-\beta}.$$

One may add another assumption that w_ε is asymptotically normal. Suppose that for any function $\ell(x) \in L_2(V)$, $\ell \not\equiv 0$, the random variable

$$\varepsilon^{-1/2} \int_V \int_V \int_0^t l(x)\, \Gamma(x, t, \xi, \tau)\, a^{ij}(\xi, \tau/\varepsilon, \omega)\, U_{\xi_i \xi_j}(\xi, \tau)\, d\tau\, d\xi\, dx \tag{9}$$

is asymptotically normal with parameters

$$(0,\ \sigma^2(l,\ \Gamma,\ U)). \tag{10}$$

In (9), $\Gamma(x,t,\xi,\tau)$ is a fundamental solution of problem (8) (see [1]).

Condition (10) is satisfied, for instance, if the $a^{ij}(x,\ t/\varepsilon,\ \omega)$ are homogeneous fields, m-dependent in the time arguments, such that $a^{ij}(x,t,\omega)$, $a^{ks}(x,t,\omega)$ are independent for $i \neq k$, $j \neq s$. Then

$$\sigma^2(l,\Gamma,U) = t\sum_{i,j=1}^{n}\int_{|\tau|<m} R_{ij}(\tau)\,d\tau,$$

where $R_{ij}(\tau) = M\tilde{a}^{ij}(s)\,\tilde{a}^{ij}(s+\tau)$ is the correlation function of $\tilde{a}^{ij}(\tau)$. For a fixed pair (i,j),

$$\tilde{a}^{ij}(\tau) = \int_V\int_V l(x)\,\Gamma(x,t,\xi,\tau)\,a^{ij}(\xi,\tau/\varepsilon)\,U_{\xi_i\xi_j}(\xi,\tau)\,d\xi\,dx.$$

Let (cf. [2]) $\overset{\circ}{W}{}_2^k(V)$ denote the closure of the set of smooth finite functions on V with respect to the norm

$$\|u\|_{2,V}^{k} = \left[\int_V\left(|u|^2 + \sum_{|\alpha|<k}|D^\alpha u|^2\right)dx\right]^{1/2},$$

and $\alpha = \{\alpha_1,\dots,\alpha_n\}$, $|\alpha| = \alpha_1+\dots+\alpha_n$, $D^\alpha = D_{x_1}^{\alpha_1}\cdots D_{x_n}^{\alpha_n}$, and W_2^{-k} is the space of generalized functions dual to $\overset{\circ}{W}{}_2^k$.

Our main result is the following.

T h e o r e m 1.1. As $\varepsilon \to 0$ the solution w_ε of problem (1) with coefficients obeying the conditions (2)-(7), (10), is asymptotically normal in the sense that the distribution generated in the space $W_2^{-3}(V)$ of generalized functions by the random field

$$v_\varepsilon = \varepsilon^{-1/2}(w_\varepsilon - U)$$

converges weakly to a centered Gaussian distribution in $W_2^{-3}(V)$ with characteristic functional

$$\varphi(\Phi) = \exp(-\sigma^2(\Phi,\ \Gamma,\ U)/2),$$

where $\sigma^2(\Phi,\Gamma,U)$ is defined in (10).

REMARK 1.1. We shall omit the ε, ω to simplify the notation for random functions.

2. Estimate of the error in averaging in the norm of $L_2(V)$

Let $u = w - U$. Then for u we have the problem

$$\left.\begin{aligned} Lu = a^{ij}(x,t)\,u_{x_i x_j} + a^{ij}(x,t)\,U_{x_i x_j}, \\ u(0,x) = 0. \end{aligned}\right\} \qquad (11)$$

Theorem 2.1. For β, $0 < \beta < \frac{1}{2}$, we can find a constant $C(\beta)$ such that

$$\mathbf{M}\|w - U\|^2_{L_2(V)} \leqslant C(\beta)\,\varepsilon^{1/2-\beta}.$$

Proof. Let us reduce problem (11) to an integral equation by using the nonrandom fundamental solution $\Gamma(x,t,\xi,\tau)$ of problem (8):

$$u = \Gamma * a^{ij} u_{\xi_i\xi_j} + \Gamma * a^{ij} U_{\xi_i\xi_j}, \qquad (12)$$

where the operator Γ is defined by the equality

$$\Gamma * f = \int_0^t \int_V \Gamma(x,t,\xi,\tau)\, f(\xi,\tau)\, d\xi d\tau.$$

In (12), iterating and integrating by parts with respect to the variable ξ, we obtain the relation (the boundary terms vanish because the field $a(x,t)$ is finite):

$$\begin{aligned} u = &-\Gamma_{\xi_i} * a^{ij} u_{\xi_j} - \Gamma * a^{ij}_{\xi_i} u_{\xi_j} + \Gamma * a^{ij}_{\xi_j} U_{\xi_i} \\ = &-\Gamma_{\xi_i} * a^{ij} \Gamma_{\xi_j} * a^{ks} u_{y_k y_s} - \Gamma_{\xi_i} * a^{ij} \Gamma_{\xi_j} * a^{ks} U_{y_k y_s} - \Gamma * a^{ij}_{\xi_i} \Gamma_{\xi_j} * a^{ks} u_{y_k y_s} \\ &- \Gamma * a^{ij}_{\xi_i} \Gamma_{\xi_j} * a^{ks} U_{y_k y_s} + \Gamma * a^{ij} U_{\xi_i \xi_j}. \end{aligned}$$

We estimate each term of the last equality individually.

Lemma 2.1.

$$F \equiv \mathbf{M}\left\| \Gamma * a^{ij} U_{\xi_i\xi_j} \right\|^2_{L_2(V)} \leqslant C\varepsilon.$$

Proof. For a fixed pair (i,j), let $U_{\xi_i\xi_j} = \tilde{U}$, $a^{ij} = a$. Then F is estimated by a finite sum of expressions of the form:

$$\mathbf{M}\int_V \int_0^t \int_0^t \int_V \int_V \Gamma(x,t,\xi,\tau)\, a(\xi,\tau)\, \widetilde{U}(\xi,\tau)\, \Gamma(x,t,y,s)\, a(y,s)\, \widetilde{U}(y,s)\, dy d\xi$$

$$\times\, ds d\tau dx = \mathbf{M}\int_V \int_0^t \int_0^t a_1(x,\tau)\, a_1(x,s)\, ds d\tau dx,$$

where

$$a_1(x,\tau) = \int_V \Gamma(x,t,\xi,\tau)\, a(\xi,\tau)\, \widetilde{U}(\xi,\tau)\, d\xi.$$

Note that $|a_1(x,\tau)| < C$ (see [3]) and $Ma_1(x,\tau)a_1(x,s) = 0$ for $|s-\tau| > \varepsilon$. Hence the proof of the lemma reduces to the obvious estimate

$$F \leqslant C \iint_{|s-\tau|<\varepsilon} ds\,d\tau \leqslant C\varepsilon.$$

L e m m a 2.2. For any β, $0 < \beta < 1$, we can find a constant $C(\beta)$ such that

$$H \equiv \mathbf{M}\left\| \Gamma_{\xi_i} * a^{ij}\Gamma_{\xi_j} * a^{ks} U_{\eta_k\eta_s} \right\|^2_{L_2(V)} \leqslant C(\beta)\,\varepsilon^{1-\beta}.$$

Proof. Let the parts (i,j), (k,s) be fixed. Then H is estimated by a finite sum of expressions of the form

$$\mathbf{M}\int_V \int_0^t \int_0^\tau q(x,\tau,\tau_1) \int_0^t \int_0^s q(x,s,s_1)\, d\bar{s}\, d\bar{\tau}\, dx,$$

where

$$q(x,\tau,\tau_1) = \int_V \int_V \Gamma_{\xi_i}(x,t,\xi,\tau)\, a(\xi,\tau)\, \Gamma_{\xi_j}(\xi,\tau,\eta,\tau_1)\, a(\eta,\tau_1)\, \widetilde{U}(\eta,\tau_1)\, d\eta\, d\xi,$$

$$d\bar{\tau} = d\tau_1 d\tau,\quad d\bar{s} = ds_1 ds.$$

In deriving the estimate of Lemma 2.2., we shall use the inequality (see [1])

$$|\Gamma_{\xi_i}(x,t,\xi,\tau)| \leqslant C(t-\tau)^{-(1+\beta)/2}|x-\xi|^{\beta-n}. \tag{13}$$

Finally, we use the notation $I\{A\}$ to indicate that the integral I is considered over the region A of variation of the time arguments. Then, due to the m-dependence of the field $a(x,t)$, the centering condition and inequality (13), we obtain

$$H\{\tau - s > 2\varepsilon\} = H\{\tau - s > 2\varepsilon \cap \tau - \tau_1 < \varepsilon\} = H\{\tau - s > 2\varepsilon$$

$$\cap\, \tau - \tau_1 < \varepsilon \cap s - s_1 < \varepsilon\} \leqslant C \int_V \int_0^t \int_V |x-\xi|^{\beta-n}(t-\tau)^{-(1+\beta)/2}$$

$$\times \int_V \int_{\tau-\varepsilon}^{\tau} |\xi-\eta|^{\beta-n}(\tau-\tau_1)^{-(1+\beta)/2} \int_V \int_0^t |x-\xi_1|^{\beta-n}(t-s)^{-(1+\beta)/2}$$

$$\times \int_V \int_{s-\varepsilon}^{s} |\xi_1-\eta_1|^{\beta-n}(s-s_1)^{-(1+\beta)/2} ds_1\, d\eta_1\, ds\, d\xi_1\, d\tau_1\, d\eta\, d\tau\, d\xi\, dx$$

$$\leqslant C(\beta)\int_0^t\int_0^t ((t-\tau)(t-s))^{-(1+\beta)/2}\int_{\tau-\varepsilon}^{\tau}\int_{s-\varepsilon}^{s}((\tau-\tau_1)(s-s_1))^{-(1+\beta)/2}\,d\bar{s}\,d\bar{\tau}$$

$$\leqslant C(\beta)\,\varepsilon^{1-\beta}.$$

The integral $H\{s-\tau > 2\varepsilon\}$ is estimated in the similar way. For $H\{|s-\tau| \leqq 2\varepsilon\}$ we have the representation

$$\begin{aligned} H\{|s-\tau|\leqslant 2\varepsilon\} = H\{|s-\tau|\leqslant 2\varepsilon \cap \tau-6\varepsilon<\tau_1<\tau\} \\ + H\{|s-\tau|\leqslant 2\varepsilon \cap s-6\varepsilon<s_1<s\} \\ + H\{|s-\tau|\leqslant 2\varepsilon \cap 0<\tau_1<\tau-6\varepsilon \cap 0<s_1<s-6\varepsilon\} \\ + H\{|s-\tau|\leqslant 2\varepsilon \cap \tau-6\varepsilon<\tau_1<\tau \cap s-6\varepsilon<s_1<s\}. \end{aligned}$$

Each of the above integrals is estimated in the same way as $H\{\tau-s > 2\varepsilon\}$. Lemma 2.2 is proved.

REMARK 2.1. It follows from the proof of Lemma 2.2 that for any β, $0 < \beta < 1$, one has

$$\mathbf{M}\left\|\Gamma * a^{ij}_{\xi_i}\Gamma_{\xi_j} * a^{hs}U_{\eta_k\eta_s}\right\|^2_{L_2(V)}\leqslant C(\beta)\,\varepsilon^{1-\beta}.$$

Indeed, the function $\Gamma(x,t,\xi,\tau)$ *a priori* satisfies inequality (13) used in the proof.

L e m m a 2.3. For any β, $0 < \beta < \frac{1}{2}$, we can find a constant $C(\beta)$ such that

$$\Psi\equiv\mathbf{M}\left\|\Gamma_{\xi_i} * a^{ij}\Gamma_{\xi_j} * a^{ks}u_{y_ky_s}\right\|^2_{L_2(V)}\leqslant C(\beta)\,\varepsilon^{1/2-\beta}.$$

Proof. Interchanging the order of integration and taking the mathematical expectation, one can obtain the representation

$$\Psi=\int_{V^5}\int_{\Lambda}\psi\,dx_0dx_1dx_2dy_1dy_2d\tau d\tau_1d\sigma d\sigma_1,$$

where all the spatial variables in the integration run through the domain V and all the time variables run through the domain $\Lambda = \{0 < \tau_1 < \tau < t,\ \ 0 < \sigma_1 < \sigma < t\}$. The integrand has the form

$$\begin{aligned}\psi=\mathbf{M}\Big\{&\Gamma_{x_{1i}}(x_0,t,x_1,\tau)\,a^{ij}_{\varepsilon}(x_1,\tau/\varepsilon,\omega)\\ &\times\Gamma_{x_{1j}}(x_1,\tau,x_2,\tau_1)\,a^{ks}_{\varepsilon}(x_2,\tau_1/\varepsilon,\omega)\,u_{ks}\\ &(x_2,\tau_1,\omega)\,\Gamma_{y_{1p}}(x_0,t,y_1,\sigma)\,a^{pq}_{\varepsilon}(y_1,\sigma/\varepsilon,\omega)\,\Gamma_{y_{1q}}(y_1,\sigma,y_2,\sigma_1)\,a^{rl}_{\varepsilon}\\ &\times(y_2,\sigma_1/\varepsilon,\omega)\,u_{rl}(y_2,\sigma_1,\omega)\Big\},\quad u_{ks}(z,\tau)=\partial^2u/\partial z_k\partial z_s.\end{aligned}$$

Note that $\psi = 0$ for $\tau > \max [\tau_1, \tau, \sigma_1] + \varepsilon$. Indeed, the solution $u_\varepsilon(x,\tau,\omega)$ is determined by the initial data and the values of the equation's coefficients in the layer $R^n \times [0,\tau]$. Hence in our case, the random variables $a_\varepsilon^{ij}(x_1, \tau/\varepsilon, \omega)$ do not depend on $a_\varepsilon^{ks}(x_2, \tau_1/\varepsilon, \omega)$, $u_{ks}(x_2, \tau_1, \omega)$, $a_\varepsilon^{pq}(y_1, \sigma/\varepsilon, \omega)$, $a_\varepsilon^{r\ell}(y_2, \sigma_1/\varepsilon, \omega)$, $u_{r\ell}(y_2, \sigma_1, \omega)$ and one can use the centering condition (7). Similar arguments are valid for $\sigma > \max (\sigma_1, \tau, \tau_1) + \varepsilon$. In the rest of the region, because of the uniform estimates for the second derivatives of the solution (see Section 4) and inequality (13), we have the estimate

$$\psi \leqslant C(\beta) \prod_{j=0}^{1} |x_j - x_{j+1}|^{\beta - n} |y_j - y_{j+1}|^{\beta - n} q_1,$$

where $y_0 = x_0$, $q_1 = [(t-\tau)(\tau-\tau_1)(t-\sigma)(\sigma-\sigma_1)]^{-\frac{1}{2}-\beta}$. Integrating over spatial variables yields the estimate

$$\Psi \leqslant C(\beta) \int_{\Lambda_1 \cap [0,t]^4} q_1 d\tau_1 d\tau d\sigma_1 d\sigma,$$

where the region Λ_1 is covered by the combination

$$\Lambda_2 = \{\sigma - \varepsilon < \sigma_1 < \sigma\}, \quad \Lambda_3 = \{\tau - \varepsilon < \tau_1 < \tau\}, \ \Lambda_4 = \{|\tau - \sigma| < \varepsilon\}.$$

Elementary estimates show that the integral of q_1 with respect to each of the regions Λ_i $(i = 2,3,4)$ is estimated by the quantity $C(\beta)\varepsilon^{\frac{1}{2}-\beta}$. Lemma 2.3 is proved.

REMARK 2.2. It follows from the proof of Lemma 2.3 that for any $0 < \beta < \frac{1}{2}$

$$\mathbf{M} \left\| \Gamma * a_{\xi_i}^{ij} \Gamma_{\xi_j} * a^{ks} u_{y_k y_s} \right\|_{L_2(V)}^2 \leqslant C(\beta) \varepsilon^{1/2-\beta}.$$

The validity of Theorem 2.1 follows from Lemmas 2.1 - 2.3 and Remarks 2.1 - 2.2.

3. An asymptotically normal component of the solution

We are considering representation (12). Our further calculations are aimed at establishing that the summand $\Gamma * a^{ij} U_{\xi_i \xi_j}$ is the major contribution to the random component u_ε.

Lemma 3.1.

$$\mathbf{M}\left\| \Gamma * a^{ij} u_{\xi_i \xi_j} \right\|_{W_2^{-2}(V)} = o\left(\varepsilon^{1/2}\right).$$

Proof. Let the pair (i,j) be fixed. Using the finiteness property of the field a(x,t), we integrate by parts and obtain the inequalities

$$\begin{aligned} \Gamma * a^{ij} u_{\xi_i \xi_j} \|_{W_2^{-2}(V)} &\leqslant \left\| \Gamma_{\xi_i} * a^{ij} u_{\xi_j} \right\|_{W_2^{-2}(V)} + \left\| \Gamma * a^{ij}_{\xi_i} u_{\xi_j} \right\|_{W_2^{-2}(V)} \\ &\leqslant \left\| \Gamma_{\xi_i \xi_j} * a^{ij} u \right\|_{W_2^{-2}(V)} + \left\| \Gamma_{\xi_i} * a^{ij}_{\xi_j} u \right\|_{W_2^{-2}(V)} + \left\| \Gamma_{\xi_j} * a^{ij}_{\xi_i} u \right\|_{W_2^{-2}(V)} \\ &\quad + \left\| \Gamma * a^{ij}_{\xi_i \xi_j} u \right\|_{W_2^{-2}(V)}. \end{aligned}$$

By the results of Section 5 (see Lemmas 5.2 and 5.6 below), it follows now that $\| \Gamma * a^{ij} u_{\xi_i \xi_j} \|_{W_2^{-2}(V)}$ is estimated by a finite sum of expressions of the form $\| \Gamma * a^{ij} u \|_{L_2(V)}$.

Let

$$g(x) = \int_0^t \int_V \Gamma(x, t, \xi, \tau)\, a^{ij}(\xi, \tau)\, u(\xi, \tau)\, d\xi d\tau.$$

From the representation (12) we obtain the equality

$$g = \Gamma * a^{ij} \Gamma * a^{ks} \tilde{u} + \Gamma * a^{ij} \Gamma * a^{ks} \widetilde{U}.$$

To prove Lemma 3.1, we need to estimate

$$A \equiv \mathbf{M} \| \Gamma * a^{ij} \Gamma * a^{ks} \widetilde{U} \|^2_{L_2(V)} \quad \text{and} \quad B \equiv \mathbf{M} \| \Gamma * a^{ij} \Gamma * a^{ks} \tilde{u} \|^2_{L_2(V)}.$$

We shall check that the contribution of A is negligible.

Lemma 3.2.

$$A \leqq C\varepsilon^2 .$$

Proof. Let the pairs (i,j), (k,s) be fixed. Then A is estimated by a finite sum of expressions of the form

$$\mathbf{M} \int_V \int_0^t \int_0^\tau h(x, \tau, \tau_1) \int_0^t \int_0^{\bar{\tau}} h(x, s, s_1)\, ds d\bar{\tau} dx,$$

where

$$h(x, \tau, \tau_1) = \int_V \int_V \Gamma(x, t, \xi, \tau)\, a(\xi, \tau) \Gamma(\xi, \tau, \eta, \tau_1) \times a(\eta, \tau_1)\, \widetilde{U}(\eta, \tau_1) d\eta d\xi.$$

Note that $|H(x,\tau,\tau_1)| < C$ (see [3]). Due to the m-dependence of the field a(x,t) and the centering condition we have

$$A\{\tau - s > 2\varepsilon\} = A\{\tau - s > 2\varepsilon \cap \tau - \varepsilon < \tau_1 < \tau\}$$
$$= \int_V \int_0^t \int_{\tau-\varepsilon}^{\tau} \mathbf{M} h(x, \tau, \tau_1)\, d\bar{\tau} \int_0^t \int_0^s \mathbf{M} h(x, s, s_1)\, d\bar{s} dx \leqslant C\varepsilon^2.$$

Similarly, $A\{s - \tau > 2\varepsilon\} \leqslant C\varepsilon^2$.

Finally,

$$A\{|\tau - s| \leqslant 2\varepsilon\} = A\{|\tau - s| \leqslant 2\varepsilon \cap \tau - 6\varepsilon < \tau_1 < \tau\} +$$
$$+ A\{|\tau - s| \leqslant 2\varepsilon \cap s - 6\varepsilon < s_1 < s\} + A\{|\tau - s| \leqslant 2\varepsilon \cap$$
$$\cap 0 < s_1 < s - 6\varepsilon\} - A\{|\tau - s| \leqslant 2\varepsilon \cap \tau - 6\varepsilon < \tau_1 < \tau \cap s - 6\varepsilon < s_1 < s\}.$$

Each of these integrals is estimated in the same way as $A\{\tau\text{-}s > 2\varepsilon\}$. Lemma 3.2 is proved.

L e m m a 3.3. $B = o(\varepsilon)$.

Proof. We make use of the arguments applied in proving Lemmas 2.2 and 2.3. Let the pairs (i,j), (k,s) be fixed. Then B is estimated by a finite sum of expressions of the form

$$\mathbf{M} \int_V \int_0^t \int_0^{\tau} p(x, \tau, \tau_1) \int_0^t \int_0^s p(x, s, s_1)\, d\bar{s} d\bar{\tau} dx,$$

where

$$p(x, \tau, \tau_1) = \int_V \int_V \Gamma(x, t, \xi, \tau)\, a^{ij}(\xi, \tau/\varepsilon, \omega)$$
$$\times \Gamma(\xi, \tau, \eta, \tau_1)\, a^{ks}(\eta, \tau_1/\varepsilon, \omega)\, u_{\eta_k \eta_s}(\eta, \tau_1)\, d\xi d\eta.$$

According to Lemma 4.1 (see Section 4) and [3] we have

$$|p(x, \tau, \tau_1)| < C. \tag{14}$$

We partition the integration domain for the time variables in the following way:

$$B = B\{\tau - 6\varepsilon < \tau_1 < \tau\} + B\{s - 6\varepsilon < s_1 < s\}$$
$$+ B\{0 < \tau_1 < \tau - 6\varepsilon \cap 0 < s_1 < s - 6\varepsilon\}$$
$$- B\{\tau - 6\varepsilon < \tau_1 < \tau \cap s - 6\varepsilon < s_1 < s\}.$$

Using the Cauchy-Buniakowski inequality, we obtain

$$B\{\tau - 6\varepsilon < \tau_1 < \tau\} \leqslant \mathbf{M} \left(\int_V \left(\int_0^t \int_{\tau-6\varepsilon}^{\tau} p(x, \tau, \tau_1)\, d\bar{\tau} \right)^2 dx \right)^{1/2}$$
$$\times \left(\int_V \left(\int_0^t \int_0^s p(x, s, s_1)\, d\bar{s} \right)^2 dx \right)^{1/2} \leqslant C\varepsilon B^{1/2}. \tag{15}$$

Similarly,

$$B\{s-6\varepsilon<s_1<s\}\leqslant C\varepsilon B^{1/2}. \tag{16}$$

The next estimate is obviously by virtue of (14):

$$B\{\tau-6\varepsilon<\tau_1<\tau\cap s-6\varepsilon<s_1<s\}\leqslant C\varepsilon^2. \tag{17}$$

Let

$$r(x,\tau,\tau_1,\tau_2)=\int\limits_V\int\limits_V\int\limits_V\Gamma(x,t,\xi,\tau)\,a(\xi,\tau)\,\Gamma_{\xi_k}(\xi,\tau,\eta,\tau_1)$$
$$\times\, a(\eta,\tau_1)\,\Gamma_{\eta_s}(\eta,\tau_1,y,\tau_2)\,a(y,\tau_2)\tilde{u}(y,\tau_2)\,dyd\eta d\xi.$$

By Lemma 4.1 (see Section 4) and [1], we have

$$|r(x,\tau,\tau_1,\tau_2)|\leqslant C(\beta)(\tau_1-\tau_2)^{-1/2-\beta}(\tau-\tau_1)^{-1/2-\beta}. \tag{18}$$

Using representation (12), iterating and integrating by parts (the boundary terms vanish because the field a(x,t) is finite), it is not hard to see that to prove Lemma 3.3 it suffices to estimate the following integral:

$$K\equiv \mathbf{M}\int\limits_V\int\limits_0^t\int\limits_0^{\tau-6\varepsilon}\int\limits_0^{\tau_1} r(x,\tau,\tau_1,\tau_2)\,d\bar{\tau}\int\limits_0^t\int\limits_0^{s-6\varepsilon}\int\limits_0^{s_1} r(x,s,s_1,s_2)\,dsdx.$$

Note that $K\{\tau-s>\varepsilon\}=K\{s-\tau>\varepsilon\}=0.$

By (18),

$$K\{|s-\tau|\leqslant\varepsilon\cap|\tau_1-s_1|\leqslant 2\varepsilon\}\leqslant C(\beta)\iint\limits_{|s-\tau|<\varepsilon}$$
$$\times\left(\iint\limits_{|\tau_1-s_1|<2\varepsilon}(\tau-\tau_1)^{-1/2-\beta}(s-s_1)^{-1/2-\beta}ds_1d\tau_1\right)d\tau ds\leqslant C\varepsilon^{1.5-\beta}. \tag{19}$$

The integrals

$$K\{|\tau-s|\leqslant\varepsilon\cap\tau_1-s_1>2\varepsilon\},\quad K\{|\tau-s|\leqslant\varepsilon\cap s_1-\tau_1>2\varepsilon\}$$

are estimated analogously. The validity of Lemma 3.3 follows from (15) - (17) and (19) and that of Lemma 3.1 from Lemmas 3.2 and 3.3.

4. A bound for the maximum of the second derivative for the problem with random coefficients

We shall consider problem (1) under the conditions imposed in Section 1.

L e m m a 4.1. There exists a nonrandom constant $C > 0$ such that

$$\sup_{x\in V, t\in[0,T]} |w_{x_i x_j}(x, t)| < C.$$

Proof. Let $Z(x,t,\xi,\tau)$ be the random fundamental solution of equation (1) considered in the deterministic case in [4]. It is well known ([4]) that

$$Z(x, t, \xi, \tau) = G(x-\xi, t, \xi, \tau) + \int_\tau^t \int G(x-y, t, y, \beta)$$
$$\times \varphi(y, \beta, \xi, \tau)\, dy d\beta \equiv G(x-\xi, t, \xi, \tau) + W(x, t, \xi, \tau),$$

where

$$G(x, t, \xi, \tau) = (2\pi)^{-n} \int e^{i(x,\sigma)} \exp\left\{-\int_\tau^t (b^{kj}(s, \xi) + a^{kj}(s, \xi))\, ds\sigma_k\sigma_j\right\} d\sigma,$$

and $\phi(x,t,\xi,\tau)$ is the solution of the integral equation

$$\varphi(x, t, \xi, \tau) = -K(x, t, \xi, \tau) + \int_\tau^t d\beta \int -K(x, t, y, \beta)\, \varphi(y, \beta, \xi, \tau)\, dy. \tag{20}$$

Here

$$K(x, t, \xi, \tau) = Q(G(x-\xi, t, \xi, \tau))$$

and Q is the operator in (1).

The integral equation (20) can be solved by the sequential approximation method, $\phi(x,t,\xi,\tau)$ being its resolvent:

$$\varphi(x, t, \xi, \tau) = \sum_{m=1}^{\infty} (-1)^m K_m(x, t, \xi, \tau),$$
$$K_1(x, t, \xi, \tau) = K(x, t, \xi, \tau),$$
$$K_m(x, t, \xi, \tau) = \int_\tau^t d\beta \int K(x, t, y, \beta)\, K_{m-1}(y, \beta, \xi, \tau)\, dy.$$

The definition of the fundametnal solution implies the equality

$$\omega = Z * f = G * f + W * f,$$

where, as before,

$$Z * f = \int_0^t \int_V Z(x, t, \xi, \tau) f(\xi, \tau)\, d\xi d\tau.$$

Using inequalities (6), we can easily see that there exists a nonrandom constant C such that

$$G_{x_i}(x,t,\xi,\tau) + G_{\xi_i}(x,t,\xi,\tau) = \tilde{G}(x,t,\xi,\tau),$$
$$\left|\tilde{G}_{x_k}(x,t,\xi,\tau)\right| \leqslant C(t-\tau)^{-(n+1)/2}\exp(-C|x-\xi|^2/(t-\tau)).$$

Using the formula for integration by parts, we then obtain

$$G_{x_i} * f = -G_{\xi_i} * f + \tilde{G} * f = G * f_{\xi_i} + \tilde{G} * f.$$

Then

$$G_{x_i x_j} * f = G_{x_j} f_{\xi_i} + \tilde{G}_{x_j} * f. \tag{21}$$

From (21), due to the integrability of the singularities of the functions $G_{xj}(x,t,\xi,\tau)$, $G_{xj}(x,t,\xi,\tau)$ and inequality (6), we obtain

$$\| G_{x_i x_j} * f \|_{C(V\times[0,T])} < C$$

with nonrandom constant C. An analogous bound for $W_{x_i x_j} * f$ may be obtained by applying inequalities (5), (6) and a verbatim repetition of the arguments of [4]. Lemma 4.1 is proved.

5. Representation of the second derivatives of the fundamental solution

In this section we again examine the nonrandom fundamental solution $\Gamma(x,t,\xi,\tau)$ of equation (8) as done in [1].

It is known [1] that

$$\Gamma(x,t,\xi,\tau) = \Gamma^*(\xi,\tau,x,t),$$

where $\Gamma^*(x,t,\xi,\tau)$ is the fundamental solution of the equation

$$L^* v = v_t + b^{ij}(x,t) v_{x_i x_j} + 2b^{ij}_{x_j} v_{x_i} + b^{ij}_{x_i x_j} v = f(x,t)$$

conjugate to (8). According to [1],

$$\Gamma^*(\xi,\tau,x,t) = Z^*(\xi,\tau,x,t) + R(\xi,\tau,x,t),$$
$$Z^*(\xi,\tau,x,t) = c(x,t,\tau)\exp\{-(b(\xi-x),(\xi-x))/4\},$$

where $b = \|b_{ij}(x,t,\tau)\|$ is the inverse of the matrix $(t-\tau)\|b^{ij}(x,t)\|$, $c(x,t,\tau) = (\det(b/2\pi))^{n/2}$, and

$$R(\xi,\tau,x,t) = \int_\tau^t \int Z^*(\xi,\tau,\eta,\sigma)\Phi^*(\eta,\sigma,x,t)\,d\eta d\sigma. \tag{22}$$

The function $\Phi^*(\eta,\sigma,x,t,)$ is the solution of the integral equation

$$\Phi^*(\eta,\sigma,x,t) = L^* Z^*(\eta,\sigma,x,t) + \int_\sigma^t \int L^* Z^*(\eta,\sigma,\xi,\tau)\Phi^*(\xi,\tau,x,t)\,d\xi d\tau,$$

admitting the representation

$$\Phi^*(\eta, \sigma, x, t) = \sum_{m=1}^{\infty} (L^*Z^*)_m(\eta, \sigma, x, t),$$

where

$$(L^*Z^*)_1(\eta, \sigma, x, t) = L^*Z^*(\eta, \sigma, x, t)$$
$$=(b^{ij}(\eta, \sigma) - b^{ij}(x, t))\, Z^*_{\eta_i\eta_j}(\eta, \sigma, x, t) + 2b^{ij}_{\eta_j} Z^*_{\eta_i}(\eta, \sigma, x, t)$$
$$+ b^{ij}_{\eta_i\eta_j} Z^*(\eta, \sigma, x, t), \quad (L^*Z^*)_{m+1}(\eta, \sigma, x, t) = \int_\sigma^t \int L^*Z^*(\eta, \sigma, y, s)$$
$$\times (L^*Z^*)_m(y, s, x, t)\, dyds\,.$$

Let $n_\beta(x) = \beta^{-n/2}\exp(-|x|^2/\beta)$.

The next lemmas are verified by straightforward calculation.

Lemma 5.1.

$$\left| Z^*_{x_i}(\xi, \tau, x, t) + Z^*_{\xi_i}(\xi, \tau, x, t) \right| \leqslant C n_{C(t-\tau)}(x - \xi).$$

Lemma 5.2.

$$\left|\left(Z^*_{x_i}(\xi, \tau, x, t) + Z^*_{\xi_i}(\xi, \tau, x, t)\right)_{x_j} + \left(Z^*_{x_i}(\xi, \tau, x, t)\right.\right.$$
$$\left.\left. + Z^*_{\xi_i}(\xi, \tau, x, t)\right)_{\xi_j}\right| \leqslant C n_{C(t-\tau)}(x - \xi).$$

We proceed to estimate the derivatives of $R(\xi,\tau,x,t,)$. A direct calculation of derivatives also shows that

Lemma 5.3.

$$L^*Z^*(\eta, \sigma, x, t) = N^{(1)}_{x_ix_j}(\eta, \sigma, x, t) + N^{(2)}_{x_i}(\eta, \sigma, x, t) + N^{(3)}(\eta, \sigma, x, t),$$

where

$$\left| N^{(i)}(\eta, \sigma, x, t) \right| \leqslant C n_{C(t-\sigma)}(x - \eta), \quad i = 1, 2, 3.$$

In particular,

$$N^{(1)}(\eta, \sigma, x, t) = (b^{ij}(\eta, \sigma) - b^{ij}(x, t))Z^*(\eta, \sigma, x, t).$$

Expressions for $N^{(2)}$, $N^{(3)}$ have a similar, but less cumbersome, form.

The validity of the representation follows form Lemma 5.3 and Lemma 3 of [1].

Lemma 5.4.

$$\Phi^*(\eta, \sigma, x, t) = M^{(1)}_{x_ix_j}(\eta, \sigma, x, t) + M^{(2)}_{x_i}(\eta, \sigma, x, t) + M^{(3)}(\eta, \sigma, x, t).$$

Also,

$$|M^{(i)}(\eta, \sigma, x, t)| \leqslant C n_{C(t-\sigma)}(x - \eta),\ i = 1, 2, 3.$$

Lemma 5.5.

$$\Phi^*(\eta,\sigma,x,t| - \Phi^*(\xi,\sigma,x,t)| = K^{(1)}_{x_ix_j}(\eta,\sigma,x,t,\xi) + K^{(2)}_{x_i}(\eta,\sigma,x,t,\xi) + K^{(3)}(\eta,\sigma,x,t,\xi),$$

where

$$|K^{(i)}(\eta,\sigma,x,t,\xi)| \leqslant C|\eta-\xi|^{\alpha}(t-\sigma)^{-\alpha/2}(n_{C(t-\sigma)}(x-\eta)+n_{C(t-\sigma)}(x-\xi)).$$

Proof. First we estimate the difference $|Z^*(\eta,\sigma,x,t) - Z^*(\xi,\sigma,x,t)|$. For $t-\sigma \leqq |\eta-\xi|^2$, the estimate

$$|Z^*(\eta,\sigma,x,t) - Z^*(\xi,\sigma,x,t)| \leqslant C(t-\sigma)^{-\alpha/2} \times |\eta-\xi|^{\alpha}(n_{C(t-\sigma)}(x-\eta)+n_{C(t-\sigma)}(x-\xi)) \tag{23}$$

is obvious, since $|Z^*(\eta,\sigma,x,t)| \leqq Cn_{C(t-\sigma)}(x-\eta)$. For $t-\sigma > |\eta-\xi|^2$, by the mean value theorem, using an explicit expression for $Z^*_{x_i}(\xi,\sigma,x,t)$, we obtain

$$|Z^*(\eta,\sigma,x,t) - Z^*(\xi,\sigma,x,t)| \leqslant C|\eta-\xi|(t-\sigma)^{-1/2}\times n_{C(t-\sigma)}(x-\theta),$$

where θ is a point in (η,ξ). Using the triangle inequality, we have

$$C|\theta-x|^2 + C|\xi-\eta|^2 \geqslant |x-\eta|^2 + |\xi-x|^2.$$

Therefore

$$-\frac{C|\theta-x|^2}{t-\sigma} \leqslant -\frac{|x-\eta|^2}{t-\sigma} - \frac{|\xi-x|^2}{t-\sigma} + C.$$

Whence inequality (23) holds for $t-\sigma > |\eta-\xi|^2$. But then it follows from Lemma 5.3 that the assertion of Lemma 5.5 holds for the difference $L^*Z^*(\eta,\sigma,x,t) - L^*Z^*(\xi,\sigma,x,t,)$.
Lemma 5.5 is finally proved by applying Lemma 3 of [1] and Theorem 7 of [1].

We shall next estimate $R_{\xi_i\xi_j}(\xi,\tau,x,t)$, where the function $R(\xi,\tau,x,t)$ is given by (22).

Lemma 5.6.

$$R_{\xi_i\xi_j}(\xi,\tau,x,t) = S^{(1)}_{x_ix_j}(\xi,\tau,x,t) + S^{(2)}_{x_i}(\xi,\tau,x,t) + S^{(3)}(\xi,\tau,x,t),$$

where

$$|S^{(i)}(\xi,\tau,x,t)| \leqslant Cn_{C(t-\tau)}(x-\xi),\ i=1,2,3.$$

Proof.

$$R_{\xi_i\xi_j}(\xi,\tau,x,t) = \int_{\frac{t+\tau}{2}}^{t}\int Z^*_{\xi_i\xi_j}(\xi,\tau,\eta,\sigma)\,\Phi^*(\eta,\sigma,x,t)\,d\eta d\sigma$$

$$+\int_{\tau}^{\frac{t+\tau}{2}}\int Z^*_{\xi_i\xi_j}(\xi,\tau,\eta,\sigma)\,(\Phi^*(\eta,\sigma,x,t)-\Phi^*(\xi,\sigma,x,t))\,d\eta d\sigma$$

$$+\int_{\tau}^{\frac{t+\tau}{2}}\int \Phi^*(\xi,\sigma,x,t)\,d\sigma Z^*_{\xi_i\xi_j}(\xi,\tau,\eta,\sigma)\,d\eta.$$

For brevity, we limit ourselves to the principal terms in the $\Phi^*(\eta,\sigma,x,t)$:

$$\left|\int_{\frac{t+\tau}{2}}^{t}\int Z^*_{\xi_i\xi_j}(\xi,\tau,\eta,\sigma)\,M^{(1)}(\eta,\sigma,x,t,)\,d\eta d\sigma\right.$$

$$+\int_{\tau}^{\frac{t+\tau}{2}}\int Z^*_{\xi_i\xi_j}(\xi,\tau,\eta,\sigma)\,K^{(1)}(\eta,\sigma,x,t,\xi)\,d\eta d\sigma+\int_{\tau}^{\frac{t+\tau}{2}}\int M^{(1)}(\xi,\sigma,x,t)$$

$$\left.\times Z^*_{\xi_i\xi_j}(\xi,\tau,\eta,\sigma)\,d\eta d\sigma\right|\leqslant C\int_{\frac{t+\tau}{2}}^{t}(\sigma-\tau)^{-1}\int n_{C(\sigma-\tau)}(\xi-\eta)\,n_{C(t-\sigma)}(\eta-x)d\eta d\sigma$$

$$+C\int_{\tau}^{\frac{t+\tau}{2}}\int(\sigma-\tau)^{-1}|\xi-\eta|^{\alpha}(t-\sigma)^{-\alpha/2}n_{(\sigma-\tau)}(\xi-\eta)\,(n_{(t-\sigma)}(x-\eta)$$

$$+n_{(t-\sigma)}(x-\xi))\,d\eta d\sigma+C\int_{\tau}^{\frac{t+\tau}{2}}n_{(t-\sigma)}(x-\xi)\,(\sigma-\tau)^{(\alpha-2)/2}d\sigma\leqslant Cn_{(t-\tau)}(x-\xi).$$

Lemma 5.6 is proved.

6. Proof of the main result

It is well known (see [5]) that the embedding operator of $W_2^{-2}(V)$ into $W_2^{-3}(V)$ is completely continuous, being the conjugate of a completely continuous one. Therefore the ball in $W_2^{-2}(V)$ is compact in $W_2^{-3}(V)$. By the Chebyshev inequality

$$\mathbf{P}\left\{\|v_\varepsilon\|_{W_2^{-2}(V)}>s\right\}\leqslant\mathbf{M}\|v_\varepsilon\|_{W_2^{-2}(V)}s^{-1}. \tag{24}$$

By Lemmas 2.1 and 3.1 we have

$$\sup\mathbf{M}\|v_\varepsilon\|_{W_2^{-2}(V)}<C$$

and $s^{-2} \to 0$ as $s \to \infty$, which together with (24) yield

$$\lim_{s\to\infty} \sup \mathbf{P} \left\{ \| v_\varepsilon \|_{W_2^{-2}(V)} > s \right\} = 0,$$

i.e., the family of distributions generated in the space $W_2^{-3}(V)$ of generalized functions by the random field v_ε is weakly compact. Finally, it follows from Lemams 2.1, 3.1 and condition (10) that for every fucntion $\Phi \in L_2(V)$, $\langle \Phi, v_\varepsilon \rangle$ is asymptotically normal as $\varepsilon \to 0$ with parameters $(0, \sigma^2(\Phi,\Gamma,U))$, where $\sigma^2(\Phi,\Gamma,U)$ is defined in condition (10).

REFERENCES

[1] Friedman, Avner. *Partial Differential Equations of Parabolic Type*. Englewood Cliffs, N.J.: Prentice-Hall, 1964.

[2] Ladyzhenskaya, O.A., and Ural'tseva, N. *Linear and Quasi-linear Elliptic Equations*. New York: Academic Press, 1968. (English transl.)

[3] Ladyzenskaja, O.A., Solonnikov, V.A., and Ural'ceva, N.N. *Linear and Quasi-linear Equations of Parabolic Type*. Providence, Rhode Island: American Mathematical Society, 1968. (English transl.)

[4] Eidel'man. Samuel D. *Parabolic Systems*. Amsterdam: North-Holland, 1969. (English transl.)

[5] Maurin, Krzysztof. *Methods of Hilbert Spaces*. Warszawa, Poland: Polish Scientific Publishers, 1967. (2nd ed. rev., 1972).

V.A. Topchij

ASYMPTOTICS OF THE PROBABILITY OF CONTINUATION FOR CRITICAL GENERAL BRANCHING PROCESSES WITH NO SECOND MOMENT FOR THE NUMBER OF DESCENDANTS

1. Introduction

We describe a general branching process $\xi(t)$, $t \geqq 0$, in terms of the evolution of some population (t is time). The population begins to evolve at the $0th$ instant with one particle, which is labelled (0). It can produce descendants labelled $(0,i)$, where i denotes the cardinal number of a new descendant. The lifetime of (0) and the times of birth of the particles $(0,i)$ are given by a random process $\{\eta, N(t)\}$, where η is the lifespan and $N(t)$ is the number of descendants generated by the particle (0) over time t. We require that this particle have no descendants after its death, i.e., $N = \lim_{t\to\infty} N(t) = N(\eta)$.

Assume that the moment of birth of the particle $(0,i_1,i_2,\ldots,i_n)$ is defined and equal to $\bar{t}$. With this particle we associate the independent process $\{\bar{\eta}, \bar{N}(t)\}$ having the distribution of $\{\eta, N(t)\}$. Then we label its ith descendant by $(0,i_1,\ldots,i_n,i)$, and the moment of birth with $t + \inf\{t \mid \bar{N}(t) \geqq i\}$. Note that if $\inf\{t \mid \bar{N}(t) \geqq i\}$ is not defined, there will be no ith descendant, i.e., there is no particle labelled $(0,i_1,\ldots,i_n,i)$. The initial particle $(0,i_1,\ldots,i_n)$ dies at the instant $\bar{t} + \bar{\eta}$.

Finally, we set $\xi(t)$ equal to the number of particles existing but not dying at the instant t. The full list of labels of particles of the realization of $\xi(t)$, including times of their birth and death can be taken as a probability space $\xi(t)$ (see [1], [2]).

The probability space $\xi(t)$ can be viewed as a direct product of spaces of independent processes $\{\eta, N(t)\}$ corresponding to particles labelled $(0,i_1,\ldots,i_n)$. Hence, defining new processes on the spaces

$\{\eta, N(t)\}$ as to correspond to the lists of labels for $\xi(t)$, we construct thereby a new branching process on the space $\xi(t)$. More specifically, let there be a procedure Φ for constructing a random process $\{\eta_1, N_1(t)\} = \Phi\{\eta, N(t)\}$ on the space of the process $\{\eta, N(t)\}$. The new processes have the above-described properties of the initial processes and $N_1 = N$ everywhere. The latter makes it possible to associate each particle $(0,i_1,\dots,i_n)$ of the realization of $\xi(n)$ with its conversions defined by $\{\bar{\eta},N(t)\}$ with the particle $(0,i_1,\dots,i_n)_1$ with its conversions defined by $\Phi\{\bar{\eta},\bar{N}(t)\}$. Again the new labels make up a branching process $\xi_1(t)$, which we call the process constructed on the space $\xi(t)$ with the aid of the procedure Φ.

We shall be using the same notation for all processes. However, when it is necessary, we equip a particular process with an additional index.

We do not exclude the possibility that $P\{\eta = 0\}$ and $P\{dN(0) > 0\} > 0$. Hence, $P\{\xi(0) = 1\} \leqq 1$, although only one particle generates the entire process.

This model covers conventional processes, such as Sevast'yanov, Bellman-Harris, and Markov processes (see [3]).

Introduce the notation:

$$N = \lim_{t\to\infty} N(t),\ G(t) = \mathbf{P}\{\eta \leqslant t\},\quad A(t) = \mathbf{M}N(t),$$

$$A = \mathbf{M}N,\ a = \int_0^\infty t\,dA(t),\ h(x) = \mathbf{M}x^N = h_{(1)}(x) = \sum_{n=0}^\infty h_n x^n,$$

$$h_{(n)}(x) = h(h_{(n-1)}(x)) \text{ for } n > 1,\ Q(t) = \mathbf{P}\{\xi(t) > 0\}.$$

We investigate the critical case, i.e., $A = h'(1) = 1$. Moreover, we assume that

$$h(x) = x + (1-x)^{1+\alpha} L(1-x), \tag{1}$$

where $\alpha \in (0,1]$, and $L(x)$ is a slowly varying function as $x \to 0$ (see the definition in [4]).

We consider the case where the sequences $1 - A(n)$ and $n(1 - G(n))$ decrease no more slowly than $1 - h_{(n)}(0)$ for which we have (see [5])

$$1 - h_{(n)}(0) \sim n^{-\beta} L_1(n) \quad \text{for} \quad \alpha^{-1} = \beta, \tag{2}$$

where $L_1(t)$ is a slowly varying function as $t \to \infty$ (the relationship of $L_1(t)$ and $L(x)$ is described in Section 3).

By the rapid decrease of the quantities listed in the title we mean the relations

$$\varlimsup_{n\to\infty} n\,(1-G\,(n))\,(1-h_{(n)}\,(0))^{-1} < \infty, \tag{3}$$

$$\varlimsup_{n\to\infty} (1-A\,(n))\,(1-h_{(n)}\,(0))^{-1} < \infty. \tag{4}$$

Theorem 1. Let $A = 1$, $0 < a < \infty$, and let (1), (4) and (3) hold for $0 < \alpha < 1$. Then, as $t \to \infty$,

$$Q(t) \asymp t^{-\beta} L_1(t), \tag{5}$$

where the symbol $\asymp$ means that the upper limits of the ratios of each of the expressions to the other is finite.

To obtain the precise asymptotics of $Q(t)$, conditions (3) and (4) need to be strengthened. We shall impose the following conditions:

$$\lim_{n\to\infty} n\,(1-G\,(n))\,(1-h_{(n)}\,(0))^{-1} = c_1 < \infty, \tag{6}$$

$$\lim_{n\to\infty} (1-A\,(n))\,(1-h_{(n)}\,(0))^{-1} = c_2 < \infty, \tag{7}$$

$$\lim_{t\to\infty} (\mathbf{P}\{\eta \geqslant t\} - \mathbf{M}\{x^N;\, \eta \geqslant t\})\, t^{\beta+1} L_1^{-1}\,(t) = g\,(y) \tag{8}$$

for $1 - x \sim yt^{-\beta}L_1(t)$. We discuss (8) in detail in Section 7.

We generalize the result proved by Vatutin in [6] for the Bellman-Harris processes.

Theorem 2. Let $A = 1$, $0 < a < \infty$, and let (1), (6), (4) and (8) hold for $0 < \alpha < 1$. Then, as $t \to \infty$,

$$Q(t) \sim \rho t^{-\beta} L_1(t), \tag{9}$$

where ρ is the unique nonnegative root of the equation

$$y^{1+\alpha} - ay - \alpha c_1 + \alpha g(y) = 0. \tag{10}$$

For Bellman-Harris processes, $g(y) \equiv 0$.

In Theorems 1 and 2 we excluded the possibility of $\alpha = 1$, because in proving the theorems we shall essentially employ the condition $\alpha < 1$.

Note that Theorem 1 can be sharpened by means of Theorem 2, i.e., explicit estimates for the ratios of the left and right sides of (5) can be obtained. They are expressed in terms of the upper and lower limits of the expressions under the limit signs in (3) and (8), starting from an analog of (10).

Rougher estimates obtain if the expression under the limit sign in formula (8) is estimated from below by 0 and from above by $P\{\eta \geqq t\} t^{\beta+1} L_1^{-1}(t)$.

The numbering of formulas in each section is independent. When referring to formulas from other sections we use the number of that section as well. $\chi\{A\}$ is the indicator of the set A. For processes $\{\eta, N(t)\}$ with conversions on a lattice, we assume that condition (P) is satisfied if the smallest lattice-support of the distribution of $N(t)$ (i.e., the one generated by $\{t \mid dA(t) \neq 0\}$) contains the support of η.

2. The reduction. A representation for generating functions

We note first that if either the process $\xi_1(t)$ or $\xi_2(t)$ is constructed on the space of the other one, so that everywhere $\eta_1 \leqq \eta_2$ and $N_1(t) \geqq N_2(t)$ for all $t \in [0,\infty)$, then

$$Q_1(t) \leqslant Q_2(t) \ . \tag{1}$$

Indeed, these assumptions imply that on each elementary outcome the conversions (birth and death) of the respective particles occur for $\xi_1(t)$ no later than for $\xi_2(t)$, i.e., (1) is true.

On the space of the initial process $\xi(t)$, let us define two auxiliary processes $\xi_H(t)$ and $\xi_B(t)$ by the relations

$$N_H(t) = H([(t\Delta^{-1}+1)\gamma^{-1}]\Delta) \ ,$$

$$\eta_H = [\eta\Delta^{-1}]\Delta \ ,$$

$$\eta_B = [(\eta\Delta^{-1}+2)\gamma^{-1}]\Delta \ ,$$

$$N_B(t) = N([t\Delta^{-1}]\Delta) \ ,$$

where $\Delta > 0$ and $0.5 < \gamma < 1$ are fixed scalars.

By construction, $h_B(x) = h_H(x) = h(x)$, and the prerequisites for (1) are true, whence

$$Q_B(t) \geqslant Q(t) \geqslant Q_H(t) \ . \tag{2}$$

The conditions of Theorems 1 and 2 remain satisfied for $\xi_B(t)$ and $\xi_H(t)$, although the constants a and c_1 and the function $g(y)$ from (1.8) for $\xi(t)$ will possibly be different for the processes constructed. In this case there exist arbitrarily small $\Delta > 0$ for which condition (P) holds for $\xi_B(t)$ and $\xi_H(t)$. To prove Theorem 2, we have to observe that the characteristics a, c_1 and $g(y)$ with the appropriate indices in the definition of ρ_H and ρ_B converge to a, c_1 and $g(y)$ as $\Delta \to 0$ and $\gamma \to 1$, i.e., $\rho_H \to \rho$ and $\rho_B \to \rho$.

Indeed, the convergence of a_H to a follows from the relations

$$a_H = \int_0^\infty t dA_H(t) = \int_0^\infty t dA\left(\left[t\Delta^{-1} + 1\right)\gamma^{-1}\right]\Delta\right) = O(\Delta) + \int_0^\infty t dA\left(t\gamma^{-1} + \Delta\gamma^{-1}\right) = O(\Delta) + \gamma a.$$

Next, let $x\Delta^{-1}$ be an integer and

$$G_B(x) = \mathbf{P}\{[(\eta\Delta^{-1} + 2)\gamma^{-1}]\Delta \leqslant x\} = \mathbf{P}\{(\eta\Delta^{-1} + 2)\gamma^{-1} < x\Delta^{-1} + 1\} = \mathbf{P}\{\eta < -2\Delta + \gamma\Delta + x\gamma\}.$$

The latter, due to the monotonicity of the distribution function, (1.2) and (1.6), implies that $c_{1B} \to c_1$ as $\Delta \to 0$ and $\gamma \to 1$.

Checking the convergence of a_B and c_{1H} is even simpler, and therefore we omit it.

For t multiple of Δ, $P\{\eta \geqq t\} = P\{\eta_H \geqq t\}$ and $M\{x^n;\ \eta \geqq t\} = M\{x^{N_H};\ \eta_H \geqq t\}$, i.e., $g_H(y) = g(y)$. By (1.2) and (1.6) $P\{\eta_B \geqq t\} - M\{x^{N_B};\ \eta_B \geqq t\} = p\{\eta \geqq \gamma t\} - M\{x^N;\ \eta \geqq \gamma t\} + o(t^{-\beta-1}L_1(t))$, yielding $g_B(y) = (1-\gamma)O(1) + g(y)$.

The proof of the convergence of ρ_H and ρ_B to ρ is completed.

It is easy to verify that the assertions of the theorems are invariant under a linear time substitution, i.e., one can assume without loss of generality that the lattice on which the conversions of the processes take place is an integer-valued one.

For $\xi_B(t)$ and $\xi_H(t)$, sufficiently long-lived particles do not produce descendants during the $1-\gamma$ part of their life before death, and in the sequel we will assume that all the particles produce no descendants during the $1-\gamma$ part of their life before death. On the one hand, this can easily be achieved by a trivial modification of $\xi_B(t)$ and $\xi_H(t)$; on the other hand, the relation $h_n(z_0,\dots,z_n) \equiv h_n(z_0,\dots,z_{\gamma n})$ following from this assumption can be changed to $h_n(z_0,\dots,z_n) \equiv h_n(z_0,\dots,z_{\gamma(n)})$, where $\gamma(n) \sim \gamma n$, and the arguments set forth in this work can be repeated almost verbatim with more elaborate notation.

To sum up, by virtue of (2) it suffices to prove the theorem for processes with conversions on an integer-valued lattice satisfying condition (P), when the particles produce no descendants during the last $1-\gamma$ part of their life. We call such processes discrete-time processes.

Introduce the notation: $\xi(n)$, $n \in N$, is the process considered

$$P_n = \mathbf{P}\{\xi(n)=0\}, \quad Q_n = 1 - P_n,$$

$$h_n(z_0, \ldots, z_n) = \mathbf{M}\left(\prod_{i=0}^{n} z_i^{dN(i)} \,\middle|\, \eta = n\right),$$

$$f_n = \mathbf{P}\{\eta = n\}, \quad h_n(i) = h_n(P_i, \ldots, P_{i-n}), \quad P_i = 1 \text{ for } i < 0,$$

$$q_n = \sum_{i=n+1}^{\infty} f_i, \qquad a_{in} = \frac{\partial}{\partial z_i} h_n(1, \ldots, 1), \; a_i = \sum_{n=i}^{\infty} a_{in} f_n,$$

$$f_a(z) = \sum_{n=0}^{\infty} a_n z^n, \quad \Phi_n = \sum_{i=0}^{n} Q_{n-i} \sum_{j=n+1}^{\infty} a_{ij} f_j,$$

$$H_j(n) = 1 - h_j(n) - \sum_{i=0}^{j} a_{ij} Q_{n-i},$$

$$(1 - f_a(z))^{-1} - a^{-1}(1-z)^{-1} = \sum_{n=0}^{\infty} \beta_n z^n.$$

We shall frequently wind up with fractions in our calculations, but we of course mean the largest integral part. Moreover, instead of $n=[t]$ we will sometimes write $n-t$. This ambiguity in interpreting indices will be allowed in expressions where it has no essential influence on the asymptotic results.

We shall derive the basic representation for Q_n, which is in fact a slight modification of relation (13) in [4], and therefore we shall only sketch the arguments.

Applying the total probability formula to the generating function of $\xi(n)$ with respect to the conversions of the initial particle, upon simple substitutions we obtain

$$Q_n = \sum_{i=0}^{n} Q_{n-i} a_i + \sum_{i=0}^{n} H_i(n) f_i - \Phi_n + q_n. \tag{3}$$

Recall that we have assumed the conditions

$$A = \sum_{n=0}^{\infty} \sum_{i=0}^{n} a_{in} f_n = \sum_{i=0}^{\infty} a_i = f_a(1) = 1, \tag{4}$$

$$0 < a = f_a(1) < \infty , \tag{5}$$

to be satisfied.

Relations (3), (4) and (5), together with the generating functions, make it possible to obtain

$$Q_n = \sum_{i=0}^{n} \left(q_i - \Phi_i + \sum_{j=0}^{i} H_j(i) f_j \right) (a^{-1} + \beta_{n-i}), \tag{6}$$

which upon identical transformations of the difference of Q_n and Q_{n-1} yields

$$Q_n - Q_{n-1} = \sum_{i=1}^{8} I_i, \tag{7}$$

where

$$I_1 = a^{-1}(q_n - \Phi_n),\ I_2 = a^{-1} \sum_{i=0}^{n} H_i(n) f_i \quad I_3 = \sum_{i=0}^{np} (q_i - \Phi_i)(\beta_{n-i} - \beta_{n-i-1}),$$

$$I_4 = \sum_{i=np+2}^{n} (\Phi_{i-1} - \Phi_i - f_i) \beta_{n-i},\ I_5 = (q_{np+1} - \Phi_{np+1}) \beta_{n-[np]-1},$$

$$I_6 = \sum_{i=0}^{np} \times \sum_{j=0}^{i} f_j H_j(i) (\beta_{n-i} - \beta_{n-i-1}),$$

$$I_7 = \sum_{i=np+2}^{n} \beta_{n-i} \left(\sum_{j=0}^{i} H_j(i) f_j - \sum_{j=0}^{i-1} H_j(i-1) f_j \right),$$

$$I_8 = \sum_{j=0}^{np+1} H_j(np+1) f_j \beta_{n-[np]-1},$$

and $0 < p < 1$ is any fixed number.

As will be shown in subsequent sections, it suffices to prove the theorems only in two cases when conditions (1.6) and (1.7) are satisfied instead of (1.3) and (1.4), respectively. If $c_1 \neq 0$ or $c_2 \neq 0$, we have $q_n = n^{-\beta-1} L_2(n)$ or $\sum_{i=n+1}^{\infty} a_i = n^{-\beta} L_3(n)$, where $L_i(t)$ is a slowly varying function and $L_2(t) \sim c_1 L_1(t)$, $L_3(t) \sim c_2 L_1(t)$. We shall show later that, without loss of generality, in the Karamata re-representation for a slowly varying function one has

$$L_i(x) = c_i(x) \exp \int_{x_0}^{x} \varepsilon_i(y)\, y^{-1} dy,$$

where $c_i(x) \to \bar{c}_i \neq 0$, and $\varepsilon_i(x) \to 0$ as $x \to \infty$ (see [4], [5]); for large x the function $c_i(x)$ can be taken constant. This assumption simplifies the proofs of the theorems.

We begin the verification from the relation

$$q_n = n^{-\beta-1} L_2(n). \tag{8}$$

For sufficiently small $\varepsilon > 0$ it is easy to construct a slowly varying function $L_4(t)$ with $c_4(t)$, constant for large t such that $L_4(t) \geqq L_2(t)$ $(L_4(t) \leqq L_2(t))$, $L_4(t)L_2^{-1}(t) \geqq 1 - \varepsilon$ $(L_4(t)L_2^{-1}(t) \leqq 1 + \varepsilon)$, $L_4(t) \sim \bar{c}L(t)$, and (8) conserves the probabilistic sense with $L_2(n)$ replaced by $L_4(n)$. Roughly speaking, this can be achieved by replacing $c_2(x)$ by $\bar{c}_2 \pm 0.5\varepsilon$ for large x.

We define the process $\xi_1(n)$ by the relations $N_1(n) \equiv N(n)$ and $\eta = G^{-1}(\xi)$, $\eta_1 = G_1^{-1}(\xi)$, where ξ is uniformly distributed on $[0,1]$, and $G(t)$ and $G_1(t)$ are the distribution functions of η and η_1 defined by (8), and $q_{n1} = n^{-\beta-1}L_4(n)$. The functions G^{-1} and G_1^{-1} denote the inverse functions to the closures of the graphs of $G(t)$ and $G_1(t)$, as is customary in defining arbitrary random variables in terms of uniformly distributed ones on $[0,1]$.

It is obvious that for fixed γ and sufficiently small ε, the definition is correct, and the particles produce no descendants during some time before they die.

Depending on the inequality sign between $L_2(t)$ and $L_4(t)$, $Q_1(t)$ estimates $Q(t)$ from above or from below; since $q_{n1} \sim \bar{c}\, n^{-\beta-1} L_2(n)$ where $\bar{c} \to 1$ as $\varepsilon \to 0$, the theorems hold true for $\xi(n)$ if they are true for $\xi_1(n)$ and the assumption on $L_2(n)$ is justified.

An analogous transformation will be made in detail in Section 6. Hence we show briefly how to regularize the second relation $\sum_{i=n+1}^{\infty} a_i = n^{-\beta} L_3(n)$.

The new processes $\xi_2(n)$ are constructed from $\xi(n)$ without changing η but changing the time scale for $N(t)$, i.e., $N_2(\tau) = N(t)$, where τ and t are linked by the relation described above for η and η_1 in replacing (8) by $\sum_{i=n+1}^{\infty} a_i = n^{-\beta} L_3(n)$, which makes sense due to the possibility of interpreting a_i as the probability of an integer-valued random variable, since $A = 1$. For small $\varepsilon > 0$ the definition will be correct, and by analogy with what has been said, the upper and lower bounds converge to each other as $\varepsilon \to 0$ if the assertions of the theorems are true for them. In this case, for small $\varepsilon > 0$, the earlier condition that the particles produce no descendants before they die is satisfied.

The slowly varying functions $L_i(t)$, $i \geqq 2$, introduced in this section, will not be used explicitly later. To apply the foregoing assumptions, we write $c_j L_1(t)$ $(j = 1,2)$ instead of $L_i(t)$ and invoke the properties which follow from $c_i(x)$ being constant: for $\nu > 0$, $\Delta > 0$ and for large t,

$$t^{\nu} L_i(t) - (t+\Delta)^{\nu} L_i(t+\Delta) \sim -\nu \Delta t^{\nu-1} L_i(t), \tag{9}$$

$$L_i(t) - L_i(t+\Delta) = o(L_i(t)) \Delta t^{-1}. \tag{10}$$

3. A lower bound for $Q(t)$

Consider an arbitrary process $\xi(t)$ satisfying the conditions of Theorem 1. On its space, we construct $\xi_1(t)$, starting out from the relations $\eta_1 = t_0 \chi\{\eta \geqq 2t_0\}$, $dN_1(0) = N(2t_0)$, $dN_1(t_0) = N - N(2t_0)$, where $t_0 > 0$ is an arbitrary scalar ensuring that $\xi_1(t)$ be nontrivial. Assuming that $\xi_1(t)$ is a discrete-time process (see Section 2) and making note of the properties following from the definition of $\xi_1(n)$: $1 - f_a(z) = a_{11} f_1(1-z)$, $a_{ij} = 0$ for $j > 1$ or $i > 1$, $f_i \equiv 0$ for $i > 1$ and so on, we rewrite (2.7) for $n > 2$:

$$a_{11} f_1(Q_n - Q_{n-1}) = (1 - h_0(P_n) - a_{00} Q_n) f_0 + (1 - h_1(P_n, P_{n-1}) - a_{01} Q_n - a_{11} Q_{n-1}) f_1.$$

By Theorem 6.2.3 of [2], $Q_n \to 0$. Applying Lagrange's mean-value formula to $h_1(P_n, x) - a_{11} x$, the latter can be brought to the form

$$a_{11} f_1(Q_n - Q_{n-1})(1 + o(1)) = 1 - h(P_n) - Q_n,$$

which in view of (1.1) implies

$$Q_n - Q_{n-1} = -Q_n^{1+\alpha} L(Q_n)(1 + o(1)) a^{-1}, \tag{1}$$

which in turn implies (see [5])

$$Q_n \sim n^{-\beta} L_{01}(n), \tag{2}$$

$L_{01}(t)$, being equivalent to $L_1(t)$ of (1.2) to within a^β, since the asymptotic behavior of $1 - h_{(n)}(0)$ obtains from relations different from (1) only by the factor $a^{-1}(1+o(1))$. Furthermore, one may easily derive from [5] that

$$L^\beta(n^{-\beta} L_1(n)) \cdot L_1(n) \alpha^\beta \sim 1. \tag{3}$$

By virtue of (2.1) and the construction of $\xi_1(t)$, relation (2) implies

$$\underline{\lim}_{t\to\infty} Q(t) L_1^{-1}(t) t^\beta > 0. \tag{4}$$

4. Results from renewal theory

Throughout below we assume that the lattice for the indices with $a_n \neq 0$ has unit step.

If $c_2 \neq 0$ in (1.7), then

$$\sum_{i=n+1}^{\infty} a_i = c_2 n^{-\beta} L_1(n) \tag{1}$$

to within an equivalent slowly varying function.

Our assumption on the form of $L_1(n)$ (see Section 2), by virtue of (2.9) implies

$$a_n \sim c_2 \beta n^{-\beta-1} L_1(n). \tag{2}$$

The required assertions are the following.

L e m m a 1. Let (1) hold, where $\beta = \alpha^{-1}$ and $\alpha \in (0,1)$. Then

$$\beta_n = a^{-2} c_2 \alpha (1-\alpha)^{-1} n^{-\beta+1} L_1(n), \tag{3}$$

$$\beta_{n-1} - \beta_n \sim a^{-2} c_2 n^{-\beta} L_1(n). \tag{4}$$

L e m m a 2. Let (2) hold, where $\beta = \alpha^{-1}$ and $\alpha \in (0,1)$. Then

$$\beta_n - \beta_{n-1} - \beta_{n-i} + \beta_{n-i-1} = a^{-2} \Theta(i) a_n, \tag{5}$$

where $\Theta(i) \sim i$ for $i = O(1)$ and $\Theta(i) \sim O(i)$ for $i \leq n\Theta$ if $0 < \Theta < 1$.

L e m m a 3. Let $a_i = 0$ for $i > n_0$. Then for any $k > 0$, $\beta_n = o(n^{-k})$.

Relation (4) follows from [9], and (3) is easily derived from it by summation (see [4]).

Lemma 3 is a particular case of Theorem 1 in [8].

Lemma 2 can be proved by the methods of [9] and [8], if it is noted that $\beta_n - 2\beta_{n-1} + \beta_{n-2}$ is the coefficient of z^n in the MacLaurin series of $(1-z)^2(1-f_a(z))^{-1}$ (the derivative of the last expression has to be investigated).

REMARK. Under conditions (1) and (2), if the constants $c_2\beta$ and c_2 are replaced by $O(1)$ and the relations $a_n - a_{n+1} = o(n^{-\beta-1}L_1(n))$ and $\sum_{i=n}^{\infty} a_i \sim \sum_{i=n+1}^{\infty} a_i$ are required to hold, then the factor $O(1)$ appears on the right sides of (3) and (4), while in (5) one has to add $o(n^{-\beta-1}L_1(n))$ to a_n.

Proof of this is analogous to an earlier proof. The difference is in the following: in estimating sums of the form $\sum_{i=0.5n}^{\infty} a_i g_{n-i}$, where $\sum_{i=0}^{\infty} |g_i| < \infty$, the principal term $a_n \sum_{i=0}^{\infty} g_i$ turns into

$$\left(a_n + o\left(n^{-\beta-1}L_1(n)\right)\right) \sum_{i=0}^{\infty} g_i.$$

Note that in proving the analog of Lemma 2, one obtains

$$\beta_{n-1} - \beta_n \sim a^{-2} \sum_{i=n+1}^{\infty} a_i + o\left(n^{-\beta}L_1(n)\right),$$

which sharpens the formulation of the Remark.

5. Proof of Theorem 1 when conditions (1.6) and (1.7) are satisfied with $c_2 \neq 0$

In view of (1.2), (1.6), (2.9) and the assumption on the form of the slowly varying function in (2.8), we have

$$q_n = c_1 n^{-\beta-1} L_1(n), \tag{1}$$

$$f_n \sim (\beta+1) c_1 n^{-\beta-2} L_1(n) \tag{2}$$

or for $c_1 = 0$

$$q_n = o(n^{-\beta-1} L_1(n)). \tag{3}$$

Recall that in the processes considered the particles produce no descendants during the last $1-\gamma$ part of their life, i.e., $a_{ij}=0$ for $i>\gamma j$.

Let

$$h_i(s,\ldots,s)=\sum_{j=0}^{\infty} h_{ij}s^j,$$

$$q_{(1)j}=\sum_{i=j+1}^{\infty} h_i,\quad q_j(h_i)=\sum_{s=j+1}^{\infty} h_{is},\quad q_{(2)j}=\sum_{i=j+1}^{\infty} q_{(1)i},$$

$$T^{-1}=n^{-\beta}L_1(n)\varepsilon,\qquad \varepsilon>0.$$

By Theorem 5 of [4], it follows from (1.1) that

$$q_{(2)n}\sim\Gamma^{-1}(\alpha)n^{-\alpha}L(n^{-1}). \tag{4}$$

Noting (3.3), we have $q_{(2)T}\sim\Gamma^{-1}(\alpha)\,\varepsilon^{\alpha}n^{-1}\beta$. Hence

$$\begin{aligned}\sum_{j=n}^{\infty}\sum_{i=0}^{\gamma j} a_{ij}f_j=\sum_{i=0}^{\infty}\sum_{j=n}^{\infty} q_i(h_j)f_j\leqslant q_{(2)T}+\sum_{i=0}^{T}\sum_{j=n}^{\infty} q_i(h_j)f_j\\ \leqslant\varepsilon^{\alpha}n^{-1}\Gamma^{-1}(\alpha)\beta+O(T)\sum_{j=n}^{\infty} f_j.\end{aligned} \tag{5}$$

We will assume that (3) is satisfied. Since ε is arbitrary, (5) imply

$$\sum_{j=n}^{\infty}\sum_{i=0}^{\gamma j} a_{ij}f_j=o(n^{-1}). \tag{6}$$

The case when (1) and (2) are satisfied will be considered at the end of this section, but for now we note that these conditions, instead of (6), imply

$$\sum_{j=n}^{\infty}\sum_{i=0}^{\gamma j} a_{ij}f_j=O(n^{-1}). \tag{7}$$

Noting the monotonicity of Q_n and $H_j(n)$ as the arguments increase, we obtain

$$\left|\sum_{j=0}^{n} H_j(n)f_j\right|\leqslant Q_{n(1-\gamma)}^{1+\alpha}L(Q_{n(1-\gamma)}). \tag{8}$$

The next assertion may be proved, using natural identity transformations and the triangle inequality.

L e m m a 4. Let $|c_{in}|<\tau\omega_n$ for $np\leqq i\leqq n$. Then

$$\left|\sum_{i=np+2}^{n}\beta_{n-i}\times(c_{in}-c_{i-1n})\right|<C_0\tau\omega_n,$$

where

$$C_0 = |\beta_0| + \max_{i\in N} |\beta_i| + \sum_{i=1}^{\infty} |\beta_i - \beta_{i-1}| < \infty. \tag{9}$$

Inequality (9) holds by Lemma 1.

To shorten the notation, we denote by ϕ_n the elements of a finite number of sequences $\{b_{jn}\}_{n\in N}$, if the b_{jn} can be estimated by a linear function of $Q_i^{1+\alpha}L(Q_i)$ and Q_i, $i \leqq n$, and upon substitution of $i^{-\beta}\rho_i^{\beta}L_1(i\rho_i^{-1})$ for Q_i, where $L_1(t)$ has constant $c_1(x)$ in the Karamata representation (see Section 2), the estimate is $o(n^{-\beta-1})\rho_n^{\beta+1}L_1(n\rho_n^{-1})$ in two cases: a) $\rho_n > 0$, $\rho_n \asymp 1$ and 2) for n such that $\rho_n \geqq \rho_i > 0$ for $i < n$, if $\sup \rho_j = \infty$ and $j\rho_j^{-1} \to \infty$. The monotonicity of Q_i and the validity of (2.9) for the substitute expression make it sufficient to verify these conditions for $\rho_n \equiv 1$.

If the foregoing estimates yield $O(\cdot)$ instead of $o(\cdot)$, we replace b_{jn} by Ψ_n.

By (6) and (4.2) for $p > \gamma$, in the sequel as well, we have

$$\Phi_n = \sum_{i=0}^{n(1-p)} Q_i a_{n-i} + Q_{n(1-p)} o(n^{-1}) = a_n \sum_{i=0}^{np} Q_i + \varphi_n. \tag{10}$$

Hence in view of (3),

$$I_1 = -a^{-1}a_n \sum_{i=0}^{np} Q_i + \varphi_n, \tag{11}$$

which together with (4.3) imply

$$I_5 = \phi_n . \tag{12}$$

Subtracting from Φ_n the principal term on the right side of (10), using Lemma 4 and formula (3), we obtain

$$I_4 = \varphi_n + \sum_{i=np+2}^{n} a^{-1}c_2\beta\beta_{n-i}\Big((i-1)^{-\beta-1}$$

$$\times L_1(i-1)\sum_{j=0}^{(i-1)p} Q_j - i^{-\beta-1}L_1(i)\sum_{j=0}^{ip} Q_j\Big),$$

which in view of (2.9) immediately implies

$$I_4 = \phi_n . \tag{13}$$

By (8) and (4.3), we have

$$I_8 = \phi_n . \tag{14}$$

Noting (2.3), we have

$$I_3 + I_6 = \sum_{i=0}^{np}\left(Q_i - \sum_{j=0}^{i} Q_{i-j}a_j\right)(\beta_{n-i} - \beta_{n-i-1}) = \sum_{i=0}^{np}\left(Q_i - \sum_{j=0}^{i} Q_{i-j}a_j\right)$$
$$(\beta_n - \beta_{n-1}) + \sum_{i=0}^{np}\left(Q_i - \sum_{j=0}^{i} Q_{i-j}a_j\right)(\beta_{n-1} - \beta_{n-i-1} - \beta_n + \beta_{n-1}).$$

Let $s > 0$ be some fixed scalar. Then by Lemma 2,

$$\sum_{i=0}^{s}\left(Q_i - \sum_{j=0}^{i} Q_{i-j}a_j\right)(\beta_{n-i} - \beta_{n-i-1} - \beta_n + \beta_{n-1})$$
$$\sim - \sum_{i=0}^{s}\left(Q_i - \sum_{j=0}^{i} Q_{i-j}a_j\right) ia^{-2}a_n.$$

Set

$$I_9 = \sum_{i=0}^{np}\left(Q_i - \sum_{j=0}^{i} Q_{i-j}a_j\right) \quad \text{and} \quad I_{10}(s) = \sum_{i=0}^{s} i\left(Q_i - \sum_{j=0}^{i} Q_{i-j}a_j\right) + a\sum_{i=0}^{s} Q_i.$$

If upon substitution of $i^{-\beta}L_1(i)$, for Q_i, the estimates $I_{10}(s) \to 0$ and $I_9 = o(n^{-1})$, as $s,n \to \infty$, are true, the last relations, (4.4), Lemma 2, (2.3), (8), (10) and the definition of ϕ_n yield

$$I_3 + I_6 = \varphi_n + a^{-1}a_n \sum_{i=0}^{np} Q_i. \tag{15}$$

Let us make the routine substitutions

$$|I_9| = \sum_{i=0}^{np} Q_i - \sum_{j=0}^{np} a_j \sum_{i=0}^{np-j} Q_i = \sum_{i=0}^{np} Q_i\left(1 - \sum_{j=0}^{np} a_j\right)$$
$$+ \sum_{j=0}^{np} a_j \sum_{i=np-j+1}^{np} Q_i \leqslant \sum_{i=0}^{np} Q_i \sum_{j=np+1}^{\infty} a_j + \sum_{j=0}^{0.5np} ja_jQ_{0.5np} + \sum_{j=0.5np}^{np} a_j \sum_{i=0}^{0.5np} Q_i.$$

Analogously,

$$|I_{10}(s)| = \sum_{i=0}^{s} iQ_i \sum_{j=s+1}^{\infty} a_j + \sum_{j=0}^{s} a_j \sum_{i=s-j+1}^{s} iQ_i - \sum_{i=0}^{s}\sum_{j=0}^{i} ia_jQ_{i-j}$$
$$+ a\sum_{i=0}^{s} Q_i \leqslant \sum_{i=0}^{s} iQ_i \sum_{j=s}^{\infty} a_j + \sum_{j=0}^{s} ja_jsQ_{0.5s} + \sum_{j=0.5s}^{s} a_j \sum_{i=0}^{s} iQ_i$$
$$+ \sum_{j=s+1}^{\infty} ja_j \sum_{i=0}^{s} Q_i + \sum_{j=0}^{0.5s} ja_jQ_{0.5s}s + \sum_{j=0.5s}^{s} ja_j \sum_{i=0}^{s} Q_i.$$

In view of (2.5) and (4.1), the required estimates for I_9 and $I_{10}(s)$ are obtained trivially, i.e., (15) is proved.

It follows from (2.7), (8), (11)-(15), Lemma 4 and the definition of Ψ_n that

$$Q_n - Q_{n-1} = \Psi_n. \tag{16}$$

Now we investigate I_2. First, for any $V > 0$, by Lagrange's mean-value formula

$$I_{11} \equiv \sum_{m=0}^{i} H_m(i) f_m - \sum_{m=0}^{i} \Big(1 - h_m(P_i, \ldots, P_{i-V}, P_i, \ldots, P_i) - \sum_{j=0}^{V} Q_{i-j}a_{jm} - \sum_{j=V+1}^{mp} Q_i a_{jm}\Big) f_m = O\Big(\sum_{j=V+1}^{ip} a_j(Q_{i-j} - Q_i)\Big).$$

Similarly, using the mean-value theorem and the monotonicity of the generating functions, we have

$$I_{12} \equiv \sum_{m=0}^{i} \Big(1 - h_m(P_i, \ldots, P_{i-V}, P_i, \ldots, P_i) - \sum_{j=0}^{V} Q_{i-j}a_{jm} + h_m(P_i, \ldots, P_i) - 1 + \sum_{j=0}^{V} Q_i a_{jm}\Big) f_m = O(Q_{i-V} - Q_i) \sum_{m=0}^{i} f_m \sum_{j=0}^{V} (a_{jm} - D_j h_m(P_{i-V}, \ldots, P_{i-V})) f_m = O(Q_{i-V} - Q_i)\Big[1 - h'(P_{i-V}) + \sum_{j=V}^{\infty} a_j + \sum_{m=i+1}^{\infty} \sum_{j=0}^{m} a_{jm} f_m\Big],$$

where $D_j h_m = \frac{\partial}{\partial x_j} h_m(x_0, x_1, \ldots, x_m)$.

It follows from (16), (4.1), (2.4) and the continuity of $h'(x)$ that for all $V > 0$ as $i \to \infty$, $I_{12} = \Psi_i(V_0(1) + V^{-\beta+1}L_1(V))$. On the other hand, by (2.5) and (16), for all $\varepsilon > 0$ for sufficiently large V, $|I_{11}| < \varepsilon\Psi_{n(1-p)}$. Since ε is arbitrary, the latter implies $I_{11} + I_{12} = o(\Psi_i)$, i.e., $I_{11} + I_{12} = \phi_i$.

We have thus proved that

$$I_2 = a^{-1} \sum_{i=0}^{n} \Big((1 - h_i(P_n, \ldots, P_n) - \sum_{j=0}^{i\gamma} a_{ji} Q_n\Big) f_i + \varphi_n,$$

which by (1.1), (3) and (6) implies

$$I_2 = -a^{-1} Q_n^{1+\alpha} L(Q_n) + \varphi_n, \tag{17}$$

which, by Lemma 4, implies in turn

$$I_7 = \varphi_n + \sum_{i=np+2}^{n} \beta_{n-i}\left(Q_i^{1+\alpha}L(Q_i) - Q_{i-1}^{1+\alpha}L(Q_{i-1})\right). \tag{18}$$

Let $s_1 = P_n$, $s_2 = P_{n-1}$. Then

$$\begin{aligned} Q_n^{1+\alpha}L(Q_n) - Q_{n-1}^{1+\alpha}L(Q_{n-1}) &= h(s_1) - s_1 - h(s_2) + s_2 \\ &= \sum_{k=2}^{\infty} h_k(s_1 - s_2)\left(\sum_{i=0}^{k-1} s_1^{k-1-i}s_2^i - k\right). \end{aligned} \tag{19}$$

Next, $(1-x)^i \geqq 1 - ix$ for $x \geqq 0$. Hence,

$$\left|\sum_{i=0}^{k-1} s_1^{k-1-i}s_2^i - k\right| \leqslant \min(k, k^2Q_{n-1}). \tag{20}$$

From (19), (20), (4.3) and the fact that Q_n tends to zero, we find that for any fixed m,

$$\sum_{i=n-m}^{n} \beta_{n-i}\left(Q_i^{1+\alpha}L(Q_i) - Q_{i-1}^{1+\alpha}L(Q_{i-1})\right) = o(1) \sum_{i=n-m}^{n} |Q_i - Q_{i-1}|. \tag{21}$$

Note that if in Lemma 4 the summation is carried out up to $n - m$, then (9) becomes

$$c_{0m} = |\beta_m| + \max_{i\in N}|\beta_{i+m}| + \sum_{i=1}^{\infty}|\beta_{i+m} - \beta_{i+m-1}| < \infty, \tag{22}$$

which makes it possible to obtain

$$\left|\sum_{i=np+2}^{n-m} \beta_{n-i}\left(Q_i^{1+\alpha}L(Q_i) - Q_{i-1}^{1+\alpha}L(Q_{i-1})\right)\right| \leqslant c_{0m}Q_{np}^{1+\alpha}L(Q_{np}). \tag{23}$$

By choosing $m > 0$, the c_{0m} in (22) can be made arbitrarily small. This together with (18), (16), (21) and (23) imply

$$I_7 = \phi_n . \tag{24}$$

Thus, noting (11)-(15), (17) and (24), we obtain from (2.7) that

$$Q_n - Q_{n-1} = -a^{-1}Q_n^{1+\alpha}L(Q_n) + \varphi_n. \tag{25}$$

The idea of estimating Q_n from above, applied in [10] to Bellman-Harris processes and developed for general processes in [7], can also be used in this situation, starting out from (25).

Set

$$Q_n = n^{-\beta}\rho_n^{\beta}L_1(n\rho_n^{-1}),$$

where the slowly varying function $L_1(t)$ is taken with constant in Karamata's representation for large values of the argument (Section 2).

We prove that ρ_n is bounded from above. To this end, we assume the converse: $\overline{\lim_{n\to\infty}} \rho_n = \infty$. By assumption, $\max_{0\le k\le n} \rho_k = \rho_{k(n)} \to \infty$ as $n \to \infty$. In the sequel, by $k(n)$ we shall mean a strictly monotonic sequence such that

$$\rho_{k(n)} = \max_{0\leqslant k\leqslant k(n)} \rho_k .$$

Then $\rho_{k(n)}$ monotonically increases and

$$\lim_{n\to\infty} \rho_{k(n)} = \infty . \tag{26}$$

Since $Q \downarrow 0$ as $n \to \infty$, by the definition of ρ_n for large n, $n\rho_n^{-1} \uparrow \infty$, and hence for any fixed i,

$$\lim_{n\to\infty} \rho_{k(n)-i}/\rho_{k(n)} = 1 . \tag{27}$$

Set $v_n = \rho_n^{\beta}$. Then, dividing (25) by $v_{n-1} n^{-\beta} v_n L_1(n/\rho_n)$, applying (2.10) to $L_1(n/\rho_n)$ and setting $n = k(n)$, using (3.3), (26) and (27), the definitions of a slowly varying function and of ϕ_n, we have

$$\begin{aligned} v_{k(n)}^{-1} - v_{k(n)-1}^{-1} &= -\beta\left(\frac{k(n)}{k(n)-1}\right)^{\beta} k^{-1}(n)\, v_{k(n)}^{-1}(1+o(1)) \\ &+ k^{-1}(n)\, v_{k(n)}^{-1} o\left(\frac{k(n)}{\rho_{k(n)}} - \frac{k(n)-1}{\rho_{k(n)-1}}\right) + a^{-1}k^{-1}(n)\, \rho_{k(n)} v_{k(n)-1}^{-1}\beta(1+o(1)). \end{aligned} \tag{28}$$

Let us rewrite (16) in the form

$$\begin{aligned} &k^{-\beta}(n)\rho_{k(n)}^{\beta} L_1(k(n)/\rho_{k(n)}) - (k(n)-1)^{-\beta} \\ &\quad\times \rho_{k(n)-1}^{\beta} L_1((k(n)-1)/\rho_{k(n)}) \\ &= O(k^{-\beta-1}(n))\rho_{k(n)}^{\beta+1} L_1(k(n)/\rho_{k(n)}) . \end{aligned}$$

Applying (2.9) to the latter, we have

$$\begin{aligned} &(k(n)/\rho_{k(n)} - (k(n)-1)/\rho_{k(n)-1}) L_1(k(n)/\rho_{k(n)}) \\ &\times k^{-\beta-1}(n)\rho_{k(n)}^{\beta+1} = O(k^{-\beta-1}(n))\rho_{k(n)} L_1(k(n)/\rho_{k(n)}) \end{aligned}$$

yielding

$$k(n)/\rho_{k(n)} - (k(n)-1)/\rho_{k(n)-1} = O(1) ,$$

which together with formula (27) enable us to write (28) in the form

$$v_{k(n)}^{-1} - v_{k(n)-1}^{-1} = (-\beta + a^{-1}\beta\rho_{k(n)}(1 + o(1)) + o(1))k^{-1}(n)v_{k(n)-1}^{-1} . \tag{29}$$

By definition of $k(n)$, $v_{k(n)}^{-1} - v_{k(n)-1}^{-1} \leqq 0$. Consequently, $-\beta + a^{-1}\beta\rho_{k(n)}(1+o(1)) \leqq 0$. The resulting relation contradicts (26). We have thus proved

$$\overline{\lim_{n\to\infty}}\, Q_n n^{\beta} L_1^{-1}(n) < \infty. \tag{30}$$

Therefore, (25) becomes

$$Q_n - Q_{n-1} = -a^{-1}Q_n^{1+\alpha}L(Q_n)(1 + o(1)).$$

As was noted in Section 3 in deriving (3.2), the latter implies that

$$Q_n \sim a^{\beta}n^{-\beta}L_1(n). \tag{31}$$

Thus, for $c_2 > 0$ and $c_1 = 0$, Theorems 1 and 2 have been proved. Note that in this case condition (1.8) stated in terms of Theorem 2 is superfluous, since by (6) is is always satisfied for $g(y) \equiv 0$.

Let us note the changes that have to be made in the preceding text upon substitution of (1) and (2) for (3). These changes are necessary because (6) is replaced by (7). Instead of (10), we obtain

$$\Phi_n = a_n \sum_{i=0}^{np} Q_i + \varphi_n + Q_{n(1-p)}O(n^{-1}) ,$$

which together with (1) and (3.4) yield

$$I_1 = -a^{-1}a_n \sum_{i=0}^{np} Q_i + \varphi_n + Q_{n(1-p)}O(n^{-1}).$$

In estimating I_4, taking (2) into account simplifies the calculations. On the other hand, the change of the estimate for I_1 entails an addition of the summand $O(n^{-1})Q_{np(1-p)}$ to the formula (13).

All the estimates up to (17) remain unchanged, while $O(n^{-1})Q_n$ and $O(n^{-1})Q_{np}$ have to be added to (17) and (18), which, in the end, instead of (25), leads to

$$Q_n - Q_{n-1} = -a^{-1}Q_n^{1+\alpha}L(Q_n) + O(n^{-1})Q_{n(1-p)p} + \varphi_n. \tag{32}$$

In the arguments following (25), we have to add $O(k^{-1}(n))\rho_{k(n)} \times v^{-1}_{k(n)-1}$; which is inessential, however, and the relation obtained from (29) contradicts (26), i.e., for $c_1 > 0$, (30) holds true. By (3.4), this completes the proof of Theorem 1 for (1.6) and (1.7) with $c_2 > 0$.

6. Proof of Theorem 1 in the general case

To complete the proof of Theorem 1, we need to obtain an upper bound for $Q(t)$, since the lower bound was derived in Section 3.

In this section, we construct several processes on probability spaces already existing with probabilities of continuation majorizing these same probabilities for the initial processes. The last process satisfies conditions (1.6) and (1.7) with $c_2 \neq 0$. However, we do not require that the particles produce no descendants before they die, neither the condition (P), since the appropriate reduction was described in Section 2. We prove Theorem 1 using the results of Sections 2 and 5.

Thus, consider the process $\xi_B(n)$, $n \in N$, for which conditions (1.4) and (1.3) are satisfied. The index B will be omitted.

Let

$$\varlimsup_{n\to\infty}(1 - A(n))\, n^{\beta} L_1^{-1}(n) = c_3, \qquad (1)$$
$$\varliminf_{n\to\infty}(1 - A(n))\, n^{\beta} L_1^{-1}(n) = c_4.$$

If $c_4 > 0$, so is c_3. In this case, the constraint (1.7) is easy to get rid of. From $\xi(n)$ we construct $\xi_2(n)$ for which (1.7) holds with $c_2 > 0$.

Set $\delta > 0$ and $1 - F_2(n) = (c_3+\delta)n^{-\beta}L_1(n)$ for large n, $F_2(n)$ being the distribution function of some integer-valued random variable and $A(n) \geqq F_2(n)$. Let us describe $\{\eta_2, N_2(t)\}$. We make a randomized time substitution $\tau = F_2^{-1}(t)$ if $n = A^{-1}(t)$, where t is uniformly distributed on $[0,1]$, and $F_2^{-1}(t)$ and $A^{-1}(t)$ are viewed as inverse to the closures of the graphs of $F_2(t)$ and $A(x)$. Roughly speaking, we have thus associated to each value n the random value $\tau(n) \in F_2^{-1}(A(n))$, where $\tau(n) \leqq \tau(n+1)$, and equality can be attained. Set $\eta_2 = \tau(n)$ if $\eta = n$ and $\forall t \in [0,1]$,

$N_2(\tau) = N_2(F_2^{-1}(t)) = N(A^{-1}(t)) = N(n)$. It follows from the definition of $\xi_2(n)$ and (1) that for $c_4 > 0$, the ratios η_2/η and $\tau/(n+1)$ are uniformly bounded from above. As the upper bound one can take the larger of the quantity $(c_3 + 1 + \delta)c_4^{-1}$ and the one depending on a finite number of the values of $F_2(n)$ and $A(n)$; which is done in routine fashion, and therefore we omit the details.

This means that the conditions of Theorem 1 still hold for $\xi_2(n)$, and moreover, (1.7) is satisfied with $c_2 = c_3 + \delta$ and $Q_2(t) \geqq Q(t)$.

We shall show at the end of this section that the additional condition (1.6) is inessential, but for the time being we assume it be satisfied.

If $c_3 = 0$, then $o(1)$ will appear on the right side of the statements of Lemmas 1 and 2, and everything reduces easily to the scheme already considered. However, if $c_4 = 0$ and $c_3 \neq 0$, then the reduction gets tricky. On the one hand, $\xi_2(n)$ cannot be used since $\eta_2\eta^{-1}$ and $\tau(n+1)^{-1}$ become unbounded and may violate the conditions of Theorem 1; on the other hand, we have essentially used the regularity of a_i (see (4.2)). Let us take $F_3(n)$ suitable for a time substitution later on, as well. We assume in (1) that $c_3 < \infty$.

Let an integer-valued sequence $\{x_n\}_{n=0,1,...}$ be such that for some $\varepsilon > 0$, $x_{n+1} \sim (1+2\varepsilon)x_n$. From it we construct $\{y_n\}_{n=0,1,...}$ such that $y_n \in N$, $x_n < y_n < x_{n+1}$ and $y_n \sim (1+\varepsilon)x_n$.

Let

$$F_3(k) = A(n)\chi\{k < x_0\} + \sum_{n=1}^{\infty} ((A(x_n) - A(x_{n-1})) f((k - x_n)(y_n - x_n)^{-1}$$
$$+ A(x_n))\chi\{x_n \leqslant k \leqslant y_n\} + \sum_{n=1}^{\infty} A(x_n)\chi\{y_n < k < x_{n+1}\}$$
$$+ A(x_0)\chi\{x_0 \leqslant k < x_1\},$$

where $f(x)$ is a monotonic continuously differentiable function on $[0,1]$, $f(0) = f'(0) = f'(1) = 0$, $f(1) = 1$.

Note that under this definition, $F_3(n)$ possesses the following properties:

1) $1 - F_3(n) = O(n^{-\beta}L_1(n))$,

2) $dF_3(n) = O(n^{-\beta-1}L_1(n))$,

3) $dF_3(n) - dF_3(n-1) = o(n^{-\beta-1}L_1(n))$,

4) $1 - F_3(n) \sim 1 - F_3(n+1)$

or 1), 2) and

3') $F_3(n) = 0$ for sufficiently large n.

Fulfillment of condition 3' means that $A(n)$ possesses this same property (i.e., $a_n \equiv 0$ for large n), and in Section 5, instead of Lemmas 1 and 2, one can use Lemma 3 with $k > 2\beta + 2$, which substantially simplifies the estimates and leads immediately to the same result for $\xi(n)$.

It is seen from the construction of $F_3(n)$ that by a random time substitution $n = A^{-1}(\xi)$, $\tau = F_3^{-1}(\xi)$, where ξ is uniformly distributed on $[0,1]$, one may assume without loss of generality that $1 \leqq \tau(n+1)^{-1} \leqq 1 + 4\varepsilon$, since these inequalities hold for large n, while for small ones we can set $F_3(n) = A(n)$.

Taking into account the needs of the next section, we construct $\xi_3(n)$ starting from the relations $\eta_3 = [(1+8\varepsilon)\eta]$ for large values and $\eta = \eta_3$ for small values of $N_3(\tau) = N(n)$, respectively, the relationship of and n being as shown earlier. In that case, we assume the condition that the particles produce no descendants after their death to be satisfied. By the construction of $\xi_3(n)$, $Q_3(t) \geqq Q(t)$. Proving Theorem 1 for $\xi_3(n)$ eliminates the constraint (1.7).

To shorten the notation, we assume that $\xi_3(n) \equiv \xi(n)$ and use $\{\tau_n\}_{n\in N}$ to denote a finite number of sequences $\{b_{nj}\}_{n\in N}$ such that $b_{nj} = O(n^{-\beta}L_1(n))$, identically for all j.

Thus, we have

$$a_n = \tau_n n^{-1}, \quad a_n - a_{n+1} = o(n^{-\beta-1}L_1(n)),$$

$$\sum_{i=n+1}^{\infty} a_i = \tau_n \quad \text{and} \quad \sum_{i=n}^{\infty} a_i \sim \sum_{i=n+1}^{\infty} a_i.$$

To investigate Q_n for $c_1 = 0$ (i.e., to prove (5.3)), we need to make an almost verbatim repetition of the reasoning of Section 5 up to (5.30) inclusively, using instead of (4.1) and (4.2) the relations obtained above as well as noting the remark in Section 4.

We list next the main changes in the estimates obtained as a result of this substitution.

Instead of (5.10), (5.11) and (5.13), we will have

$$\Phi_n = \tau_n n^{-1} \sum_{i=0}^{np} Q_i + \varphi_n,$$

$$I_1 = \tau_n n^{-1} \sum_{i=0}^{np} Q_i + \varphi_n \quad \text{and} \quad I_4 = \tau_n n^{-1} \sum_{i=0}^{np} Q_i + \varphi_n.$$

Similarly, (5.15) becomes

$$I_3 + I_6 = \tau_n n^{-1} \sum_{i=0}^{np} Q_{\ i} + \varphi_n.$$

However, estimation of the sum associated with $I_{10}(s)$ is different, namely, we use the conditions on a_i, the remark in Section 4, the monotonicity of Q_n, the change in the order of summation, and (2.5):

$$\sum_{i=0}^{np} \left(Q_i - \sum_{j=0}^{i} Q_{i-j} a_j\right) (\beta_{n-i} - \beta_{n-i-1} - \beta_n + \beta_{n-1})$$

$$= \sum_{i=0}^{np} \left|Q_i - \sum_{j=0}^{i} Q_{i-j} a_j\right| i\tau_n n^{-1} = \tau_n n^{-1} \sum_{i=0}^{np} i \sum_{j=0.5i}^{\infty} a_j Q_{0.5i}$$

$$+ \tau_n n^{-1} \sum_{i=0}^{np} i \sum_{j=0}^{0.5i} (Q_{i-j} - Q_i)\, a_j = \tau_n n^{-1} \left(\sum_{i=0}^{np} Q_i + \sum_{j=0}^{np} \sum_{i=2j}^{np} Q_{i-j} j a_i \right.$$

$$\left. + \sum_{j=0}^{np} a_j \sum_{i=j}^{2j} iQ_i \right) = \tau_n n^{-1} \sum_{i=0}^{np} Q_i \,. \tag{2}$$

All the later deductions up to (5.25) remain in vigor, although the conditions on a_i have become less restrictive. The latter takes on the form

$$Q_n - Q_{n-1} = -a^{-1} Q_n^{1+\alpha} L(Q_n) + \varphi_n + \tau_n n^{-1} \sum_{i=0}^{np} Q_i. \tag{3}$$

Now we verify that (3) implies an addition of the summand $k^{-1}(n) \times o(1) v_{k(n)-1}^{-1}$ (in (5.28)), i.e., (5.28) does not change and (5.30) still holds.

We need to investigate the contribution of the last summand in (3) to (5.28). Instead of $k(n)$ we write n, assume that $0 < \delta < 0.5(\beta-1)$ is fixed, use the monotonicity of $x^{-\beta} L_1(x)$, $x^{\delta} L_1(x)$ and $n\rho_n^{-1}$ for large x (it follows from the assumptions of Section 2 on the form of $L_1(x)$ and on the decreasing Q_n), (5.26), (5.27), $\beta > 1$, and the fact that $i\rho_n^{-1} \sim k$ for $k\rho_n < i < (k+1)\rho_n$,

$$n^{\beta}v_n^{-1}v_{n-1}^{-1}\tau_n L_1^{-1}(n\rho_n^{-1})\,n^{-1}\sum_{i=0}^{n} i^{-\beta}\rho_i^{\beta}L_1(i\rho_i^{-1}) = O(n^{-1}v_n^{-2})\,L_1^{-1}(n\rho_n^{-1})\,n^{\delta}n^{-\delta}$$

$$\times L_1(n)\sum_{i=0}^{n} i^{-\beta}\rho_n^{\beta}L_1(i\rho_n^{-1}) = O(n^{-1})\,v_n^{-1}\rho_n^{(-\beta+1)0,5} = o(n^{-1})\,v_n^{-1}.$$

This proves Theorem 1 under the additional condition (5.3). In passing from (5.3) to (5.1) and (5.2), the arguments of Section 5 remain valid where an analogous transition is made for (1.7) with $c_2 \neq 0$.

To complete the proof of Theorem 1, we need to get rid of the constraint (1.6) for $\xi(n)$. For this, we use the procedure of constructing $\xi_1(n)$, described at the end of Section 2. Let

$$\overline{\lim_{n\to\infty}}\,(1 - G(n))\,n(1 - h_{(n)}(0))^{-1} = c_5$$

and let $G_1(n)$ of an integer-valued random variable be such that

$$G(n) \geqq G_1(n) \quad \text{and} \quad \lim_{n\to\infty}(1 - G_1(n)) \times n(1 - h_{(n)}(0))^{-1} = c_6 \geqq c_5.$$

Then, letting $N_1(n) \equiv N(n)$, $\eta = G^{-1}(\xi)$ and $\eta_1 = G_1^{-1}(\xi)$, where ξ is uniformly distributed on $[0,1]$, we construct $\xi_1(n)$ satisfying the conditions of Theorem 1 and (1.6), with $Q_n \leqq Q_{n1}$. This completes the proof of Theorem 1.

7. Proof of Theorem 2

In Section 5, Theorem 2 was proved for the case $c_1 = 0$ and $c_2 > 0$. We shall get rid of the first restriction at the beginning and of the second restriction at the end of this section.

We use the results of Section 5. By Theorem 1, (5.30) is true. Hence, in the estimates one can write $o(n^{-\beta-1}L_1(n))$ instead of ϕ_n, which we shall be taking into account without replacing ϕ_n by a cumbersome explicit expression. The same applies to Ψ_n. In Section 5 we showed which changes had to be made in the previous arguments to prove Theorem 1 for $c_1 > 0$. These changes are only for I_1, I_2, I_4 and I_7. We obtain Theorem 2 for $c_1 > 0$, if we impose additional restrictions on its hypothesis and estimate I_1, I_2, I_4 and I_7 more precisely than was done at the end of Section 5. Let us set

$$A_n = \sum_{j=n+1}^{\infty}\sum_{i=0}^{\nu j} a_{ij}f_j$$

and investigate Φ_n.

By the definition of Φ_n and (5.16), one has

$$\Phi_n = \sum_{i=0}^{n(1-p)} Q_i a_{n-i} + Q_n A_n - \sum_{i=0}^{np} (Q_n - Q_{n-i}) \sum_{j=n+1}^{\infty} a_{ij} f_j$$

$$= \sum_{i=0}^{n(1-p)} Q_i a_{n-i} + Q_n A_n + \Psi_n \sum_{i=0}^{np} \sum_{j=n+1}^{\infty} i a_{ij} f_j,$$

which by virtue of (2.5) and (4.2) yields

$$\Phi_n = a_n \sum_{i=0}^{n(1-p)} Q_i + Q_n A_n + \varphi_n. \tag{1}$$

It follows from (5.1) and (1) that

$$I_1 = -a^{-1}\left(a_n \sum_{i=0}^{n(1-p)} Q_i + Q_n A_n - c_1 n^{-\beta-1} L_1(n)\right) + \varphi_n. \tag{2}$$

The presence of the summand $Q_n A_n$ in (1) and the derivation of (5.13) we had earlier imply

$$I_4 = \varphi_n + \sum_{i=np+2}^{n} \beta_{n-i} (Q_{i-1} A_{i-1} - Q_i A_i). \tag{3}$$

We can no longer get (5.17), rather only what precedes it:

$$I_2 = a^{-1} \sum_{i=0}^{n} \left(1 - h_i(P_n, \ldots, P_n) - \sum_{j=0}^{i\gamma} a_{ji} Q_n\right) f_j + \varphi_n. \tag{4}$$

Noting this, we have to add

$$\sum_{i=np+2}^{n} \beta_{n-i} \left(Q_i A_i - Q_{i-1} A_{i-1} + \sum_{j=i}^{\infty} (1 - h_j(P_{i-1}, \ldots, P_{i-1})) f_j \right.$$

$$\left. - \sum_{j=i+1}^{\infty} (1 - h_j(P_i, \ldots, P_i)) f_j \right)$$

to the right side of (5.18), which will entail this addition to (5.24), too. Hence, in view of (3) one has:

$$I_4 + I_7 = \varphi_n + \sum_{i=np+2}^{n} \beta_{n-i} \left(f_i O(1) \right.$$

$$\left. + \sum_{j=i+1}^{\infty} (h_j(P_{i-1}, \ldots, P_{i-1}) - h_j(P_i, \ldots, P_i)) f_j \right),$$

which, by virtue of (5.16), Lagrange's mean-value theorem, (5.7), (4.3) and (5.2), implies

$$I_4 + I_7 = \varphi_n \, . \tag{5}$$

It follows from (2), (4) and (5.15) that

$$I_1 + I_2 + I_3 + I_6 = c_1 a^{-1} n^{-\beta-1} L_1(n) - a^{-1} Q_n^{1+\alpha} L(Q_n) - a^{-1} \sum_{j=n+1}^{\infty} (1 - h_j(P_n, \ldots, P_n)) f_j + \varphi_n$$

or, noting additionally (2.7), (5), (5.14), (5.12) and (1.8), we obtain

$$Q_n - Q_{n-1} = c_1 a^{-1} n^{-\beta-1} L_1(n) - a^{-1} g\left(Q_n n^{\beta} L_1^{-1}(n)\right) n^{-\beta-1} L_1(n) - a^{-1} Q_n^{1+\alpha} L(Q_n) + \varphi_n. \tag{6}$$

The derivation of an explicit expression for Q_n in (6) resembles that of (5.29) from (5.25). Thus, set $Q_n = n^{-\beta} \rho_n^{\beta} L_1(n/\rho_n)$ and assume that $\lim_{n\to\infty} \rho_n$ is not defined, i.e.,

$$0 < \underline{\rho} = \varliminf_{n\to\infty} \rho_n < \varlimsup_{n\to\infty} \rho_n = \bar{\rho} < \infty,$$

which together with the relation $\rho_n \sim \rho_{n-1}$ easily obtainable from (6) enable one to choose two sequences $\tau(k)$ and $\nu(k)$ such that $\rho_{\tau(k)} \to \underline{\rho}$, $\rho_{\nu(k)} \to \bar{\rho}$ as $k \to \infty$ and $\rho_{\tau(k)} \leqq \rho_{\tau(k)\pm 1}$, $\rho_{\nu(k)} \geqq \rho_{\nu(k)\pm 1}$.

Applying the procedure of deriving (5.29) from (5.25) to (6) for $n = \tau(k)$, $\tau(k)+1$, $\nu(k)$, $\nu(k)+1$ (the condition $\rho_n \asymp 1$ makes it possible to simplify a great deal), and using the foregoing definitions and properties, it is easy to obtain that

$$a^{-1}\beta\rho_{\tau(k)} - \beta + o(1) - c_1 a^{-1} \rho_{\tau(k)}^{-\beta} + g\left(\rho_{\tau(k)}^{\beta}\right) a^{-1} \rho_{\tau(k)}^{-\beta} = 0,$$

$$a^{-1}\beta\rho_{\nu(k)} - \beta + o(1) - c_1 a^{-1} \rho_{\nu(k)}^{-\beta} + g\left(\rho_{\nu(k)}^{\beta}\right) a^{-1} \rho_{\nu(k)}^{-\beta} = 0.$$

Passing to the limit in these relations, we have that $\underline{\rho}^{\beta}$ and $\bar{\rho}^{\beta}$ satisfy the same equation (1.10). We now show that it has a unique nonnegative solution. Consider the function $z = y^{1+\alpha} - ay - c_1\alpha$. It decreases for $0 \leqq y < (1+\alpha)^{-\beta} a^{\beta}$ and increases for $y \geqq a^{\beta}(1+\alpha)^{-\beta}$, and in the second region has a unique zero. The nonnegativity following from (5.7) together with the definitions of continuity and growth of $g(y) \leqq c_1$ imply the uniqueness of a solution for (1.10).

We have thus arrived at a contradiction with the assumption $\underline{\rho} < \bar{\rho}$, i.e., Theorem 2 is proved for $c_2 > 0$ (the quantity $\lim_{n\to\infty} \rho_n$ is found similarly).

What makes it possible to get rid of condition (1.7) with $c_2 > 0$ is the fact that with (1.7) replaced by properties 1) - 4) from Section 6 for $F_3(n)$, with the $dF_3(n)$ regarded as a_n and the remark from Section 4 instead of Lemmas 1 and 2, the proof remains almost intact, if one does not replace a_n by the exact asymptotic expression (4.2). The estimates I_4 and $I_3 + I_6$ are the only exception. The correction in the case $c_1 > 0$ (at the beginning of this section) remains unchanged.

Estimation of $I_3 + I_6$ is different in investigating the sum connected with the determination of $I_{10}(s)$, since we add the summand $io(n^{-\beta-1}L_1(n))$ to the right-hand side of (4.5). In addition, one needs to prove the boundedness of

$$\sum_{i=0}^{np} i \left| Q_i - \sum_{j=0}^{i} Q_{i-j} a_j \right|,$$

but this easily follows from (6.2). This and the previous derivation of (5.15) prove its validity for $\xi_3(n)$, too.

By Theorem 1 and the properties of a_n, (11) can be written in the form $I_1 = a_n\mu + \phi_n$, where $\mu = -a^{-1} \sum_{i=0}^{\infty} Q_i$. Then, as before,

$$I_4 = \varphi_n + O(1) \times \sum_{i=np+2}^{n} \beta_{n-i}(a_i - a_{i-1}).$$

Let us estimate

$$\sum_{i=np+2}^{n} \beta_{n-i}(a_i - a_{i-1}) = \sum_{i=np+2}^{n-m} \beta_{n-i}(a_i - a_{i-1}) + \sum_{i=n-m+1}^{n} \beta_{n-i}(a_i - a_{i-1}).$$

One can apply to the first sum a natural generalization of Lemma 4 with (5.9) replaced by

$$C_0 = 2 \max_{i \in N} |\beta_{m+i}| + \sum_{i=m}^{\infty} |\beta_i - \beta_{i-1}|,$$

which is small for large m. The second sum, for mixed m, is equal to ϕ_n, whence $I_4 = \phi_n$, i.e., all the estimates we had previously when (1.7) held have been preserved, and Theorem 2 is proved for $\xi_3(n)$.

Note that Theorem 1, as the estimate of I_1 illustrates, simplifies the arguments of Section 5 while proving Theorem 2. However, we often referred to the discussion of Section 5 and therefore made no simplifications.

Using condition (1.4), from $\xi(n)$ one can construct $\xi_3(n)$ in Section 6, which estimates Q_n from above. In this case, if $\varepsilon \to 0$, the parameters of $\xi_3(n)$ in the definition of ρ_3 converge to the corresponding parameters for $\xi(n)$, i.e., the validity of Theorem 2 for $\xi_3(n)$ yields an upper estimate for Q_n by a function that satisfies (1.9).

For a lower bound we construct a process $\xi_0(n)$. Fix $n_0 > 0$ and set

$$\eta_0 \equiv \eta, \ N_0(n) = (N - N(n))\chi\{n > n_0\} + N(n).$$

Then by (2.10, $Q_{n0} \leqq Q_n$.

In this case, $f_0(x)$ has only a finite number of nonzero coefficients, which makes it possible to use Lemma 3 instead of Lemmas 1 and 2 in investigating (2.7). More precisely, the summands containing $\beta_{0n} - \beta_{0n-1}$ are estimated immediately by any power function n^{-k}, and in Φ_n the previosuly present term $a_n \sum_{i=0}^{np} Q_i$ (see (5.10) and (1)) disappears.

This makes the proof of Theorem 2 for $c_2 > 0$ appropriate as well as simple for investigating $\xi_0(n)$. But $h_0(z,\ldots,z) = h(z,\ldots,z)$, and $a_0 \to a$ as $n_0 \to \infty$. We hereby conclude that the lower bound for Q_n also satisfies (1.9).

Thus, we have proved Theorem 2 for $c_2 = 0$, i.e., we have removed all additional constraints used in the proof.

To conclude, we consider condition (1.8). If (5.6) is satisfied, then $g(y) \equiv 0$. Here is an example with $g(y) \not\equiv 0$. Let $\xi(n)$ be defined by $\{\eta, N(n)\}$, given by the relations $f_i = i^{-\beta-2}$, $h_i(z_0,\ldots,z_i) = z_1^{\gamma(i)}$, where $\gamma(i) \sim i^{\beta}$ for large i, and for small i are such that

$$\sum_{i=1}^{\infty} f_i = 1, \ \sum_{i=1}^{\infty} h_i'(1,\ldots,1) f_i = 1.$$

Using Theorem 5 (see [4]) and (3.3), it is easily seen that

$$(h(s)-s)(1-s)^2 \sim (1-s)^{\alpha-1} \times \Gamma(1-\alpha)(1+\alpha)^{-1} ,$$

i.e., (1.1) holds with

$$L(1-s) \sim \Gamma(1-\alpha)(1+\alpha)^{-1} .$$

Obviously,

$$\sum_{k=n+1}^{\infty} h_k(z_0, 1 - yn^{-\beta}, z_2 \dots, z_h) f_k$$

$$\sim \sum_{k=n+1}^{\infty} \exp\{-yk^{\beta}n^{-\beta}\}\ k^{-\beta-2} \sim n^{-\beta-1} \int_1^{\infty} \exp\{yx^{\beta}\}\, x^{-\beta-2}dx.$$

It follows from (3.3) that

$$L_1(n) \sim (\beta + 1)^{\beta}\Gamma^{-\beta}(1 - \alpha) \equiv C$$

and

$$g(y) = C^{-1}\left((\beta + 1)^{-1} - \int_1^{\infty} \exp\{-yCx^{\beta}\}\, x^{-\beta-2}dx\right).$$

REFERENCES

[1] Crump, Kenny S., and Mode, Charles J. "A General Age-Dependent Branching Process. I; II." *Journal of Math. Analysis and Applications* 24 (1968): 494-508; 25 (1969): 8-17.

[2] Jagers, P. *Branching Processes With Biological Applications*. New York: John Wiley & Sons Publishers, 1975.

[3] Sevast'yanov, B.A. *Vetvyashchiesya protsessy* (Branching Processes). Moscow: Nauka, 1971. (German transl.: Sewastjanow, B.A. *Verzweigungsprozesse*. München Wien: R. Oldenbourg Verlag, 1975.)

[4] Feller, William. *An Introduction to Probability Theory and Its Applications*, vol.II. New York: John Wiley & Sons Publishers, 1966.

[5] Bojanic, R., and Seneta, E. "Slowly Varying Functions and Asymptotic Relations." *Journal of Math. Analysis and Applications* 34 (1971): 302-15.

[6] Vatutin, V.A. "A New Limit Theorem for the Critical Bellman-Harris Branching Process." *Math. USSR Sbornik*, vol.37, no.3 (1980): 411-23. (English transl.)

[7] Topchij, V.A. *Integral'naya predel'naya teorema dlya kriticheskikh vetvyashchikhsya protessov Krampa-Yagersa s diskretnym vremenem* (The Integral Limit Theorem for Critical Crump-Jagers Discrete-Time Branching Processes). Novosibirsk: The Institute of Mathematics of the Siberian Branch of the Academy of Sciences of the USSR, 1977. Preprint.

[8] Rogozin, B.A. "An Estimate of the Remainder Term in Limit Theorems of Renewal Theory." *Theory Probab. Applications*, vol.18, no.4 (1973): 662-77. (English transl.)

[9] Borovkov, A.A. "Remarks on Wiener's and Blackwell's Theorems." *Theory Probab. Applications*, vol.10, no.2 (1964): 303-12. (English transl.)

[10] Nagaev, S.V. "Transition Phenomena for Age-Dependent Branching Processes with Discrete Time. I." *Siberian Math. Journal*, vol.15, no.2 (1974): 261-81; "Transfer Effects for Age-Dependent Discrete-Time Branching Processes. II." *Siberian Math. Journal*, vol.15, no.3 (1974): 408-15. (English transl.)

V.F. Dem'yanov, and L.V. Vasil'ev
Nondifferentiable Optimization
1985, xvii + 455 pp.
ISBN 0-911575-09-X Optimization Software, Inc.
ISBN 0-387-90951-6 Springer-Verlag New York Berlin Heidelberg Tokyo
ISBN 3-540-90951-6 Springer-Verlag Berlin Heidelberg New York Tokyo

A.A. Borovkov, Ed.
Advances in Probability Theory: Limit Theorems For Sums of Random Variables
1985, XII + 301 pp.
ISBN 0-911575-17-0 Optimization Software, Inc.
ISBN 0-387-96100-3 Springer-Verlag New York Berlin Heidelberg Tokyo
ISBN 3-540-96100-3 Springer-Verlag Berlin Heidelberg New York Tokyo

V.F. Kolchin
Random Mappings
1986, approx. 250 pp.
ISBN 0-911575-16-2 Optimization Software, Inc.

B.T. Polyak
Introduction to Optimization
1986, approx. 450 pp.
ISBN 0-911575-14-6 Optimization Software, Inc.

V.P. Chistyakov, B.A. Sevast'yanov, and V.K. Zakharov
Probability Theory For Engineers
1986, approx. 200 pp.
ISBN 0-911575-13-8 Optimization Software, Inc.

V.F. Dem'yanov, and A.M. Rubinov
Quasidifferential Calculus
1986, approx. 300 pp.
ISBN 0-911575-35-9 Optimization Software, Inc.

N.I. Nisevich, G.I. Marchuk, I.I. Zubikova, and I.B. Pogozhev
Mathematical Modeling of Viral Diseases
1986, approx. 400 pp.
ISBN 0-911575-06-5 Optimization Software, Inc.

Continued on page 298

L. Telksnys, Ed.
Change Detection in Random Processes
1986, approx. 250 pp.
ISBN 0-911575-20-0 Optimization Software, Inc.

V.A. Vasilenko
Spline Functions: Theory, Algorithms, Programs
1986, approx. 280 pp.
ISBN 0-911575-12-X Optimization Software, Inc.

I.A. Boguslavskij
Filtering and Control
1986, approx. 400 pp.
ISBN 0-911575-21-9 Optimization Software, Inc.

R.F. Gabasov, and F.M. Kirillova
Methods of Optimization
1986, approx. 350 pp.
ISBN 0-911575-02-2 Optimization Software, Inc.

V.V. Ivanishchev, and A.D. Krasnoshchekov
Control of Variable Structure Networks
1986, approx. 200 pp.
ISBN 0-911575-05-7 Optimization Software, Inc.

V.G. Lazarev, Ed.
Processes and Systems in Communication Networks
1986, approx. 250 pp.
ISBN 0-911575-08-1 Optimization Software, Inc.

A.N. Tikhonov, Ed.
Problems in Mathematical Physics and Computational Mathematics
1986, approx. 500 pp.
ISBN 0-911575-10-3 Optimization Software, Inc.

G.I. Marchuk, Ed.
Computational Processes and Systems
1986, approx. 350 pp.
ISBN 0-911575-19-7 Optimization Software, Inc.

Continued on page 301

TRANSLITERATION TABLE

R	E	R	E
а А	a	р Р	r
б Б	b	с С	s
в В	v	т Т	t
г Г	g	у У	u
д Д	d	ф Ф	f
е Е	e	х Х	kh
ё Ё	e	ц Ц	ts
ж Ж	zh	ч Ч	ch
з З	z	ш Ш	sh
и И	i	щ Щ	shch
й Й	j	ъ Ъ	"
к К	k	ы Ы	y
л Л	l	ь Ь	'
м М	m	э Э	eh
н Н	n	ю Ю	yu
о О	o	я Я	ya
п П	p		

I.E. Kazakov, and S.V. Mal'chikov
State Space Theory
1986, approx. 350 pp.
ISBN 0-911575-15-4 Optimization Software, Inc.

B.A. Berezovskij, et al.
Multicriteria Optimization: Theory and Applications
1986, approx. 200 pp.
ISBN 0-911575-11-1 Optimization Software, Inc.

Yu.I. Merzlyakov, Ed.
Advances in Probability Theory: Groups and Algebraic Systems With Endpoint Constraints
1986, approx. 300 pp.
ISBN 0-911575-33-2 Optimization Software, Inc.

V.M. Glushkov, V.V. Ivanov, and V.M. Yanenko
Modelling of Evolution Systems
1986, approx. 400 pp.
ISBN 0-911575-32-4 Optimization Software, Inc.